Nanoscale
Communication Networks

Artech House Series
Nanoscale Science and Engineering

Series Editors
Xiang Zhang and Kang L. Wang

Nanoscale Communication Networks

Stephen F. Bush

ARTECH
HOUSE

BOSTON | LONDON
artechhouse.com

Library of Congress Cataloging-in-Publication Data
A catalog record for this book is available from the U.S. Library of Congress.

British Library Cataloguing in Publication Data
A catalogue record for this book is available from the British Library.

ISBN-13: 978-1-60807-003-9

Cover design by Vicki Kane

© 2010 ARTECH HOUSE
685 Canton Street
Norwood, MA 02062

10 9 8 7 6 5 4 3 2 1

Contents

Preface

OBJECTIVE

Richard Feynman presciently stated in his 1959 talk that "There's Plenty of Room at the Bottom." I believe that within this vast room at the bottom there will be a requirement for communication. Think of entire networks, or at least network nodes, that are the size of today's bits. I am *not* referring to nanotechnology being applied to today's macroscale communication, but rather to entire network systems that exist at the nanoscale, with nanoscale "users" such as nanorobots and molecular machines. The relatively new field of nanoscale communication networking seeks to shrink network communication in size to the scale of individual molecules; this topic includes the goal of transmitting information where traditional communication has never gone, including inside the cells of living organisms. Physics, biology, and chemistry will play a much greater role in this form of networking compared to the traditional electronics to which most people are accustomed today. This book will prepare the reader for the convergence of nanotechnology and networking. My goal is to provide students and professionals with a rapid, easy way to become proficient without having to personally read the 1,000+ papers on this topic; a selected set of 271 comprise the bibliography.

GENESIS

The idea for nanonetworks was inspired by a variety of sources, including early work on the effect of misalignment of carbon nanotubes within field effect transistors. Carbon nanotubes are a high mobility (rapidly switching) semiconductor. The idea was that the more we could reduce the requirement for perfect alignment among nanotubes, the cheaper and easier it would be to produce such transistors, assuming we could understand the impact of misalignment upon transistor performance. Once a certain degree of randomness in nanotube orientation was allowed, it was very compelling to think of the nanotubes as ad hoc Internet connections on a very small scale. Of course, this is only one of many possible approaches to nanonetworking as you will see in this book.

Through early conferences on this topic, such as the Nano-Networks conference series, other researchers thinking about diverse aspects of nanonetworking came together and formed the IEEE Emerging Technologies Committee on Nano-Scale, Molecular, and Quantum Networking.[1] I also introduced this topic as part of a general computer networking course at the State University of New York (SUNY) University at Albany campus; the students were intrigued by the concept and I owe them a debt of gratitude for providing the questions and enthusiasm to help build upon this topic.

---

1   http://www.comsoc.org/nano

APPROACH AND CONTENT

Nanoscale communication takes many forms, and includes the concept also known as molecular communication. The term "molecular communication" is often used when the transmission media is biological in nature. I take a much broader view of nanonetworking as shown in the overview of the contents of the book in Figure 1. Dashed lines show alternative coherent paths through the chapters. I recommend reading Chapter 1 and then either following the chapters in numerical order, or choosing among Chapters 2, 3, 4, and 5, as they relate to the underlying media that most interests you. Chapters 6, 7, and 8 follow with material that is common to all nanoscale media. Note that Chapter 7 includes architectural concerns that can be media-specific.

This book is written for both the student and the professional. Both can benefit by exercising their knowledge with the problem sets at the end of each chapter. This facilitates use as a supplementary course textbook or as a book for professionals to quickly become acclimated to this new field. There is an extensive bibliography with more than 270 papers that were researched in the writing of this book; a bibliography is provided for each chapter. Solutions to the exercises, slides, and other addenda to the book will be provided at www.comsoc.org/nano.

We begin the book by defining the nascent field of nanoscale networking, looking at how it has come into existence, and examining the driving forces behind this new field within Chapter 1. The following four chapters concentrate on specific nanoscale media. Thus, Chapters 2 and 3 describe biologically oriented media; they focus upon molecular motor communication as well as gap junction communication and cell signaling, respectively. These areas are biologically inspired; it is a very diverse research field with collaborative research from the biological fields. The goal of this book is not to convert network researchers into biologists and quantum physicists, but rather, to provide enough conceptual background to allow network researchers to understand how biological systems can be leveraged to transport information. Chapter 4 focuses on carbon nanotube networks; it can be divided into two parts: a brief introduction to research into nanonetworks used as chip interconnects and then a stronger focus on research into nanonetworks (largely random nanotube networks), which goes beyond application to chip fabrication. This topic, like the biological nanonetworks just mentioned, is a very diverse and collaborative field, which includes not only computer science but computer engineering, physics, and many other fields. Chapter 5 addresses what is truly the ultimate in small-scale phenomena, quantum computation and communication. Quantum networking and quantum computation is a vast topic in its own right; I have had to make some difficult decisions in narrowing down the topic and introducing the concepts as concisely as possible.

By this point in the book, the topic of nanonetworking will have been well defined and various specific nanoscale media will have been explained. The next step is to discuss general techniques for analyzing such systems. First, in Chapter 6, I address information-theoretic aspects of nanonetworks. The information-theoretic aspects focus not only on traditional information and entropy applied to nanoscale systems, but also to relatively recent work on more unique aspects of nanoscale networking such as communication through diffusion and a brief introduction to quantum information. Then, in Chapter 7, architectural issues are discussed. Most readers are familiar with macroscale Internet layered communication protocol stacks and commonly accepted principles of network architecture. We consider how these principles change when we shrink entire networks down to the nanoscale. For example, the active network architecture [1] has been suggested by multiple researchers as a potentially more efficient architectural approach.

Because nanoscale networking is still a revolutionary field, there are very limited resources for simulation and analysis of nanoscale networks. The appendix attempts to address this problem by highlighting relevant research centers and tools that are likely to aid nanoscale network researchers. The appendix includes a review of several molecular simulation platforms in which simulation software can be used to test hypotheses about molecular communication.

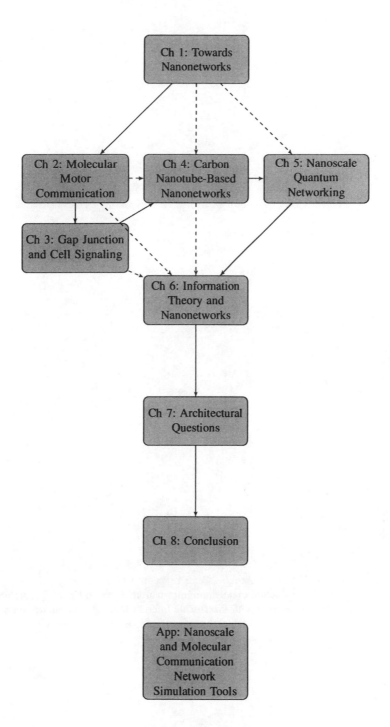

Figure 1 Chapter dependencies for nanoscale communication networks.

Acknowledgments

This book could not have come together without the expertise and mutual support of many dedicated researchers. I extend thanks to Imrich Chlamtac and Gian Maggio, and in particular to the folks involved in the International Conference on Nano-Networks, particularly in the Third conference (Nano-Net 2008). Alexandre Schmid did an outstanding job presiding over the review process used to select the papers to be presented at this conference. I am deeply grateful to the members of the Technical Program Committee for their tremendous effort in evaluating (along with many anonymous reviewers) the submissions and for organizing the papers into sessions, all on a tight schedule. I extend my deepest thanks to the organizing committee, in particular, Alexandre Schmid, as well as the active organizing members, namely, Sasitharan Balasubramaniam, Alexander Sergienko, Nikolaus Correll, Kaustav Banerjee, Radu Marculescu, and Tatsuya Suda. There were three outstanding plenary speakers in Tatsuya, Sylvain Martel, and Neil Gershenfeld, who gave inspirational speeches on a variety of truly fascinating topics regarding networking at the nanoscale. In the panel discussion on "Using Advanced Micro/Nano-electronic Technology to Establish Neuromorphic Systems," panelists Prof. Garrett Rose, Dr. Vladimir Gorelik, Prof. Eugenio Culurciello, Shih-Chii Liu, and Dr. Matthew Hynd shared their perspectives on this growing technology related to the formation of the ultimate nanoscale network, a brain-like system. Yun Li organized an excellent tutorial session for Nano-Networks 2008. I would like to extend my warm thanks to Wei Lu for stepping up to share his ideas and expertise with the community in the form of the first Nano-Network tutorial. Maggie X. Cheng has done an excellent job as publications chair. I also owe a special debt of gratitude to Sanjay Goel and Damira Pon of SUNY for an outstanding and highly professional job in organizing the Nano-Network Workshop and hosting the symposium web site. They responded to literally hundreds of requests for changes and updates to the Web site—all accomplished in a timely and efficient manner. I also extend thanks to Andrew Eckford who has done outstanding work in furthering the information theory aspect of nanonetworking.

Chapter 1

Towards Nanonetworks

Progress, far from consisting in change, depends on retentiveness. When change is absolute there remains no being to improve and no direction is set for possible improvement: and when experience is not retained, as among savages, infancy is perpetual. Those who cannot remember the past are condemned to repeat it.

George Santayana

The goal of this chapter is to examine how the concept of nanonetworking has come into existence, to provide a proper definition for nanoscale networking, to outline some of the driving forces behind this new field, and finally to provide an introduction to the techniques required to advance this exciting field that will be used in the remainder of the book. Having defined the concept of nanoscale networking, it is important to understand the status and trajectory of today's technology and the challenges that will need to be overcome to implement the broader vision of nanoscale networking. As a preview of what is to come, Chapters 2 and 3 describe biologically oriented nanonetwork media. They will require basic knowledge of thermodynamics provided in this chapter and will focus on molecular motor communication as well as gap junction communication and cell signaling, respectively. Chapter 4 focuses on carbon nanotube networks; this will require basic facility with electronics, matrices, and eigenvalues. Chapter 5 addresses quantum communication and networking, which will assume the background provided in this chapter. Chapter 6 addresses information-theoretic aspects of nanonetworks; a basic introduction to the origin of information theory from physics is presented in this chapter. Chapter 7 addresses architectural issues including the notion of self-assembly; consideration is given to potential changes in today's commonly accepted networking principles when entire networks shrink to the nanoscale. All of the remaining chapters will assume at least a cursory understanding of the concepts introduced in this chapter. The reader is urged to attempt the exercises at the end of each chapter; many formulae and equations are explained throughout the text with the goal of mastering the exercises and ultimately being able to transition the material from a cursory understanding to actual analysis and application.

1.1 BRIEF HISTORICAL CONTEXT

An intersection of two worlds, emerging nanotechnologies and network and communication theory, is poised to change the nature of networking. New communication paradigms will be derived from the transition from micro- to nanoscale devices. The related degrees of freedom and constraints associated with these new technologies will change our notions about efficient networks, system design, and the nature of networking. Work is ongoing on a multidisciplinary front towards new techniques in modeling, design,

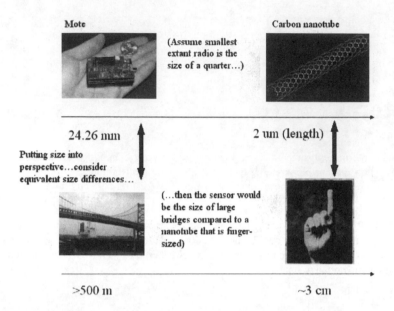

Figure 1.1 Comparison of macro- and nanoscale networking. The size of a wireless mote sensor is to a nanotube as the length of a large bridge (or an Ethernet segment) is to a finger on the human hand.

simulation, and fabrication of network and communication systems at the nanoscale. This section reviews the state of the art and considers the challenges and implications for networking. As specific examples, consider three fundamental manifestations of nanoscale networks: (1) biological networking, (2) nanotube interconnections, and (3) quantum communication.

Networks communicating information already exist on a nanoscale [2]. Interconnected carbon nanotubes, micrometers in length, and nanometers in diameter, convey signals across areas of tens of square micrometers [3]. Wireless transmission and reception among components on a single chip have been designed in [4] and patented in [5]. Consider the impact of the extreme difference in scale between today's networks and nanoscale networks. In Figure 1.1 the size of a wireless mote sensor is to a nanotube as the length of a large bridge (or approximately an Ethernet segment) is to a finger on the human hand. Thus, it is clearly much easier to manipulate and replace components in today's Internet than at the nanoscale.

In terms of nanoscale sensor networks, the network components are on the same scale as the individual molecules of the sensed elements. Management of the complexity of such systems becomes significantly more difficult. The ability to detect and mitigate malicious behavior is thus more difficult. The problems are twofold: (1) the significant increase in the complexity of nanoscale systems due to their larger number of components within a compact space; (2) the mismatch in the size of the networking components, making them individually more difficult to detect and handle.

Solutions from macroscale wide-area networking are being proposed for use in on-chip networks. The implementations for the routers vary widely using techniques of packet or circuit switching, dynamic or static scheduling, wormhole or virtual-cut through routing. The majority of the current router implementations for network-on-chip (NoC) are based on packet-switched, synchronous networks. Some research has proposed an NoC topology and architecture that injects data into the network using multiple sub-NICs (network interface controllers), rather than one NIC, per node. This scheme achieves significant improvements in nanonetwork latency and energy consumption with only negligible area overhead and complexity over existing architectures. In fact, in the case of MESH network topologies, the proposed scheme provides substantial savings in area as well, because it requires fewer on-chip routers.

Another theme that drives research in on-chip networks is the likelihood that production of chips with massive numbers of processing elements and interconnections will increase uncertainty with respect to on-chip properties. Researchers following this theme begin to address issues that will also be of concern in the long-term for self-assembled systems. For example, some links might be so long that communications between processing elements cannot occur in a single clock cycle [6]. In other cases, chip properties might lead to transient, intermittent, or permanent communication errors [7]. Other research considers how to operate a chip when dimensions are so small as to preclude distribution of a reliable clock [8]. Such uncertainty leads researchers to propose various schemes for robust on-chip communications [9–11].

1.2   NANOROBOTICS

Another driver for nanonetworks has been nanorobotics, in which there are two major research thrust areas [12]. The first area deals with design, simulation, control, and coordination of robots with nanoscale dimensions. The second research area focuses on overall miniaturization of mobile robots down to micrometer overall sizes. Nanorobots, nanomachines, and other nanosystems are objects with overall dimensions at or below the micrometer range and are made of assemblies of nanoscale components with individual dimensions ranging approximately between 1 to 100 nm. In these mobile robotic systems, overall system size is very limited, which induces severe constraints in actuators, sensors, and motion mechanisms; power sources, computing power, and wireless communication capability. When scaling down, the surface-to-volume ratio increases and surface forces dominate volume-based forces. At nanometer scales, interatomic forces or surface chemistry plays a significant role in robot mechanics. Thus, inertial forces and weight are almost negligible and micro/nanoscale surface interatomic forces, fluid dynamics, heat transfer, surface chemistry, and adhesion-based contact mechanics and friction dominate robot mechanics. These micro/-nanoscale forces have many different characteristics compared to macroscale ones [13]. Our focus is on information transmission among such nanomachines [14] and whether nanoscale forces have an impact upon the fundamentals of communication. Research into nanorobotics is well under way [15].

1.3   DEFINITION OF NANONETWORKS

Nanonetworks are communication networks that exist mostly or entirely at the nanometer scale. The vision of nanoscale networking achieves the functionality and performance of today's Internet with the exception that node size is measured in nanometers and channels are physically separated by up to hundreds or thousands of nanometers. In addition, nodes are assumed to be mobile and rapidly deployable. Nodes are assumed to be either self-powered or "sprinkled" onto their target location. Individual nodes of such small size may seem too constrained to have any significant functionality. However, quite the opposite is true. The operation of molecular motors has appeared in a wide variety of contexts and with amazingly complex operation. Molecular motors will be discussed in great detail in Chapter 2. Most useful applications of nanoscale networking will require communication in environments where the nodes are in motion, for example, within the complex dynamic environment of living organisms or as nanoscale sensors or robots inspecting sensitive parts in an automated process. Communication should leverage the natural environment with as little disruption as possible.

Nanonetworks are communication networks. As such, they have all features analogous to wireless communication networks and ultimately, the Internet. Information must be collected, coded, transported, forwarded, received, decoded, and delivered to the appropriate application. This entails all the traditional concepts of information theory including compression, bandwidth, error correction, and network architecture. Network architecture includes the manner in which data is physically realized, the construction of channels, access control, the implementation, and allocation of functionality within the system in order to implement a network. A challenging issue involves not only interfacing among nanoscale components, but also between nanoscale and macroscale networks. It can be enlightening to consider the impact that the difference in scale has upon space as illustrated in Exercise 2 and bit rate in Exercise 3.

Humans are extending the notion of traditional networking from both ends, towards the deep space and interplanetary scale [16] and now to the nanoscale. Even within the nanoscale, as will become apparent later, different techniques perform better at different transmission lengths. Thus, it's important to clearly define the terms for the order of magnitude of the scales involved. As humans, we tend to define scale from our own anthropomorphic perspective, so we'll call anything from 10^6m (e.g., Mount Everest) to 10^{-6}m (e.g., bacterium) *human scale*. Anything larger is the *astronomical scale*. From 10^{-6}m down to 10^{-9}m (e.g., the size of a DNA helix or virus), or in other words, down to the length of a single nanometer, is the *cellular* scale, where cellular refers to the lengths typical to biological cells. From 10^{-9}m down to 10^{-15}m (e.g., the size of protons and neutrons) is the *atomic scale*. Finally, below 10^{-15}m (e.g., the size of electrons and quarks) is the *subatomic scale*. It is no coincidence that the nanometer is at the cellular scale and many nanonetworking approaches are attempting to leverage biologically inspired cellular techniques as we will see.

1.3.1   Requirements to be a nanonetwork

Wireless transmission as well as nanoscale transport have existed for at least 3.5 billion years; ever since magnetic storms and lightning have lit the sky, ever since van der Waals forces have existed, and ever since the first living cells transported nucleotide units such as RNA and DNA. The point is that humans do not really "create" anything new; we leverage and fine-tune what nature provides. Wireless communication existed in nature long before man existed. For example, the discovery of magnetic storms may be attributed to Baron Alexander von Humboldt who reported the observation of the deflection in the local magnetic declination at Berlin on December 21, 1806. He also noted simultaneous auroral activity overhead with the magnetic deflection. The general description of a geomagnetic storm in the early days is through the time history of the horizontal or H-component of the magnetic field [17], where the Earth's magnetic field is a vector quantity; at each point in space it has a strength and a direction (to completely describe it we need three quantities: two strength components and an angle (H, Z, D)). In fact, we will see in Chapter 7 how bacteria that are sensitive to the Earth's magnetic field are being harnessed for in vivo nanoscale communication. With regard to the nanoscale, recent research indicates that all prokaryotes (more primitive nonnucleated cellular organisms) have cytoskeletons, albeit more primitive than those of eukaryotes. Eukaryotes differ from prokaryotes in that they have a distinct nucleus. Molecular motors, potentially carrying information, ride along the microtubules that compose the cytoskeleton. Besides homologues of actin and tubulin (MreB and FtsZ), the helically arranged building block of the flagellum, flagellin, is one of the most significant cytoskeletal proteins of bacteria as it provides structural backgrounds of chemotaxis, the basic cellular physiological response of bacteria. The oldest known fossilized prokaryotes were laid down approximately 3.5 billion years ago. So again, we see the ancient workings of complex communication activity. The tremendous complexity of a flagellum is on an amazingly small scale; the diameter of a rotor of a flagellar motor is about 30 nm and its length is about 50 nm [18].

These natural analogs of wireless and nanoscale communication: magnetic storms, lightning, cytoskeletal and flagellar motors, have propagated information before humans existed. One can argue whether what is transported by these phenomena is really "information," but that leads to long philosophical debates about the nature of information and from whose perspective it is defined, which is outside the scope of this book. Humans cannot change the laws of nature, but they can begin to understand them well enough to harness them, as demonstrated by human society's pervasive digital wireless media. From one perspective, human-scale digital wireless communication is most advanced (from a human perspective), because natural concepts have been understood well enough to be transformed to engineered solutions that people are willing to purchase.

The question for nanonetworks can then be framed similarly; at what point does our understanding allow us to transform the natural phenomena of nanoscale communication and networking to an engineered solution that solves useful problems at the human scale? Nanoscale transport already solves the useful problem of keeping us alive at the cellular level, but that is not "engineered" nor is its information transport immediately and fully accessible or controllable at the human scale. More specifically, when engineered information flow can be sent, retrieved, and routed at the cellular level with full control at the human level, a nanoscale network will be born.

1.3.2 Driving forces behind nanoscale networking

The definition of nanoscale networks will also be shaped by its applications; that is, the forces that are driving its development, as well as the techniques to implement it that are being envisioned. Nanoscale communication networks are an inevitable convergence from the human scale above and the cellular and atomic scale below. From above, there is a drive to make sensors and sensor networks smaller. One driver for this is noninvasive or minimally invasive sensing. Sensors inside the body must be small enough not to disrupt organ function. Sensors inspecting equipment must be able to fit inside small corners and crevices.

From the microscopic scale world below, there are many drivers. An early driver has been chip manufacturers trying to fit ever more transistors on on a single chip. Nanorobotics researchers are challenged with coordination and communication among micro- and nanoscale robots. Because there is "plenty of room at the bottom," there is a need for communication and control at this level. Early advances in the sensing component of sensors have led to nanoscale sensors, but a similar reduction in scale of the communications components has been more challenging. Simultaneously, quantum networking is developing at a rapid pace. Quantum networking is currently revolutionizing communications in the macroscopic world with advanced security. Quantum behavior is inherently a small scale phenomenon, and communication among small elements such as quantum dots is an enabler for nanonetworking. Systems biology has been both a driver and an enabler of nanonetworking. As a driver, experimental validation in systems biology requires that information from very small sensors be transmitted and collected. As an enabler of nanonetworking, results from systems biology and network science are benefiting our understanding of very large scale network behavior as well as providing insight into molecular communication.

Nanoscale networking has been driven by several factors, a significant one being the fact that industry is reaching limits regarding the speed of processors that can be placed onto an integrated circuit chip with acceptable properties of power consumption, current leakage, and heat dissipation. This is leading to new multicore architectures, where a multicore processor is an integrated circuit (IC) to which many, sometimes in the hundreds, of processors have been attached for enhanced performance, reduced power consumption, and more efficient simultaneous processing of multiple tasks. A multiple core setup is somewhat comparable to having multiple, separate processors in the same chip. Multicore processing is a growing industry trend as single-core processors rapidly reach the physical limits of possible complexity and speed.

The current means of connecting elements on a chip will prove insufficient as chips advance to include many independent processing elements. This motivates research into various forms of networks on chip (NoC) to connect the processing elements. Another term often used is system on chip (SoC), which refers to the integration of an entire macroscopic device, such as a general-purpose computer, on a single chip. In the short term, current lithography-based approaches will continue to evolve to fabricate chips and only the architecture of the chips will change. However, longer-term at the scale of 22 nanometers and less, current techniques simply cannot be used to produce large-scale integrated circuits. Here, nanonetworking workshops and conferences are providing venues for novel ideas for fabricating computing devices, such as combining self-assembled DNA structures with processing and communication elements based on carbon nanotubes (CNTs).

Source and channel coding as well as cryptography require computational overhead that grow very rapidly with the large scale of nanonetworks, and network processing power is reduced at the nanoscale, because there is limited processing that can be packed into ever-smaller volumes. Given this limitation, more of the computation will have to be done by nontraditional means, perhaps by utilizing network topology as part of the computation.

1.3.2.1   Carbon nanotube networks

Current computer chips are fabricated with lithographic techniques operating at 65 nm with predictions for 45-nm scale chips in 2008 [19]. The industry roadmap predicts that in 2018 feature size will reach 16 nm; however, no currently known process can reliably produce this scale of interconnects in mass quantity [19]. Researchers are now looking towards carbon nanotubes to achieve this objective. Currently, the resulting population of carbon nanotubes (CNTs) is highly variable. This is a basis for considering long-term approaches based on self-assembly of DNA and integration with CNTs.

A CNT is a sequence of carbon atoms (C60), which are arranged into a long thin cylinder with a diameter of approximately one nanometer. The atomic structure of CNTs makes them mechanically strong and the atomic properties lead them to be conductors of electric current. Researchers have used CNTs to construct various electronic components, such as resistors, capacitors, inductors, diodes, and transistors. CNTs, which appear as rolled tubes of graphite (graphene) with walls constructed from hexagonal carbon rings, can be formed into large bundles (much as typical electronic wires can be bundled). CNTs come in two general forms: single-walled (SWNTs) and multiwalled (MWNTs). SWNTs have only a single tube, while MWNTs consist of concentric tubes.

1.3.2.2   Molecular communication

Due to their small size, nanotubes can reach deep into their environment without affecting their natural behavior. For example, a single CNT is small enough to penetrate a cell without triggering the cell's defensive responses. Individual nanotubes can be used to construct a network of sensing elements [20, 21]. The depth and coverage provided by such a network of sensing elements is greater than today's sensor networks.

From a medical standpoint, the use of wireless implants using current techniques is unacceptable for many reasons, including their bulky size, inability to use magnetic resonance imaging after implantation, potential radiation damage, surgical invasiveness, need to recharge/replace power, postoperative pain and long recovery times, and the reduced quality of life for the patient [22–24]. Better, more humane implant communication is needed. Development of both biological and engineered nanomachines is progressing as well; such machines will need to communicate [25, 26]. Unfortunately, networking vast collections of nanoscale sensors and robots using traditional communication techniques, including wireless techniques, is not possible without the communication mechanisms exceeding the nanoscale. One solution is to use randomly aligned nanotubes, as discussed in [27], as the communications media, thus bringing the scale of the communications network down to the scale of the sensing elements. The biological approach has been well established in [28–36].

Sensor coverage will benefit from finding better ways to communicate among smaller sensors. Also, as development in nanotechnology progresses, the need for low-cost, robust, reliable communication among nanomachines will become apparent. Communication and signaling within newly engineered inorganic and biological nanosystems will allow for extremely dense and efficient distributed operation. One may imagine small nanotube networks with functionalized tubes sensing multiple targets inserted into a cell in vivo. Information from each nanotube sensor can be fused within the network. This is clearly distinct from traditional, potentially less efficient, approaches of using CNT networks to construct transistors for wireless devices that would then perform the same task.

Molecular communication enables nanoscale machines to communicate using molecules as a carrier to convey information. Within biological systems communication is typically done through the transport of molecules. This can include a broad spectrum of approaches including intracellular communication through vesicle transport, intercellular communication through neurotransmitters, and interorgan communication through hormones. There are many fundamental research issues including controlling propagation of carrier molecules, encoding and decoding information onto molecules, and achieving transmission and reception of carrier and information molecules. Simply put, the goal is to transmit over tens of micrometers using carrier molecules, such as molecular motors, hormones, or neurotransmitters. Information is encoded as proteins, ions, or DNA molecules. The environment is typically assumed to be an aqueous solution as found within and between biological cells. From an information assurance perspective in such networks, analogy will become reality; computer and network viruses may literally be biological viruses.

1.3.2.3 Solid-state quantum devices

Another approach to nanoscale electronics is to exploit devices based on quantum effects. These include tunneling diodes, single-electron transistors, and quantum dots. It is well-known that quantum devices are sensitive to noise and, if one assumes lithographic techniques for interconnection, would be highly sensitive to lithographic accuracy since quantum devices operate on the scale of one or a few electrons.

Quantum devices will enable the use of quantum cryptographic techniques to improve information assurance at the nanoscale level. The BB84 [37] quantum key distribution scheme developed by Charles Bennett and Gilles Brassard in 1984 is a well-known example. The protocol is provably secure, relying on the quantum property that information gain is only possible at the expense of disturbing the signal if the two states we are trying to distinguish are not orthogonal. Quantum approaches to information assurance are growing rapidly in both macroscale and nanoscale networks.

1.3.3 Defined in relation to sensor networks

To further our definition of nanonetworking, consider that one of the closest familiar engineered systems to a nanonetwork is a sensor network. This is because sensor nodes are small, potentially mobile, devices. From the perspective of traditional computer networking, sensor nodes are resource-limited. This is primarily due to the fact that sensor networks have been designed as evolutionary offshoots of traditional computer networking devices. As of the date this is being written, the smallest wireless sensor has a volume of approximately 1 cm^3 [38]. However, an important point to note is that the sensing element within a wireless sensor node is easily implemented at the nanoscale; carbon nanotubes are often proposed as sensing elements. Thus, the bulk of the volume in a wireless sensor is the communication component size, namely, the radio. Let us compare and contrast what has been learned from sensor networks and how it may apply to nanoscale networks. By our definition of nanonetworking, the entire nanonetwork node will shrink to the same scale as the today's actual sensing element. While the definition of nanonetworks does not require nodes to be sensors, it appears likely that this will be one of the initial primary uses of such devices. The entire network will be at the same scale as the targets being sensed, which will have significant implications to be discussed later.

What can we learn from sensor networks that may apply to nanonetworks? Here a few general concepts are discussed; more details are provided in Section 1.5. First, a few simple observations. When moving from the classical Internet to sensor networks, the size of the host nodes are reduced and more nodes tend to be concentrated into a smaller area; the node density increases [39]. The increase in node density is greater than linear with the change in scale of the size of the nodes. Second, the physical topology of the network tends to be different. The classical Internet assumes an architecture in which nodes are of roughly equal capability, and the protocol is designed such that each node has the ability to communicate with any other node. A sensor network assumes an asymmetric flow, namely, that information is primarily directed from the sensors toward a collection point, although provision is generally included for control and management information to reach the sensors. The Internet topology roughly forms a scale-free structure in which clusters naturally form around important nodes; a few nodes are highly connected, and many more nodes have much fewer connections. The number of connections drops off with a power law distribution (Equation 1.1), where $P(k)$ is the probability of a node having k connections, $k = m, \ldots, K$, m and K are the minimum and maximum connections, c is a normalization factor $c \approx (\lambda - 1)m^{(\lambda-1)}$, and λ is the power-law exponent. A log-scale plot of $P(k)$ versus k often nicely fits a negatively sloping straight line.

$$P(k) = ck^{-\lambda} \tag{1.1}$$

A sensor network topology, on the other hand, will generally look like a star. All paths lead to either one, or perhaps a few, data collector nodes. For nanoscale networks, we must anticipate their use for both node-to-node communication and as sensor networks. However, recall that the definition of nanonetworks requires the ability to handle mobile nanoscale nodes. Thus, the added dimension of motion dynamics needs to be included. The physical properties at the cellular scale differ from that at the human scale and motion patterns may be significantly different as will be discussed in more detail later.

One of the biggest challenges and perhaps one of the most important lessons in scaling down from the classical Internet to the scale of today's sensor nodes was the constraint imposed by power consumption. Power could no longer be taken for granted; batteries are relatively large energy storage devices and difficult to scale. Thus, much research has gone into increasing the energy density of batteries and reducing power consumption by the network nodes. Energy scavenging is a possibility for some of today's sensor networks; it is much more natural in nanoscale networks. Biological systems are masters of energy efficiency, including when applied to information transport, relative to human engineered systems.

Another important lesson from wireless sensor networks, in terms of efficient energy usage, is that there is an optimum distance between hops from the global perspective of energy usage within the sensor network. This is known as the characteristic distance [40]; attempting to transmit too far en route to the destination wastes energy that might have been better spent if the message were relayed between intervening hops. However, attempting to route through too many intervening hops also wastes the accumulated energy required by the intervening nodes. Thus, one can imagine that there is a nice convex function of network energy consumption versus information transmission distance. Given detailed energy parameters, the allocation of transmission among hops such that overall energy consumption is minimized can be determined; more detail is provided in the next section. Nanonetworks are showing an analog to characteristic distance, namely, there are subdivisions of the cellular scale that are optimal for different cellular scale techniques.

One of the biggest differences between wireless sensor networks and nanoscale networks is that there are fundamentally new types of channels. Wireless, or electromagnetic radiation, is still included as a physical layer channel. A wireless radio receiver is nothing more than an electromagnetic sensor. As mentioned previously, sensing elements are relatively easily constructed at the nanoscale. Thus, it is not surprising that a single carbon nanotube radio receiver was among one of the first human-engineered nanoscale communication devices [41]. However, as devices scale down in size, the frequency scales up. Nanoscale devices can be expected to exist at a higher frequency in the electromagnetic spectrum than today's wireless networks. Thus, infrared and optical transmission media will be a natural choice. Free-space optical communication can be used to facilitate a more energy-constrained approach. Random networks of carbon nanotubes may provide a rapidly deployable communication substrate for mobile devices using current flow. However, much is being learned from biological nanoscale transport channels that fundamentally changes the traditional nature of communication channels. Namely, molecular motors physically carry information and waves of information can be diffused through cell signaling.

1.4 REVIEW OF NANOTECHNOLOGY RELATED TO COMMUNICATIONS

Another aspect that leads us toward nanoscale networking is the increasing use of nanotechnology in today's wireless networks. Increasing advances in human-scale networking are occurring due to the application of nanotechnology to network components. It's important to note that, to date, the application of nanotechnology to networking is not focused on reducing network components, but rather improving current human-scale network efficiency.

We begin with a very basic review of nanotechnology. An approach to any new problem in research is to look for nonlinearities and to attempt to leverage them. Here we show some general trends in various properties as the length L is scaled. Mass and volume are significantly impacted by scale as shown in Equation 1.2, and mechanical frequency scales as shown in Equation 1.3. Even with a linear increase in scale, the frequency, and thus bandwidth, will have a useful increase that can be leveraged for more efficient communication. Resistance, in Equation 1.4, also rises roughly linearly. This is significant, particularly for routing on-chip, due to the high resistances involved. From Equations 1.5 and 1.6, it's clear that electromagnetic forces will remain very small.

$$\text{mass} \propto \text{volume} \propto L^3 \tag{1.2}$$

$$\text{frequency} \propto \frac{\text{speed}}{\text{length}} \propto L^{-1} \tag{1.3}$$

$$\text{resistance} \propto \frac{\text{length}}{\text{area}} \propto L^{-1} \tag{1.4}$$

$$\text{magnetic field} \propto \frac{\text{current}}{\text{distance}} \propto L \tag{1.5}$$

$$\text{magnetic force} \propto \text{area} \times \text{magnetic field}^2 \propto L^4 \tag{1.6}$$

1.4.1 Nanotechnology for high-frequency classical wireless transmission

These scale factors are currently improving existing wireless communications. Radio frequency operation at high frequencies (gigahertz frequencies and above) requires high clock speeds and fast processing for physical and media access control layer processing [42]. Nanoelectromechanical systems (NEMS) include nanoscale resonators that are ideal for high-speed applications. Providing the ability to perform efficient spectral sensing enables more advanced cognitive radio systems. Many of the required components for a nanoscale radio system already exist in some form at the nanoscale, such as sensing, processing unit, transceiver, and power unit constructed with nanoscale devices such as carbon nanotubes.

Following the trend in today's technology for reduction in transistor size, the limits will be reached in 2015. Complementary metal oxide semiconductor (CMOS) technology is expected to have a density of 10^{10} devices per cm^2 by 2020 with switching speeds of 12 THz, circuit speeds of 61 GHz, and requiring a switching energy of 3×10^{-18} J. New approaches based upon nanotechnology should meet or exceed these expected values. In addition to increased computing speed to more efficiently handle higher frequencies, denser and more energy-efficient memory will be available as well.

Many of the current wireless sensor networking challenges that are being addressed by algorithms at the network layer are also being addressed at the physical layer through device performance improvements enabled by nanotechnology [43]. For example, with respect to power storage, nanotubes have demonstrated significant improvement in power density and charge/discharge rates. Nanotubes can replace the typical graphite within lithiumgraphite electrodes. Due to their high surface area, nanotubes incorporate more lithium than graphite. With open, single-walled nanotubes, capacities of up to 640 ampere hour per kilogram have been demonstrated. In addition, nanocrystalline materials can be used as separator plates in batteries because of their foamlike structure, which can hold considerably more energy than conventional plates. These batteries will require less frequent recharging and last longer because of their large grain boundary (surface) area. Related to energy storage are supercapacitors. These are electrical storage devices that have a relatively high energy storage and power density. They can provide short term energy storage in power electronics. A supercapacitor is formed by exponentially increasing the surface area using carbon nanotubes. Carbon nanotubes have a narrow distribution of mesopore sizes, highly accessible surface area, low resistivity, and high stability.

As previously mentioned, the sensing element of a sensor node is one of the most easily reduced components. There are many nanotechnology-based sensing elements. Here only a few are given as examples. The world's smallest "balance" has been created by taking advantage of the unique electrical and mechanical properties of carbon nanotubes. A single particle is mounted on the end of a nanotube and an electrical charge is applied to it. The nanotube oscillates without breaking and the mass of the particle is calculated from changes in the resonance vibrational frequency with and without the particle. This approach may allow the mass of individual biological molecules to be measured. Many nanoscale biological sensors have been developed. Nanotubes are coated with analyte-specific entities such as antibodies for selective detection of complex molecules. DNA detection of nanoscale particles is being detected. A wide variety of nanoscale chemical sensors are begin developed. Chemical sensors based on individual single-walled nanotubes have been demonstrated. When exposed to gaseous molecules such as NO_2 or NH_3, the electrical resistance of a semiconducting nanotube will dramatically increase or decrease depending upon the target. This has served as the basis for many nanotube molecular sensors.

Sensor networks will benefit from increased processing capability enabled by nanoscale advances. IBM researchers have successfully created transistors from nanotubes since 1998. Ten carbon atoms formed into a cylinder are 500 times smaller than today's smallest silicon-based transistors. These new approaches will keep Moore's law in effect beyond its 10- to 20-year remaining life span that scientists have been predicting. Nanotube transistors use individual multiwalled nanotubes and single-walled carbon nanotubes as the channel of a field-effect transistor (FET). The change of current flowing through the nanotube can be changed by a factor of 100,000 by changing the voltage applied to a gate. The new nanotube FETs reported as far back as 2002 work even better in terms of the switching rate and the amount of current they can carry per width of conductor. Electrons can flow through nanotubes ten times faster than they can in circuits made using silicon, and nanotubes can carry 100 times the current and dissipate 20 times less the heat of circuits made with silicon. Nanotubes in transistors can also amplify about 20 times more current than conventional silicon-based transistors.

Another sensor component to benefit from nanotechnology is the antenna. RF nanowire and nanotube antennae have outstanding properties. A nanotube can be used as an antenna to couple nanosensors by wireless interconnects foregoing the need for lithographically fabricated electronics. A nanotube antenna differs significantly from a typical thin-wire antenna because of the much larger inductance, on the order of 104 times the inductance of a thin-wire antenna. This translates into performance predictions that are substantially different from thin-wire antennas because the wavelength of the current excitation is 100 times smaller than the wavelength of the far-field radiation. An advantage of nanotube antennas is that the nanotube can serve as an excellent impedance matching circuit to receive signals from free-space propagation to high-impedance devices.

Of course, memory will benefit significantly from nanoscale components. Network queues and buffers are significant components of network devices. Scientists have demonstrated a data storage density of a trillion bits per square inch–20 times higher than the densest magnetic storage available today. The technique uses a mechanical approach employing thousands of sharp tips to punch indentations representing individual bits into a thin plastic film, seeming much like the old-fashioned card readers, except on a nanoscale, but with two significant differences, namely (1) the technology is rewriteable (meaning it can be used over and over again), and (2) it may be able to store more than three billion bits of data in the space occupied by just one hole in a standard punch card. Another recent advance in memory is from a U.S. company called Nantero. Using nanotubes sprinkled onto a silicon wafer, they have created a device using primarily standard chip production techniques, called Nano-RAM, which works by balancing nanotubes on ridges of silicon. Under an electrical charge differential, the nanotubes can be physically tilted into one of two positions representing one and zero. Because the tubes are so small, this movement is very fast and requires very little power. The tubes are a thousand times as conductive as copper, thus it is very easy to sense their position and read the data. The tubes remain in position until there until they are reset by a signal.

This section has presented only a small sample of the manner in which nanoscale advances are impacting communication and sensor networks in particular. It is important to note that all of the applications of nanotechnology in this section have been designed with the intent of increasing the performance of human-scale networks. The goal of nanonetworks is to take a radical leap into networks of nodes composed entirely at the cellular or nanoscale. In what follows, consider the processor within today's sensor node. Deep within the processing element lies a tiny, human-engineered, communication network. In fact, one can view nanonetworking as turning the sensor node inside out; that is, utilizing the tiny network that exists inside today as a general purpose communication mechanism.

1.4.2   System-on-chip

A significant driver for early research in nanoscale networking has been the development of nanoscale networks on a chip, the goal being to improve the routing of information, or interconnection between subcomponents, within a single chip. As individual components continue to move to smaller scales and more components are placed upon a single chip, the current approach of placing wires to interconnect all the components becomes ever more complex. New approaches are actively being examined, and some of them are aligned with the direction of nanoscale communication.

In the current technology market, many intellectual property (IP) cores, or proprietary subcomponents, can be integrated, as modular entities, on a single chip. As the component density increases, interconnection and testing becomes more complex. Traditional hard-wired interconnects face severe fundamental limitations. Many alternatives have been proposed, including carbon nanotube and wireless radio interconnects [44]. Tiny receivers, transmitters, and on-chip zigzag antennae are implemented in 0.18-μm CMOS technology with area consumption of 0.116, 0.215, and 0.15 mm^2, respectively [45]. For a 2.5-cm microprocessor, the total area with one transmitter, 16 receivers, and 17 antennae consumes about 1% of the total area. Unlike the traditional "passive" metal interconnect, the "active" RF/wireless interconnect is based on low-loss and dispersion-free microwave signal transmission, near-field capacitive coupling, and even typically used communication network multiple-access algorithms [4].

The decrease in minimum feature size of devices has led to a proportional decrease in interconnect cross-sectional area and pitch, where pitch is the sum of the interconnect diameter and the distance between the next interconnect. The parasitic resistance, capacitance, and inductance associated with interconnects negatively influence circuit performance and are increasingly becoming more serious in the evolution of deep submicrometer ULSI technology. Improvements in the conventional interconnect technology have occurred by reducing the resistivity of conductors (using copper) and reducing the dielectric constant of interlayer dielectric materials by using low-κ (the ratio of the field without the dielectric (E_o) to the net field (E) with the dielectric) polymers. The larger the dielectric constant, the more charge that can be stored in a capacitive manner and the greater the capacitive coupling and interference.

However, the efficient transmission and reception of RF/microwave signals in free space requires the size of antennas to be comparable with their wavelengths as we have already discussed. As the CMOS device dimensions continue to scale down, operating speeds and cutoff frequencies of CMOS devices will exceed 100 GHz in the near future. In physics and electrical engineering, a cutoff frequency, corner frequency, or break frequency is a boundary in a system's frequency response at which energy flowing through the system begins to be reduced (attenuated or reflected) rather than passing through. But, even at this frequency, the optimal aperture size of the antenna is on the order of 1 mm, which is too large to be comfortably implemented in ultra large-scale integration. As a receiver, the antenna aperture can be visualized as the area of a circle constructed broadside to the incoming radiation where all radiation passing within the circle is delivered by the antenna to a matched load. Transmitting and receiving are reciprocal, so the aperture is the same for both. Thus incoming power density (watts per square meter) multiplied by the aperture (square meters) yields the available power from the antenna (watts). Antenna gain is directly proportional to aperture. An isotropic antenna has an aperture of $\frac{\lambda^2}{4\pi}$ where λ is the wavelength. An antenna with a gain of G has an aperture of $\frac{G\lambda^2}{4\pi}$. Generally, antenna gain is increased by directing radiation in a single direction, while necessarily reducing it in all other directions since power cannot be created by the antenna. Thus a larger aperture produces a higher gain and narrower beamwidth. A technique used on chips, since the communication distance is relatively short (several centimeters apart), is to utilize much smaller "near-field" capacitive couplers rather than utilize larger "far-field" antenna.

An interesting point, from a networking perspective, is that typical LAN and Internet protocols are being implemented to do on-chip communication. Imagine that chips being used to implement the human-scale Internet contain within them a small-scale Internet. Code division multiple access (CDMA) and/or frequency division multiple access (FDMA) algorithms can be used effectively within the chip to alleviate undesired cross-channel interference within the shared medium.

This leads to the point that as devices are scaled down to nanoscale dimensions, an issue that is becoming increasingly important is that of fault tolerance. Rather than perfecting nanotechnology, where fabrication of on-chip routing becomes exorbitant, instead have inexpensive fabrication technology with potentially low yield. One can then use the architectural flexibility to program a virtual machine on the resulting hardware by first testing and followed by reconfiguration. The key underlying enabling technology is the design of a flexible interconnection network, with a flexibility approaching that of an ad hoc network.

The main takeaway from this discussion of chip-level networking, both wired and wireless, is not that this is leading to the ultimate nanonetwork technology, but rather, it provides some indication of the state-of-the-art and, more importantly, it provides some lessons learned with regard to small-scale communication.

1.5 TODAY'S SENSOR NETWORKS

Since we have identified sensor networks as being the closest analog to nanoscale networks, this section reviews sensor networking in more detail. Applications of wireless sensor networks are well-known and wide-ranging, although the transition from academia to industry has been slower than many would have expected. The current driving forces on sensor network design are, not surprisingly, low cost and low power. Reduction in power directly reduces cost, but it also has a greater secondary impact on cost reduction by reducing the need for maintenance costs. These drivers on sensor network design are due primarily to the marketing of sensing technology to grab the relatively easier and lower-hanging fruit of human scale applications such as industrial control, home automation and monitoring, military and security applications, asset tracking, and human-scale medical monitoring. Radical reduction in size will open up entirely new markets and pervasiveness.

1.5.1 Wireless sensor networks

Today's wireless sensor networks (WSN) are comprised of microsensors coupled with wireless communication capability. They are designed to be relatively cheap with the potential to be deployed in large quantities. An abstract view of the generic node components is shown in Figure 1.2. The potentially nanoscale sensing element is inside the transducer box in the diagram. The user interface contains any sensor node management control that is required. Energy scavenging is of course optional.

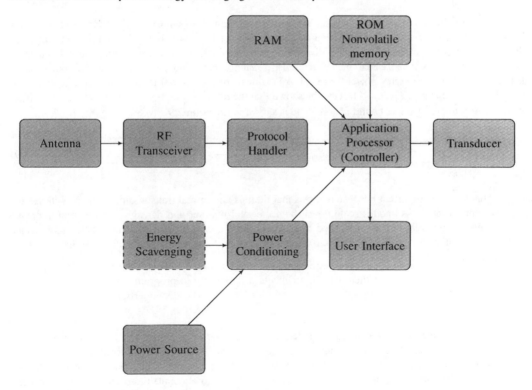

Figure 1.2 A generic wireless sensor network node.

Regarding network energy consumption, as was previously mentioned, intervening nodes can act as relays to ease the power consumption of the source node. Let us examine this in more detail. One may wish to allocate relay nodes in order to minimize the overall energy consumption. Following the results of [40], assume there are $K - 1$ relays (hops) from source to destination. The total rate of power dissipation (remember energy is power multiplied by time) is $P_{link}(D) = -\alpha + \sum_{i=1}^{K} P_{relay}(d_i)$. The negative α term adjusts for the fact that the transmitter does not need to spend any receive energy. D is the total hop distance and the d_i are the individual relay hops distances. We can estimate the total link power required P_{link} from the sum of the transmit energy, receive energy, and any energy needed to sense a bit for receiver wakeup. Transmit energy is $\alpha_{11} + \alpha_2 d^n$ where d^n assumes a $1/d^n$ path loss. α_{11} is specifically the energy/bit used in the transmitter electronics, α_2 is the energy dissipated in the transmitter operational amplifier. The receiver energy consumption is simply α_{12}, which is the energy per bit consumed by the receiver electronics and α_3 is the energy/bit for sensing a bit. The total power required by such a link is then shown in Equation 1.7, where r is the bit rate in bits per second.

$$P_{link}(d) = (\alpha_{11} + \alpha_2 d^n + \alpha_{12})r \qquad (1.7)$$

The transmit and receive power can be combined into $\alpha_1 = \alpha_{11} + \alpha_{12}$, yielding Equation 1.8.

$$P_{link}(d) = (\alpha_1 + \alpha_2 d^n)r \qquad (1.8)$$

Because $P_{relay}(D)$ is convex, Jensen's inequality can be used to show that, given the number of hops $K - 1$, the power is minimized when the hops are evenly spaced, $\frac{D}{K}$, which makes intuitive sense as well; all node parameters being equal, the relay distances should be equal. What is more interesting is that there is an optimal number of hops based upon a characteristic distance that is independent of the total transmission distance d_{char}. Since the relay power for each link is known, the total power required to transmit a distance D with $K - 1$ hops is $K P_{relay}(\frac{D}{K})$. It can be shown that the total minimum power is achieved when K_{opt} is the greatest upper or lower bound of $\frac{D}{d_{char}}$ and is shown in Equation 1.9.

$$d_{char} = \sqrt[n]{\frac{\alpha_1}{\alpha_2(n - 1)}} \qquad (1.9)$$

This result is important because it shows that there is an optimal transmission distance with regard to power consumption. As nodes scale down, both the total number and density of nodes may increase dramatically although there may be limited control over their precise placement. The question to consider is whether the other forms of nanoscale channels that we will discuss in later chapters also have energy-optimal transmission distances.

1.5.2 Antenna

A significant problem occurs regarding the antenna when the sensor is reduced in size and current radio frequencies are desired. When the antenna is reduced in size and still used at relatively low frequencies, the free-space wavelength is larger than the physical size of the antenna. This creates what is called an *electrically small* antenna. If the size mismatch is not taken into account, these antennae can be very inefficient; a significant amount of power may be dissipated as heat instead of desired electromagnetic radiation.

Using a Thevenin equivalent circuit for the antenna circuit, antenna efficiency is shown in Equation 1.10 [46].

$$\eta = \frac{R_{rad}}{R_{rad} + R_{loss}} \tag{1.10}$$

Here, R_{rad} is the radiation resistance of the antenna in ohms and R_{loss} is the loss resistance of the antenna in ohms. The radiation resistance of an antenna is the ratio of the power radiated by the antenna (watts) to the square of the rms current at the antenna feed point (amperes) as shown Equation 1.11.

$$R_{rad} = \frac{P_{rad}}{I^2} \tag{1.11}$$

Maxwell's equations can be applied to a very small length of antenna to determine the behavior of that small increment, then integrated over the length of the antenna, to obtain the behavior of the entire antenna. As a result, the derivation gives the radiation resistance of a small (less than a quarter wavelength) dipole antenna as shown in Equation 1.12.

$$R_{rad} = 20\pi^2 \left(\frac{L}{\lambda}\right)^2 \tag{1.12}$$

Here L is the dipole antenna length (meters) and λ is the free-space wavelength (meters). See Exercise 5 for further work with this concept. From Equation 1.10, the R_{loss} should be small relative to R_{rad} for better efficiency. For a given length antenna, as the frequency increases, the electrical length of the antenna increases, which raises R_{rad} relative to R_{loss}. Also, as the frequency is reduced, the input impedance of an electrically small antenna increases, which again reduces the efficiency. Thus, classical wireless sensor network designs are enticed towards higher free-space frequencies. Poor antenna efficiency reduces the transmission range of the wireless node and can only be overcome by using more power or improving receiver sensitivity, both of which consume more power.

Finally, another important consideration is the maximum instantaneous fractional bandwidth of an electrically small antenna. It is the ratio of the bandwidth to the center frequency as shown in Equation 1.13.

$$BW = \frac{2\left(\frac{2\pi d}{\lambda}\right)^3}{\eta} \tag{1.13}$$

Here BW is the maximum instantaneous fractional bandwidth, $2d$ is the maximum dimension of the electrically small antenna, where d is the diameter of the smallest sphere that can completely contain the antenna. The relevant point here is that as the efficiency η increases, the bandwidth decreases. As efficiency increases, the bandwidth narrows and the need for tuning becomes more important. Thus, as if we are to naively shrink current wireless technology down the nanoscale, there are many complex factors that need to be considered.

1.6 RELATIONSHIP BETWEEN PHYSICS AND INFORMATION

Nanonetworks are a radical leap from small-scale networking on a chip. The environment on a chip is very friendly compared to the "real world." Nodes on a chip are not in motion, the use of the chip can be rigorously specified to avoid problems, and input parameters are tightly controlled. Cellular scale and smaller sensor nodes and their unique communication channels (where other communication channels may be just as good or better than traditional wireless communication) will be at a similar scale with the smallest components of matter (individual molecules) that they may be sensing. At this scale, physics and information will become more united. Thus, this section focuses on the physics of information [47].

1.6.1   Physical entropy

Finding fundamental physical invariants and relationships that relate the human-scale world with the microscopic- and cellular-scale world has been a goal of physics since its earliest days. Reconnecting physics with information has been an ongoing effort. Beginning with some of the most elementary principles, we consider the relationship between physics at the cellular and smaller scales, physics at the human and larger scales, and how these are being related to information and ultimately, from our point of view, to the transport of information at the cellular and lower scales.

An elementary, but important invariant in relating the microscale with the macroscale is the Avogadro constant. In 1811, Amedeo Avogadro first proposed that the volume of a gas (at a given pressure and temperature) is proportional to the number of atoms or molecules. It wasn't until 1865 that the value of the Avogadro constant was first estimated by Johann Josef Loschmidt. Today, the value Avogadro's constant (N_A) is $N_A \approx 6.02 \times 10^{23}$/mol where mol is the unit for a mole. Strictly speaking, the mole today is defined as the amount of substance that contains as many "elemental entities" (e.g., atoms, molecules, ions, electrons) as there are atoms in 12g of carbon-12. Avogadro's constant gives us a sense of the number of atoms, molecules, or elementary particles that are present in a mole of substance. What is interesting here is the use of the term "elementary particle." Elementary means that they cannot be broken down further without completely changing their nature: If divided further, the physical, chemical, electrical properties would no longer be the same; they would become a different substance. Thus, as communication networks approach these small scales, they may exist at the edge of what makes a substance unique. In fact, quantum dots have been suggested as a mechanism for creating artificial atoms that can dynamically change from one substance to another [48]. However, one of the biggest challenges in creating such synthetic matter is the communication of control signals to each and every quantum dot at such small scales, thus a need for nanoscale networking.

Another early bridge between the microscopic and macroscopic scales has been the ideal gas law $pV = nRT$, where where p is the absolute pressure of the gas, V is the volume of the gas, n is the number of moles of gas, R is the universal gas constant, and T is the absolute temperature. The parameters p, V, n, and T are measurable quantities while R is an invariant that allows the equation to be universally true. $R = 8.314472 J \cdot \text{mol}^{-1} \cdot K^{-1}$, where J is energy, Joules, where mol is moles again, and K is temperature. The ideal gas law and the universal gas constant have been widely used since it was first stated by Benoit Paul Emile Clapeyron in 1834. The universal gas constant, while extremely useful in its own right, also leads us to the next fundamental tie between the microscale and the macroscale, Boltzmann's constant.

Boltzmann's constant is defined in terms of the constants previously discussed, $k_B = \frac{R}{N_A}$. It relates the energy at the microscopic scale with temperature observed at the macroscopic scale. Its value is $1.3806504 \times 10^{-23} J \cdot K^{-1}$. The Boltzmann constant transforms the ideal gas law into an equation about the microscopic properties of molecules, $pV = NkT$ where N is the number of molecules of gas. It tells us about the energy carried by each microscopic particle and its relation to the overall temperature. Notice that kT has units of Joules (energy); kT appears in many fundamental equations relating the micro- and macroscopic scales.

Boltzmann went much further in linking the microscopic and macroscopic scales. He first linked entropy and probability in 1877, which led to Boltzmann's entropy. Entropy provides a measure of the relative number of remaining ways that states can be allocated after its observable macroscopic properties have been taken into account. For a given set of macroscopic variables, such as temperature and volume, entropy measures the degree to which the probability of the system is spread out over different possible microscopic states. The more states available to the system with appreciable probability, the greater the entropy. More generally, and more to the point of nanoscale communications, entropy is as a measure of any remaining uncertainty about a system after some macroscopic properties are known. The equilibrium state of a system maximizes the entropy because we have lost all information about the initial conditions except for the conserved variables; maximizing the entropy maximizes our ignorance about the details of the system.

The function shown in Equation 1.14 quantifies this notion of uncertainty once all probabilities are known. S is the conventional symbol for entropy in statistical mechanics; information theoreticians have borrowed the concept and labeled it H as we'll discuss later. The sum runs over all microstates i consistent with the given macrostate and P_i is the probability of the ith microstate.

$$S = -k_B \sum_i P_i \ln P_i \qquad (1.14)$$

When the probability of being in any microstate is equal for all microstates, then uncertainty is greatest, and the equation reduces to that shown in Equation 1.15, where Ω is the number of microstates corresponding to the observed thermodynamic macrostate.

$$S = k \ln \Omega \qquad (1.15)$$

Another important point about entropy, which will be used in later chapters, is the relation between entropy and heat. As discussed earlier, kT shows up frequently as a measure of energy. Notice in Equation 1.14 that if we consider the portion that measures uncertainty, namely, $\sum_i P_i \ln P_i$ to be analogous to temperature, then S is analogous to energy.

When heat is added to a system at high temperature, the increase in entropy is small. When heat is added to a system at low temperature, the increase in entropy is great. This can be quantified as follows: In thermal systems, changes in the entropy can be ascertained by observing the temperature while observing changes in energy. This is restricted to situations where thermal conduction is the only form of energy transfer (in contrast to frictional heating and other dissipative processes). It is further restricted to systems at or near thermal equilibrium. We first consider systems held at constant temperature, in which case the change in entropy ΔS is given by Equation 1.16.

$$\Delta S = \frac{\Delta q}{T} \qquad (1.16)$$

In this equation, Δq is the amount of heat absorbed in an isothermal and reversible process in which the system goes from one state to another, and T is the absolute temperature at which the process is occurring.

Finally, Boltzmann's entropy plays an important role in determining the useful energy of a system. Loosely speaking, when a system's energy is divided into its "useful" energy (energy that can be used, for example, to push a piston), and its "useless energy" (that energy which cannot be used to do external work), then entropy can be used to estimate the "useless" energy, which depends on the entropy of the system. The Gibbs free energy is a measure of this quantity. It is the maximum amount of nonexpansion work that can be extracted from a closed system; this maximum can be attained only in a completely reversible process. When a system changes from a well-defined initial state to a well-defined final state, the Gibbs free energy ΔG equals the work exchanged by the system with its surroundings, minus the work of pressure forces, during a reversible transformation of the system from an initial state to the *same* final state. As we'll see later, it depends upon Boltzmann's entropy.

As network devices scale down to have the size of, and utilize the energy of, the individual microstates that we've discussed, the individual microstates themselves become more important relative to the macroscale. For example, thermal noise becomes more significant. Also, we can't help but notice the relationship between physical microstates and information and between energy and communication.

Microstates, whether defined as magnetic orientation, energy states, or quantum states, are information; coding and communicating that information requires energy. Finally, this leads us to a discussion of thermodynamics, which plays a large role in our understanding of molecular motors.

1.6.2 Thermodynamics

What follows is very brief primer on thermodynamics required for understanding the relationship with information and molecular communication. There is a small set of fundamental laws. The first law essentially states that heat is a form of energy and energy is conserved. Thus, any heat applied to a system will affect its internal energy dU by either being absorbed and increasing the heat dQ, or the heat will be converted to work dW as summarized in Equation 1.17.

$$dU = dQ - dW \qquad (1.17)$$

The second law of thermodynamics brings us back to entropy. The total entropy of any isolated thermodynamic system tends to increase over time, approaching a maximum value. Heat will flow from hot to cold reaching thermal equilibrium. The flow is irreversible; there is always an energy cost involved in reversing this direction. A simple example of this cost is the Carnot cycle, in which heat Q_1 may be absorbed expanding a gas to do work, then the gas cools, releasing an amount of heat Q_2. The work done is shown in Equation 1.18. However, the change in entropy ΔS is as shown in Equation 1.19 where the inequality is due to the second law. Thus, the work done is not precisely $W = Q_1 - Q_2$ as would be expected, but some slightly smaller amount. Thus, there is no machine that can convert heat to work without loss.

$$W = \int_{cycle} PdV \qquad (1.18)$$

$$\Delta S = \frac{-Q_1}{T} + \frac{Q_2}{T} \geq 0 \qquad (1.19)$$

This law has an information counterpart, the Landauer principle, which will be discussed in more detail later. The information concept is that there is no machine that can erase information without loss. We will see that at least $k_B T \ln 2$ energy is required to erase one bit of information.

Finally, the notion of free energy is important. This is the amount of energy available to do work. It is defined as $F \equiv U - TS$, where F is the free energy, U is the internal energy, T is the temperature, and S is the entropy. Going back to the first law in Equation 1.17, the internal energy can be defined in:

$$dU = TdS - PdV \qquad (1.20)$$

Defining free energy in differential form, we have:

$$dF = dU - TdS - SdT \qquad (1.21)$$

Using Equation 1.20, we have:

$$dF = -PdV - SdT \qquad (1.22)$$

Thus, the work that can be done is less than or equal to the negative free energy, $-dF$.

1.6.3 Physics of information

Now we can begin to relate thermodynamics and information. Although we did not specifically derive it, the Gibbs entropy (notice that there is no k_B as in the Boltzmann entropy) in Equation 1.29 can be derived from the equation for free energy as shown in Equation 1.23. This can be done by a combinatorial approach in which the number of possible microstates is used to compute the probability of each state.

$$S = \frac{U - F}{T} \tag{1.23}$$

Consider a system of spinning magnetic particles within a magnetic field; this is how electrons were modeled, which we will briefly describe. The manner in which this system has been evaluated has been applied to many applications beyond physics, from error correction in communications to image restoration, including Markov random fields. The main concept is to find the minimum energy configuration, given a large number of possible configurations and an energy function. From this we can derive the Gibbs distribution as we will show shortly. In addition, the underlying mechanics may be applicable to self-assembly, a critical component of nanoscale networks. If we assume a simple two state (spin up/spin down) system, there are 2^N states with k up and $N - k$ down. The total energy of the system is $\epsilon_m = 2m\mu H$ where H is the magnetic field in which the system resides and is assumed to be pointing upward, and μ is the magnetic moment of one of the "spinning" particles, which is a measure of the magnitude and the direction of its magnetism. If we assume k spins pointing up and $N - k$ pointing down, then $m = \frac{(N - 2k)}{2}$. The number of microscopic combinations is then shown in Equation 1.24.

$$\frac{N!}{(\frac{1}{2}N + m)!(\frac{1}{2}N - m)} \tag{1.24}$$

We can use Sterling's formula for the factorials as shown in Equation 1.25.

$$n! = \sqrt{2\pi n}\frac{n^n}{e} \tag{1.25}$$

This results in Equation 1.26.

$$2^N \left(\frac{2}{\pi N}\right)^{\frac{1}{2}} e^{\frac{-2m^2}{N}} \tag{1.26}$$

Maxwell and Boltzmann had to make some assumptions regarding the interaction among particles, namely the conservation of energy, to yield Equation 1.27.

$$p_i = \frac{e^{-\frac{\epsilon_i}{T}}}{Z} \tag{1.27}$$

In order for probability to be conserved, $\sum_i p_i = 1$. Therefore, Equation 1.28 must hold.

$$Z \equiv \sum_i e^{-\frac{\epsilon_i}{T}} \tag{1.28}$$

Z is known as the partition function. The distribution in Equation 1.27 is important in both physics and information theory. It is known by several names including the Gibbs distribution, Boltzmann distribution, Boltzmann-Maxwell distribution, and Boltzmann-Gibbs distribution. It plays in important role related to information theory, via Markov random fields, to be discussed later in this chapter.

From the partition function $F = -T \ln Z$ and from Equation 1.27, $F = \epsilon_i + T \ln p_i$. We also know that at equilibrium, the internal energy, U can be described as the mean energy, $\sum_i \epsilon_i p_i$. Inserting these values into Equation 1.23 yields the familiar Gibbs entropy shown in Equation 1.29.

$$S = -\sum_i p_i \log p_i \tag{1.29}$$

Although Gibbs entropy can be derived through physical relationships, with some careful rationalization, it is applied to information forming the bases of information theory. Define self-information as a measure of the information content associated with the outcome of a random variable. It is expressed in a unit of information, for example bits, nats, or hartleys (also known as digits, dits, bans), depending on the base of the logarithm used in its definition. By definition, the amount of self-information contained in a probabilistic event depends only on the probability of that event: The smaller its probability, the larger the self-information associated with receiving the information that the event occurred. Suppose an event C is composed of two mutually independent events A and B, then the amount of information when C is "communicated" equals the sum of the amounts of information at the communication of event A and event B, respectively. Taking into account these properties, the self-information $I(\omega_n)$ (measured in bits) associated with outcome w_n is:

$$I(\omega_n) = \log_2 \frac{1}{p(\omega_n)} \tag{1.30}$$

If the probability of the event occurring is low, then the information that it actually did occur is high. Information entropy is a measure of the uncertainty associated with a random variable. The term by itself in this context usually refers to the Shannon entropy, which quantifies, in the sense of an expected value, the information contained in a message, usually in units such as bits. Equivalently, the Shannon entropy is a measure of the average information content one is missing when one does not know the value of the random variable. Thus, in Equation 1.31, the expected value of the surprisal is shown. A slight rearrangement returns the value in Equation 1.32. Here, b is the base of the logarithm and is generally two for binary systems.

$$H(X) = E(I(X)) = \sum_1^n p(x_i) \log_b \frac{1}{p(x_i)} \tag{1.31}$$

$$H(X) = E(I(X)) = -\sum_1^n p(x_i) \log_b p(x_i) \tag{1.32}$$

Clearly, there is a strong relationship between entropy and predictability. As we scale down, individual microstates will become more significant and systems less predictable. Many of the concepts used in computing the number of microstate combinations now play a role in information theory. Markov random fields (MRF) evolved from the simplifying assumptions in computing the probabilities of microstates and is now used widely in various aspects of information theory related to both source and channel coding and image processing. Nearly direct implementations of MRFs are being researched for networking [49, 50]. Since the MRF links microstates to the macrostate, it is a natural application for nanoscale network architectures and will be discussed in more detail later. Entropy, as we have seen, is a direct crossover to information theory.

1.6.4 A brief introduction to quantum phenomena

Now we come to a brief introduction leading to some of the tools necessary to work with quantum theory and quantum networking. Only the most basic outline is presented here in order to get the reader up to speed quickly; the reader is encouraged to go deeper into other aspects on their own. The goal for this section is simply to have a general idea behind the Dirac notation, which will be further developed in Section 5.2, which is focused on quantum networking and is much simpler to work with than the content here.

We begin with the Planck constant, which is the proportionality constant between the energy (E) of a photon and the frequency of its associated electromagnetic wave (ν). The scale of the quantum regime is set by Planck's constant, which has dimensions of *energy* $\times$ *time* (or equivalently *momentum* $\times$ *length*). It is extremely small in ordinary units: $\hbar = 1.05 \times 10^{-34} Joule \cdot seconds$. This is one reason why quantum properties only manifest themselves at very small scales or very low temperatures. One has to keep in mind, however, that radically different properties at a microscopic scale (say at the level of atomic and molecular structure) will also lead to fundamentally different collective behavior on a macroscopic scale. The basic quantum of energy is $\epsilon = \hbar\nu$, where ν is frequency. The critical observation of Planck was that energy is not continuous, it can only come in discrete amounts $E = n\epsilon = nh\nu$, for $n = 0, 1, 2, \ldots$. Note that $h = \frac{\hbar}{2\pi}$, where $\hbar$ is called the reduced Planck constant.

Light appeared to have properties of both particles and waves; the Planck-Einstein equation brought together the notion of both momentum and wave frequency in Equation 1.33, where p is the photon particle momentum and k is the angular wave number, which is simply $\frac{2\pi}{\lambda}$.

$$p = \hbar k \tag{1.33}$$

In 1923, Louis de Broglie suggested that *all* matter obeys the Planck-Einstein relations. Thus, any particle with energy E has an associated wave of frequency $\frac{E}{h}$ and momentum is related to the wave number by $p = hk$. Thus all matter with momentum p has an associated wave length $\lambda = \frac{h}{p}$.

The behavior of a particle of mass m subject to a potential $V(x, t)$ is described by the Schrödinger Equation 1.34. $\Psi(x, t)$ is the wavefunction. The square of the wavefunction is a probability density regarding the location of a particle.

$$ih\frac{\partial \Psi x, t}{\partial t} = -\frac{h^2}{2m}\frac{\partial^2 \Psi(x, t)}{\partial x^2} + V(x, t)\Psi(x, t) \tag{1.34}$$

A very interesting and useful property of the solution to Equation 1.34 is that there are many possible solutions $\Psi_1(x, t), \Psi_2(x, t), \ldots, \Psi_n(x, t)$ and the linear combination of the solutions is also a solution, namely, Equation 1.35 holds. This is the reason for superposition, the ability for multiple states to exist simultaneously, the reason that there is so much computational and networking power in quantum systems.

$$\Psi = C_1\Psi_1(x, t) + C_2\Psi_2(x, t) + \ldots + C_n\Psi_n(x, t) = \sum_{i=1}^{n} C_i\Psi_i(x, t) \tag{1.35}$$

From our standpoint, it is much easier to work with states as vectors in Dirac notation. Consider a simple two-state quantum system. Here α and β are complex numbers, and thus we can think of $|\psi\rangle$ as a vector in the 2-dimensional complex vector space, denoted C^2, and we can represent the state as a column vector $\begin{pmatrix} \alpha \\ \beta \end{pmatrix}$. Superposition is captured in this representation because it indicates that the system is simultaneously in state α and state β. Thus, it appears that there is the potential to hold much more information in qubit than in a simple digital bit, which can only be in a discrete 0 or 1 state. Much more regarding quantum networking using the Dirac notation will be discussed in Section 5.2.

1.7 NEED FOR SELF-ASSEMBLY

It is difficult to discuss nanotechnology without discussing the need for self-assembly. Manipulation of such small components requires that they essentially guide and fine-tune themselves with respect to positioning and location. On the other hand, communication networks are, in fact, beautiful examples of self-assembly (although generally at the human scale today). We'll first discuss how self-assembly is being proposed and used in current nanoscale network systems, mostly geared toward chip fabrication.

1.7.1 Self-assembly for nanotube alignment

Of course, biological systems are a clear inspiration since they are the ultimate self-organizing systems on all scales. Tags are used in DNA self-assembly of nanotubes and nanowires to stimulate the construction of structures with specific properties. Once a DNA structure exists, other organic components can be attached to the structure and the attached components can be interconnected with communication links, perhaps composed of CNTs, to construct the functional equivalent of a computer chip, including large numbers of processing elements. For the short term, DNA-based self-assembly is likely to be restricted to two layers [51].

Alignment of carbon nanotubes has been the topic of vigorous research. Cost and separation of impurities, namely metallic tubes, is still an unsolved problem. In the approach proposed by [3], lower-cost, randomly oriented tubes are directly utilized as a communication media. Information flow through a CNT network may be controlled in spite of the random nature of tube alignment. The same technique used for sensing in CNT networks, namely, change in resistance of semiconducting material, may be used to effectively route information. The traditional networking protocol stack is merged in this approach because, rather than the network layer being logically positioned above the physical and link layers, the CNT network and routing of information is an integral part of the physical layer. The potential benefits of better utilizing individual nanotubes within random CNT networks to carry information is distinct from traditional, potentially less efficient and wasteful, approaches of using CNT networks to construct transistors which are then used to implement communication networks [3].

"Self-assembly is currently limited to producing small sized DNA lattices thus limiting circuit size. However, the parallel nature of self-assembly enables the construction of a large number (≈ 109–1012) of nodes that may be linked together by self-assembled conducting nanowires." [52] This implies that control over the production process (for node placement, node orientation, and internode link creation) would be quite imprecise. Resulting devices produced by the same process could differ distinctly. Systems created using such techniques would need to discover the placement, orientation, and connection among nodes and organize their run-time processes to take maximum advantage of the characteristics of the system. Different systems, created with the same processes, could yield devices with varying capabilities.

Alternatively, self-assembled systems might be considered as stochastic systems whose performance can be described only probabilistically. Ultimately, self-assembly at the nanoscale seems destined to create systems with intrinsic defects. Two types of defects have been noted: functional and positional. A functional defect corresponds to a component that does not perform its specified function and a positional defect corresponds to a potentially functionally correct component that is placed incorrectly. This implies that nanoscale systems must be designed with fault tolerance as a fundamental property. Nanotechnology provides smaller, faster, and lower-energy devices, which allow more powerful and compact circuitry; however, these benefits come with a cost, namely, nanoscale devices may be less reliable. Thermal- and shot-noise estimations alone suggest that the transient fault rate of an individual nanoscale device may be orders of magnitude higher than today's devices. As a result, one can expect combinational logic to be susceptible to transient faults, not just the storage and communication systems. Therefore, to build fault-tolerant nanoscale systems, one must protect both combinational logic and memory against transient faults. Based on these assumptions, researchers are investigating error-correcting codes that can work effectively under the higher error rates expected from nanoscale memories [53].

1.7.2 Self-assembly, complexity, and information theory

As mentioned previously, nanoscale networking will have to overcome the challenges of poor reliability. Self-assembly at the nanoscale coupled with a dynamic environment in which such networks will have to operate will yield high failure rates. Also operation will be near the thermal limit. As device sizes shrink, the energy difference between logic states will approach the thermal limit and thermal noise will be significant for communication.

However, self-assembly may be brought about by harnessing the very technique discussed regarding the Gibbs distribution earlier, which was used to connect the microstate with the macrostate of a system. This is the Markov random field (MRF). The Markov random field defines a set of random variables, $\lambda = \{\lambda_1, \lambda_2, ..., \lambda_k\}$. Each variable λ_i can take on various values. Associated with each variable, λ_i, is a neighborhood, N_i, which is a set of variables from $\{\Lambda - \lambda_i\}$. Simply put, the probability of a given variable depends only on a (typically small) neighborhood of other variables. Of course, variants of this approach are used as error correction techniques in networking today. However, utilizing this concept to work over larger, more complex network states and activities has only recently been investigated.

Recall the large number of possible states, or configurations, of the two-state spin system discussed earlier in this chapter. The MRF technique finds the most likely, or minimal energy, configuration. Variables represent states of nodes in a network. The arcs or edges in the network convey the conditional probabilities with respect to the neighboring nodes, N_i, of node i. The definition of the MRF is as follows:

$$P(\lambda) > 0 \forall \lambda \in \Lambda \text{ (Positivity)} \tag{1.36}$$

$$P(\lambda_i | \{\Lambda - \lambda_i\}) = P(\lambda_i | N_i) \text{ (Markovian)} \tag{1.37}$$

The conditional probability of a node state in terms of its neighborhood can be formulated in terms of cliques. A clique is a set of nodes, all of whom are adjacent to one another. The Hammersley-Clifford theorem enables us to show that Equation 1.38 is valid. Note the Gibbs distribution, which we have discussed previously.

$$P(\lambda_i | \{\Lambda - \lambda_i\}) = \frac{1}{Z} e^{-\frac{1}{k_b T} \sum_{c \in C} U_c(\lambda)} \tag{1.38}$$

Recall that he normalizing constant Z is called the partition function and ensures that P is in the range $[0, 1]$. The set C is the set of cliques for a given node, i. The function U_c is called the clique energy function and depends only on the nodes in the clique. Note that the probability of states depends on the ratio of clique energy of the MRF to the thermal energy $k_b T$. For instance, the probabilities are uniform at high values of $k_b T$ and become sharply peaked at low values of $k_b T$. This mimics the annealing behavior of physical systems. Thus, we have a generalization from physics that has been widely applied in information theory.

The general algorithm for finding individual states that maximize the probability of the overall network is called belief propagation (BP). It provides an efficient means of solving inference problems by propagating marginal probabilities through the network. There are three probability functions involved:

Joint probability as shown in Equation 1.39.

$$p(x_0, x_1, \ldots, x_{n-1}) \tag{1.39}$$

Marginal probability in Equation 1.40.

$$p(x_1) = \sum_{x_0} \sum_{x_1} \cdots \sum_{x_j} \cdots \sum_{x_{n-1}} p(x_0, x_1, \ldots, x_{n-1}), j \neq i \qquad (1.40)$$

Conditional probability in Equation 1.41.

$$p(x_0, x_1, \ldots, x_{i-1}, \ldots | x_i) = \frac{p(x_0, x_1, \ldots, x_{n-1})}{p(x_i)} \qquad (1.41)$$

In the general case, some states are defined and other states are undefined. The undefined states may have different reasons for being undefined in different applications. In today's networks, they may be corrupt bits that need to be corrected, in image processing they may be pixels whose color and intensity may be corrupt, and in nanoscale networks these may have more creative interpretations.

The concept behind belief propagation is that the probability of a state value at a given node in the network can be determined by marginalizing (summing) over the joint probabilities for the node state given just the probabilities for node states within the Markov neighborhood, N_i. This marginalization establishes the state probabilities for the next propagation step. In other words, given the states that are known and a particular energy function, the most likely, lowest energy global result can be found. It can be shown that this propagation algorithm will converge to the maximum probability states for the entire network system, provided there are no loops. If there are loops, more direct, but less efficient techniques can be used.

A communication network paradigm that enables the extreme flexibility required for self-assembly is active networking. The active network architecture is discussed in more detail in Chapter 7; however, it has also been applied to self-assembly in several instances. First, it has been used to self-assemble communication protocols and services [54] and it has been applied to a DNA self-assembled interconnect network [55].

1.7.3 Active networking at the nanoscale

Active networking [1, 56] at the human scale is a network paradigm in which intermediate network nodes— for example, switches, routers, hubs, bridges, gateways—perform customized computation on the packets flowing through them. The network is called "active" because new code (algorithms) are injected into nodes dynamically, thus altering the behavior of the network. Packets in an active network can carry program code in addition to data. Customized computation is embedded within the packets' code, which is executed on network nodes. By making network node computation application-specific, applications using the network can customize network behavior to suit their requirements. A specific application of this concept is software-defined radios. Active networking, because it deals with the optimal trade-off between executable code and passive data, has a strong tie to algorithmic information theory.

The active network concept is seen at in [55] where an active network architecture at the nanoscale is used to solve the problem of limited node size, which prevents the design of a single node that can perform all operations. Instead, DNA self-assembly designs different node types (e.g., add, memory, shift) based on node size constraints. A configuration phase at system startup determines defective nodes and links, organizes a memory system, and sets up routing in the network. When executed, an instruction searches for a node with the appropriate functionality (e.g., add), performs its operation, and passes its result to the next dependent instruction. In this active network execution model, the accumulator and all operands are stored within a packet, a hallmark of macroscale active networks, rather than at specific nodes, thus reducing per-node resource demands. This enables the encoding of a series of dependent instructions within a single packet. Thus, the security techniques used to assure information in macroscale active networks might be called upon to help solve nanoscale active networks.

1.8 NEED FOR NANOSCALE INFORMATION THEORY

Because nanoscale networking is still in its early stages, theoretical results on what is achievable are very important. In this section, we review a few results from traditional networking that are very relevant.

1.8.1 Capacity of networks

At the network layer in wireless networking, a significant problem is that of routing, which is exacerbated by the time-varying network interconnections, power constraints, and the characteristics of the wireless channel. There is a relation between the predictability of such dynamic networking and the benefits that prediction can provide [57].

Here we look at the fundamental capacity of wireless networks and quickly recapitulate the results from [58]. The transport capacity of a network defined such the network transports one bit-meter when one bit has been transported a distance of one meter toward its destination. This sum of the product of the bits and distances is an indicator of a network's transport capacity.

Just as in the discussion regarding minimizing energy over multiple hops, we assume that packets are sent from node to node in a multihop fashion until they reach their final destination. Given spatial reuse, that is, the ability of nodes to transmit on the same frequency simultaneously because they are physically separated, several nodes can make wireless transmissions simultaneously, provided there is no destructive interference.

X_i denotes both the location and name of a node. It is assumed that node X_i transmits over the the $m^t h$ subchannel to a node X_j successfully if Equation 1.42 is valid for every other X_k node simultaneously transmitting over the same subchannel.

$$|X_k - X_j| \geq |(1 + \Delta)X_i - X_j| \tag{1.42}$$

$\Delta > 0$ is the space taken by a guard zone that may be used to prevent a neighboring node from transmitting on the same subchannel at the same time. A physical layer view is shown in Equation 1.43. Here $\{X_k; k \in T\}$ is the subset of nodes simultaneously transmitting at some time instant over a certain subchannel and P_k is the power level chosen by node X_k for $k \in T$. The transmission from a node X_j, $i \in T$, is successfully received by a node X_j if Equation 1.43 is valid.

$$\frac{\frac{P_i}{|X_i - X_j|^n}}{N + \sum_{k \in T, k \neq i} \frac{P_k}{|X_k - X_j|^n}} \geq \beta \tag{1.43}$$

This assumes that a minimum signal-to-interference ratio (SIR) of β is necessary for successful reception, the ambient noise power level is N, and signal power decays with distance r as $\frac{1}{\alpha}$. The main result is that the transport capacity of the protocol model is $\Theta(W\sqrt{n})^1$ bit-meters per second if the nodes are optimally placed, the traffic pattern is optimally chosen, and if the range of each transmission is optimally chosen; in other words, under the best possible conditions. More specifically, an upper bound of $\sqrt{\frac{8}{\pi} \frac{W}{\Delta}} \sqrt{n}$ bit-meters per second can be achieved for all spatial and temporal scheduling strategies. The natural question to ask is what happens to the transport capacity when networks scale down to the nanoscale. In later chapters we will consider this as well as under what conditions this question makes sense.

---

1 Recall Knuths notation: $f(n) = \Theta(g(n))$ denotes that $f(n) = O(g(n))$ as well as $g(n) = O(f(n))$.

1.9 SUMMARY

Nanonetworks are ad hoc communication networks that exist mostly or entirely at the nanometer scale, achieving the functionality and performance of today's Internet with the exception that node size is measured in nanometers and channels are physically separated by up to hundreds or thousands of nanometers. Nanoscale and molecular communications is a new field [59]; many different techniques are being tested and vetted from multiple disciplines. Undoubtedly some important background topics have been overlooked in this introduction. However, the topics chosen for this chapter are primarily based upon the background necessary to understand the remainder of this book. It is no accident that they are also topics that bridge our understanding of information and physics as well as fundamental advances in our understanding of human-scale networking that will be important considerations at the cellular scale.

While nanoscale communication networks appear to be a revolutionary leap in technology, they can be traced to the drive to make devices smaller and services more ubiquitous. Nanoscale networks will provide safe, in vivo sensing and control enabling transformative new medical advances. Nanoscale and molecular channels will utilize entirely new forms of information transport that will be able to operate in harsh environments that current wireless communications cannot penetrate, for example, areas with high electromagnetic interference or harsh environmental conditions such as extremely high temperatures and pressures. A better understanding of diffusive communication may allow improved underwater communication. The new forms of communication channels will be extremely energy efficient; for in vivo applications they may utilize the body's own highly efficient energy mechanisms. While constructing nanoscale sensors is relatively easy, keeping communications from exceeding the nanoscale has been one of the most difficult challenges to overcome. Coupling nanoscale sensing with nanoscale communication will revolutionize smart materials.

A salient theme in this chapter is the fact that nanoscale and molecular communication networking is a highly interdisciplinary field. Information, physics, and biology are being coupled in an extraordinarily fundamental way, at the very foundations of matter itself. If some of these topics are challenging don't worry, there is still much to be gained by a qualitative reading of the nanonetworking techniques to be discussed in the next three chapters. Play with the exercises and think about how these topics are relevant as communication networks scale down in size; and, of course, study the relevant papers in the references via your human-scale network. Exactly how do molecular motors work? How much information can they carry? How well can they be controlled? How efficient will they be as a nanoscale communication channel? These are questions that we will address in the next chapter.

1.10 EXERCISES

Exercise 1   Nanonetwork Scale

Consider the impact of scale on size and density of matter at the nanoscale. Use the fact that silicon has an atomic density of approximately 4.995×10^{14} atom/cm^2 and the smallest wireless sensor (as of the time this book was written) is approximately 1 cm^3 to answer the questions below.

1. What is the interatomic distance of silicon assuming atoms are uniformly distributed?

2. Carbon nanotubes are well known for having a high aspect ratio (length to diameter). What is the aspect ratio if the length is 5 millimeters and the diameter is 5 nanometers?

3. A single carbon nanotube can be functionalized to act as a sensor. How many carbon nanotube sensing elements could fit within the volume of the smallest wireless sensor assuming they are 1-cm long and laid side by side?

Exercise 2   Nanonetwork Transmission Rate

Consider the impact of scale on the transmission rate within nanoscale networks. Compare the bit rate of a hypothetical channel transmitting each bit at the speed of light (299,792,458 meters/second) over a distance of one meter to a hypothetical nanoscale network in which bits move at an apparent snail's pace of 1 mile per hour over a distance of one nanometer. One mile is equivalent to 1,609.34 meters. Assume that during transmission, seconds-per-bit are the same at both scales, with the exception of the contribution due to propagation delay.

1. First, what is the bit rate in bits per second of the speed of light network?

2. Second, what is the bit rate in bits per second of the apparently slow nanoscale network?

3. How many times faster is the nanoscale network at 1 mile/hour than the speed-of-light network?

Exercise 3   Nanonetwork Bandwidth

Consider the impact of the reduction in scale on frequency. When identical randomly located nodes, each capable of transmitting at W bits per second and using a fixed range, form a wireless network, the throughput $\lambda(n)$ obtainable by each node for a randomly chosen destination is of the order of $\Theta\left(\frac{W}{\sqrt{n \log n}}\right)$ bits per second under a noninterference protocol [58].

1. Recall how mechanical frequency scales with size. Middle C on a piano is approximately 256 Hz. If its scale is shrunk by a billion, what frequency (in hertz) would we expect? Why is this useful for communication?

2. What is the density of nanoscale communication compared to today's wireless networking? (Hint: Use the formula of Kumar and Gupta.)

Exercise 4   Physics at the Nanoscale

In statistical thermodynamics, Boltzmann's equation is a probability relating the entropy S of an ideal gas to the quantity W, which is the number of microstates corresponding to a given macrostate:

$$S = k \log W \tag{1.44}$$

where k is Boltzmann's constant, which is equal to 1.38062×10^{-23} joule/kelvin, and W is the number of microstates consistent with the given macrostate. In short, the Boltzmann formula shows the relationship between entropy and the number of ways the atoms or molecules of a thermodynamic system can be arranged.

For thermodynamic systems where microstates of the system may not have equal probabilities, the appropriate generalization, called the Gibbs entropy, is:

$$S = -k \sum p_i \log p_i \tag{1.45}$$

This reduces to Equation 1.44 if the probabilities p_i are all equal. Boltzmann used this formula as early as 1866.

1. In many of these areas, the special case of thermal noise arises, which sets a fundamental lower limit to what can be measured (or signaled) and is related to basic physical processes at the molecular level described by well-established thermodynamic considerations, some of which are expressible by relatively well-known simple formulae.

2. Show how Equation 1.44 is related to Equation 1.29.

Exercise 5   Nanoscale Antenna

Suppose the dipole length of an antenna is 0.0125m and the wavelength λ is 0.125m.

1. What is the operating frequency?

2. What is R_{rad}?

3. What is the antenna efficiency?

Exercise 6   Markov Random Fields

A Markov random field attempts to find a set of states that minimizes the total energy of a system. Selecting an energy function that provides the desired global state is often a critical part of developing an MRF solution to a problem.

1. What is a simple qualitative answer to how the energy function relates to the clique energy and the global result?

Exercise 7   Quantum Introduction

Consider quantum effects, the de Broglie wavelength, and scale. Assume a particle with a mass of 1g moving at 1 cm/sec.

1. How does the de Broglie wavelength scale with particle size?

2. What is the de Broglie wavelength of the above human-scale mass?

3. Why are we not be able to detect the dual matter-wave nature of the above particle?

Exercise 8   Wireless Capacity

Consider the formula for maximum wireless capacity.

1. Which will increase the total network capacity at a faster rate: reducing node size and adding more nodes (assuming channel rate is constant) or increasing channel capacity (assuming the number of nodes is constant)?

Exercise 9   Active Networking

Algorithmic information theory and active networking are both concerned with the compression of code. Active networks seek the smallest combination of code and data within an active packet. Consider the problem of transmitting the value of π. Assume that π represents a value that is unknown to your destination (you cannot just send a pointer to π). You can use the fact that $c = 2\pi r$, $\pi = \frac{c}{2r}$. Suppose that you can send code that draws a circle, finds the circumference of the circle, and performs the required operation to compute π; the complete code is of length γ bytes and you need to send d digits of π, at two bytes per digit.

1. Find the active compression rate as a function of code length and number of digits sent, where active compression rate is the ratio of transmitting code to transmitting only passive data.

Chapter 2

Molecular Motor Communication

Without any doubt, the regularity which astronomy shows us in the movements of the comets takes place in all phenomena. The trajectory of a simple molecule of air or vapor is regulated in a manner as certain as that of the planetary orbits; the only difference between them is that which is contributed by our ignorance. Probability is relative in part to this ignorance, and in part to our knowledge.

Pierre-Simon Laplace, Philosophical Essay on Probabilities (1814), 5th edition

This chapter and the next focus upon biological mechanisms of communication. These are mechanisms operating at the nano- or cellular scale; they are systems of information transport that are currently used in, and among, living cells. This chapter describes molecular motor communication, an intracellular communication technique used to transport information and material within living cells. The next chapter explains promising intercellular communication techniques, which is communication among cells allowing for coordinated operation. We encourage the reader to review the exercises at the end of the chapter to consolidate and expand their understanding of the material.

The traditional view of a biological cell as an encased collection of fluids that operate by changing concentration levels has been transformed as we are able to peer downwards in scale. At the nanometer scale, the view changes to one in which mechanical force and movement of mass due to force become more apparent. Energy for the forces involved comes from known biochemical sources, such as adenosine-5'-triphosphate (ATP), a nucleotide that plays an important role in cell biology. A complete understanding of biological processes would take us outside the focus of network communication; however, I would like to explain enough to enable a communication researcher to understand the most important and fundamental engineering and control aspects of such a system. Nanonetworks are described as interconnections of nanomachines that will expand the capabilities of single nanomachines by allowing them to cooperate and share information. Nanomachines are devices consisting of nanoscale components, able to perform a specific task at the nanometer level. Nanomachines need not be totally mechanical; in biological communications nanomachines can be cells or cell/machine hybrids. Methods for realizing a nanomachine can be very diverse, but they are described as having a general architecture as shown in Table 2.1. Biological nanomachines are currently the most advanced nanomachines in existence and researchers are currently trying to harness these preexisting biological features to build communication networks.

Molecular communication will enable precise mechanisms for directly interacting with in vivo cells [60]. Information may be sent to and from specific cells within the body allowing detection and healing of diseases at the cellular scale. Molecular computing will also benefit from nanoscale communication. Molecular logic gates, based upon regulation of genetic activity [61] and memory [62] are being constructed from existing biological components. Efficient interconnection among such biological logic components

Table 2.1

Components of a Nanomachine

Component	Function
Control unit	Executes instructions for intended task
Communication unit	Transmit and receive messages
Reproduction unit	Fabricate other components
Power unit	Power all other components
Sensor and actuators	Interface between environment and nanomachine

will be required. A cellular communication system implemented by means of molecular diffusion [61] will be discussed in more detail in the next chapter, which deals with intercellular signaling.

This chapter begins with an explanation of the molecular motor. There are many types of molecular motors; we start with the type of motors that ride along filaments within the cell. Next an introduction to the physical aspects of molecular motor operation is presented, namely those aspects most relevant to its velocity and information-carrying capacity. After this, there is a discussion of the potential for control and routing of such motors, particularly molecular motors that reside on microtubule tracks. There is a discussion of the nature of microtubules, how they are organized, and the fact that they serve as dynamic networks for molecular motors. Microtubules are constantly dissolving and reforming, thus changing the topology of the network. Following this, there is a discussion of the ability to steer molecular motors, again focusing on motors riding on microtubules. One of the big challenges is interfacing with molecular motors, namely, the manner in which a payload is generated and attached to a motor, and how the payload is unloaded and received. It is then natural to introduce some of the information theoretic aspects of such a communication system, namely the amount of information that can be carried within a motor payload. Next, a section is devoted to other molecular motors, that is, motors that are not associated with riding along microtubules, such as motors using a flagellum. A flagellum is a tail-like structure that projects from the cell body of some prokaryotic and eukaryotic cells and provides propulsion. The flagellum is much like cilia, another tail-like projection from various organisms. However, the the flagellum operates with a spinning, propeller-like motion while cilia operate with a back and forth motion; the motor structures differ between the two types of locomotion.

2.1 MOLECULAR MOTORS ON RAILS

To a nonbiologist looking at the cell for the first time, it can be surprising to learn that the cell is more than a collection of concentrations of chemicals; there is a significant amount of mechanical activity taking place. The cell is more akin to the workings of a fine watch or a large city, depending on the scale of the analogy you wish to use. Subcellular scale molecular motors are transporting intracellular molecular shipments throughout the cell. DNA is being transported, complementary strands are zipped and unzipped, and the cells themselves move relative to one another, for example, in the contraction of a muscle. The study of the complex physical operation of the cell has attracted physicists and engineers; now it is attracting communication and network researchers.

2.1.1 Where molecular motors are found and used in nature

The molecule motors that will be the main focus of this chapter range in size from a few nanometers to tens of nanometers. As mentioned, they ride along filaments or rails; the detail of the rails themselves is a fascinating and complex topic that will be discussed later in this chapter. However, the filaments form the cytoskeleton; just like a human skeleton, the cytoskeleton helps maintain the shape and integrity of the cell. However, the cell's skeleton differs from a human skeleton in that it is both more flexible and dynamic; it does not maintain a constant shape, but changes form in complex ways. A simple way to explain it is that instead of rigid bones with a small set of joints at fixed locations, the cytoskeleton has the ability to bend at any location and is constantly dissolving and reforming. A key point is that molecular motors utilize these filaments as a road system; the motors are able to attach, and crawl along, the cytoskeletal filaments. Molecular motors can do this while also carrying a cargo; that cargo represents a packet of information.

For molecular motors, there are three major cytoskeletal motor proteins: myosin, kinesin, and dynein. Myosins either move on actin tracks or pull actin filaments. Actin is a component of the cytoskeletal rail system and part of the contractile apparatus in muscle cells. On the other hand, kinesins and dynein move on microtubules, another major component of the cytoskeletal rail system. Motors are not limited to riding along cytoskeletal rails; they can also ride along DNA. As a quick reminder, DNA is formed from amino acids and contains the blueprint for the construction of proteins. DNA encodes the proteins that are the fundamental workhorse for activities within the cell. Instead of transporting cargo, molecular motors can perform mechanical operations necessary to the proper function of DNA, such as unzipping DNA strands. Another form of molecular motors are RNA polymerases and ribosomes, which are mobile mechanical workshops that polymerize, or assemble, RNA and proteins into chains.

The molecular motor operates by attaching and detaching one of its "heads" (this terminology is often used, although they would seem more like feet) from the corresponding track, much like taking steps on a ladder. The steps on the track are equally spaced binding sites for the motor head molecule. Given that the motor head must be able to sequentially detach and reattach at the next step, it's not surprising that this is a delicate operation and there is no guarantee that the motor will not fly away from the microtubule track. The term *processivity* is used with molecular motors to measure the efficiency of movement. Here efficiency is often a trade-off between the rate of movement and the ability to remain attached to the track. There are three commonly used definitions of processivity: the average number of chemical cycles before detachment from the track; the lifetime of the motor while in a state of attachment to the track within its working cycle; and the mean distance covered by the motor on the track in a single run.

Let us call one complete motor step a cycle. Then assume that during the cycle a motor spends time τ_{on} attached to the track for its "power stroke" and τ_{off} detached from the track, during a "recovery" phase. The duty ratio r is then simply $r = \frac{\tau_{on}}{\tau_{on} + \tau_{off}}$. Thus, the duty ratio is the time spent while a head is attached to the track. The typical duty ratios of kinesins and cytoplasmic dynein are at least 0.5, whereas the ratio for conventional myosin can be as small as 0.01.

Details of the microtubule tracks will be discussed later; however, it should be noted that the tracks have a polarity. A microtubule is composed of smaller units called α and β tubulin. A microtubule's polarity is such that it has an α tubulin at its negative end and a β tubulin at its positive end. Most kinesins are positive-end-directed motors, whereas most dyneins are negative-end directed. Because only one binding site for a motor exists on each subunit of a microtubule, the minimum step size for kinesins and dyneins is 8 nm.

Consider how molecular motors get their power; they use one of the most common forms of energy found in intracellular systems, adenosine triphosphate (ATP). ATP is made from adenosine diphosphate (ADP) or adenosine monophosphate (AMP), and its use in metabolism converts it back into these precursors. ATP is therefore continuously recycled in organisms, with the human body turning over its own weight in ATP each day. ATPases are a class of enzymes that catalyze the decomposition of adenosine triphosphate into ADP and a free phosphate ion. This reaction, known as dephosphorylation, releases energy, which the enzyme harnesses to drive other chemical reactions. This process is widely used in all known forms of life. ATPase sites, where ATP binding to the filament occurs, plays a central role in molecular motors.

The motor protein acts like an enzyme and catalyzes the hydrolysis of ATP, releasing ADP and phosphate. This enzymatic process causes small changes in the conformation of the protein surrounding the ATPase site, which, in turn, propagates outward and is ultimately amplified into movement. These conformational changes generate sufficiently large forces that are responsible for the motor's unidirectional motion over long distances through repeated enzymatic cycles.

The rail-riding class of molecular motors have several common elements, namely, they have "heads," the term for a site for ATP hydrolysis, and a binding site to the cytoskeletal rail, and they have a stalk. They also have a "tail" at the opposite end of the stalk that binds with a cargo. However, there are differences within the class of rail-riding motors. For example, a kinesin has the smallest head, whereas a myosin's head is of intermediate size and a dynein's head is very large. The tail exhibits much more diversity than the head because the same motor must be able to recognize and pick up wide varieties of cargo. Some particular motors within the myosin and kinesin superfamilies are known to self-assemble into more complex structures, the most well-known among these being the myosin thick filaments in muscles.

Now we come to the small-scale effects. The motors are extremely small and thus have a very small mass. Inertial forces are small compared to viscous forces. In other words, the mass or weight of the motor has a much smaller impact than the surrounding fluid forces acting upon the motor. The Reynolds number quantifies this effect; it is a dimensionless number that provides a measure of the ratio of inertial forces to viscous forces and, consequently, it quantifies the relative importance of these two forces. The precise definition of the Reynolds number given fluid flow in a pipe is shown in Equation 2.1 where V is the mean fluid velocity (m/s), D is the diameter (m) of the flow, μ is the dynamic viscosity of the fluid ($Pa \cdot s$ or $N \cdot s/m^2$), ν is the kinematic viscosity ($\nu = \mu/\rho$) (m^2/s), ρ is the density of the fluid (kg/m^3), Q is the volumetric flow rate (m^3/s), and A is the pipe cross-sectional area (m^2).

$$Re = \frac{\rho V D}{\mu} = \frac{V D}{\nu} = \frac{Q D}{\nu A} \tag{2.1}$$

At low Reynolds numbers, laminar flow tends to occur, where viscous forces, represented by both μ and ν in Equation 2.1, are dominant. This is characterized by smooth, constant fluid motion. This is in contrast to high Reynolds numbers, dominated by inertial forces, ρ and V, where turbulent flow occurs, which tends to produce random eddies, vortices and other flow fluctuations. Intuitively, viscosity is a measure of the slipperiness of the fluid; a higher viscosity fluid will have a lower Reynolds number and more "even" flow while a low viscosity fluid, which is less slippery, will cause more random-looking flow patterns. Similarly, a high-density fluid, or a fluid injected with a higher velocity, will yield a higher Reynolds number and more random eddies and vortices. Molecular motors, being so small relative to other liquid molecules, are hit on all sides by randomly moving liquid molecules, so molecular motors experience random forces leading to noisy motor trajectories. Models of molecular motors must therefore take both the mechanics and chemistry into account. Direct application of Newton's laws of physics to each and every particle is too computationally intensive to be practical over even a single motor stepping cycle. Thus, simplifications are required. One simplification is to assume that the molecular motor behaves like a particle undergoing Brownian motion, but with a potential force caused by chemical energy. More will be said about Brownian motion in later chapters regarding cell signaling and diffusion. In short, Brownian motion is the mathematical model used to describe the apparent random movement of particles in fluids. Since diffusion will play a large role in nanoscale network transport, this is a good point to discuss some required background on diffusion and Brownian motion before continuing with the operation of the molecular motor.

2.1.2   Random walk and Brownian motion

The ancient Roman poet Lucretius in *De Rerum Natura* ("On the Nature of Things") (60 BC) helped to transfer philosophical and scientific knowledge from Greece to Rome. What is more relevant here is that his work contains a vivid description of Brownian motion in the form of dust moving, apparently at random, in the air. It presciently describes the spontaneous movement of atoms. There is a long rich history behind the development of the notion and mathematics underlying Brownian motion. While a detailed understanding of the mathematics behind Brownian motion may appear daunting, the reader will be greatly rewarded with a clear understanding of the physics impacting nanoscale communication and molecular motors in particular. Let us begin with a simple example: a random walk on a line. Steps Δz of equal length l are taken with a random direction, either left or right. The final position of a particle is the accumulation of all the steps in either direction, as shown in Equation 2.2. Assume the particle starts at position z_0 and performs N steps.

$$z_n = \Delta z_1 + \Delta z_2 + \ldots + \Delta z_{N-1} + \Delta z_N = \sum_{i=1}^{N} \Delta z_i \qquad (2.2)$$

Consider the squared length of the walk, in Equation 2.3.

$$z_n^2 = \left(\sum_{j=1}^{N} \Delta z_j \right) \left(\sum_{k=1}^{N} \Delta z_k \right) \qquad (2.3)$$

Something interesting happens when the terms are regrouped, as show in Equation 2.4.

$$z_n^2 = \sum_{j=1,k=j}^{N} \Delta z_j^2 + \sum_{j,k=1,k\neq j}^{N} \Delta z_j \Delta z_k = Nl^2 + \sum_{j,k=1,k\neq j}^{N} \Delta z_j \Delta z_k \qquad (2.4)$$

Consider the left terms in the above equations. For the indices where $j = k$, the step sizes are the same, and since they are squared, the result is positive, thus yielding Nl^2. The potentially differing steps are in the right term.

Now consider averaging over a large number of steps, shown in Equation 2.5. Note that the symbol $\langle \rangle$ is used to indicate expected value.

$$\langle z_n^2 \rangle = Nl^2 + \left\langle \sum_{j,k=1,k\neq j}^{N} \Delta z_j \Delta z_k \right\rangle \qquad (2.5)$$

The left term remains the same. The right term is the average of the remaining steps. Since the steps are assumed to be random and independent, the average of both Δz_j and Δz_k will be zero and their product will be zero. Thus, a fundamental result is derived as shown in Equation 2.6. Although this result seems simple, it is of significant importance, for it allows us to find a pattern in something that seems at first appearance to be random; it allows us to find the distance that information might spread in a random nanoscale environment.

$$\langle z_n^2 \rangle = Nl^2 \qquad (2.6)$$

The root-mean square result is shown in Equation 2.7. If a "packet" of information is buffeted by the whims of a fluid environment, each step size is l, and there are N steps, then the information can be expected to propagate a distance of R as shown in the equation.

$$R = \sqrt{\langle z_n^2 \rangle} = l\sqrt{N} \tag{2.7}$$

Suppose the particle is moving in a fluid, as we imagine the environment to be for the molecular motor. Consider for now, that the motor is one among many randomly moving particles and we wish to compute the characteristics of its random path. Assume the distance traveled between particle collisions is $l = \langle v \rangle \tau$ where $\langle v \rangle$ is the mean velocity and τ is the time between collisions. Assume the motor has radius r_m and travels a total distance L where the particle density is n. The surface area of a sphere is $4\pi r_m^2$ and it sweeps through a total path of length L. Since n is the density, or number of particles per unit area, then the motor can expect to experience $4\pi r_m^2 Ln$ collisions. The mean path length between collisions is then the distance traveled before one collision occurs, that is, $4\pi r_m^2 ln = 1$.

Given the definition of τ as the mean time between collisions, then the number of steps taken in a fluid is $N = \frac{t}{\tau}$ for a given time t. As was stated, it is assumed that $l = \langle v \rangle \tau$, therefore, the mean-square distance is shown in Equation 2.8.

$$\langle z^2 \rangle = Nl^2 = \frac{t}{\tau}(\langle v \rangle)\tau l = (\langle v \rangle l)t \tag{2.8}$$

In three dimensions, with three degrees of freedom and assuming the fluid is in equilibrium and particles are uniformly distributed, one can assume that the mean total distance traveled in each dimension will be equal, $\langle x^2 \rangle = \langle y^2 \rangle = \langle z^2 \rangle = \frac{\langle r^2 \rangle}{3}$, where r is the coordinate system for three-dimensional space. Thus, Equation 2.9 holds. Here D is the diffusion coefficient, $3\langle v \rangle l$.

$$\langle r^2 \rangle = 3\langle v \rangle lt = Dt \tag{2.9}$$

Now let's reevaluate what we have done, but in three-dimensional space, incorporating the probability of finding a particle at a particular location, and most importantly, accounting for dependence in steps, in other words, not having to assume independent steps. We now have steps $\Delta \bar{r}$ where $\bar{r}$ is a multidimensional space. The time between events Δt will be constant. The position $\bar{r}_n$ of a particle is shown in Equation 2.10.

$$\bar{r}_n = \Delta \bar{r}_0 + \ldots + \Delta \bar{r}_n \tag{2.10}$$

Everything remains analogous to the one-dimensional case discussed earlier. $\bar{r}_0$ is the starting position and each step $\Delta \bar{r}_i$ is random. We can extend the analysis by defining a probability distribution $P(\bar{r}, t_n)$ that defines the probability for the motor to be at position $\bar{r}$ at time t. Using the same technique as Equation 2.5 yields Equation 2.11.

$$\langle \bar{r}_n^2 \rangle = \sum_{j=1,k=j}^{n} \langle \Delta r^2{}_j \rangle + \sum_{j,k=1,k\neq j}^{n} \langle \Delta \bar{r}_j \Delta \bar{r}_k \rangle \tag{2.11}$$

This yields an elegant result; the left term on the right-hand side is the sum of the variances of the path lengths and the right-hand term is the sum of the covariances, as shown in Equation 2.12. This is due to the fact that $\sigma^2_{\Delta \bar{r}_j^2} = \langle \Delta \bar{r}_j^2 \rangle$ by the definition of variance, similarly for the covariance term. Comparing the above equation with the earlier result with completely independent steps, we see that the variance is l^2 and the covariance is zero.

$$\langle \bar{r}_n^2 \rangle = \sum_{j=1,k=j}^{n} \sigma^2_{\Delta \bar{r}_j^2} + \sum_{j,k=1,k\neq j}^{n} cov(\Delta \bar{r}_j \Delta \bar{r}_k) \tag{2.12}$$

It is possible to attempt to characterize Brownian motion directly by the forces causing the motion. This is slightly more complex, but can provide a better understanding of the forces at work and the corresponding motion. The approach was derived by Pierre Langevin (1908) and is derived directly from Newton's laws, in particular, $F = ma$. In this analysis, x is position, $\dot{x}$ is the velocity, and a is the damping rate, which depends on the viscosity of the fluid and the radius of the particle in motion, the molecular motor in our case. $a\dot{x}$ represents a friction force that slows, or dampens, the motion. $F(t)$ is a random, time-varying force that is caused by the collision of particles with one another. Putting the forces together results in Equation 2.13, known as the Langevin equation. From looking at this equation, it should be clear that if there were no collision force, the only other force acting on the particle is the damping force, which would slow the particle down until it eventually stopped.

$$m\dot{x} = -a\dot{x} + F(t) \tag{2.13}$$

Now, let us consider how to simplify and solve this equation. In Equation 2.14, both sides of Equation 2.13 have been multiplied by position, x.

$$mx\dot{x} = m\left[\frac{d(x\dot{x})}{dt} - \dot{x}^2\right] = -ax\frac{d}{dx} + xF(t) \tag{2.14}$$

It sometimes simplifies things to consider averages over large numbers of objects or events. We do this in Equation 2.15 where averages are taken over a large number of particles. Also, since both x and $F(t)$ are random and independent, $\langle xF(t)\rangle = 0$.

$$m\frac{d\langle x\dot{x}\rangle}{dt} = m\langle \dot{x}^2\rangle = -a\left\langle x\frac{d}{dx}\right\rangle \tag{2.15}$$

Now we can relate this equation to the physics of fluids and gases as discussed in Chapter 1. The kinetic energy of the particle is proportional to the temperature, $m\frac{\langle \dot{x}^2\rangle}{2} = \frac{kT}{2}$, where k is the Boltzmann constant and T the temperature.

$$m\frac{d\langle x\dot{x}\rangle}{dt} + a\langle x\dot{x}\rangle = kT \tag{2.16}$$

In Equation 2.17, both sides are divided by m.

$$\left(\frac{d}{dt} + \frac{a}{m}\right)\langle x\dot{x}\rangle = \frac{kT}{m} \tag{2.17}$$

In Equation 2.18, $\gamma = \frac{a}{m}$.

$$(\frac{d}{dt} + \gamma)\langle x\dot{x}\rangle = \frac{kT}{m} \tag{2.18}$$

When the above equation is solved, the following Equation 2.19 results.

$$\langle x\dot{x}\rangle = \frac{1}{2}\frac{d\langle x^2\rangle}{dt} = Ce^{-\gamma t} + \frac{kT}{m} \tag{2.19}$$

In order to solve for C in Equation 2.19, the mean square displacement at time zero must be zero, $C + \frac{kT}{a} = 0$.

$$\frac{1}{2}\frac{d\langle x^2\rangle}{dt} = \frac{kT}{a}(1 - e^{-\gamma t}) \tag{2.20}$$

Integrating to obtain the mean square distance yields Equation 2.21.

$$\langle x^2 \rangle = \frac{2kT}{a} \left[t - \frac{1}{\gamma}(1 - e^{-\gamma t}) \right] \tag{2.21}$$

If we are considering times that are much smaller than the collision time, then $t << \frac{1}{\gamma}$. In this case, the solution can be approximated by $\langle x^2 \rangle \sim t^2$. This is known as ballistic diffusion and indicates that particles are not slowed by collision forces and can diffuse relatively quickly. On the other hand, when $t >> \frac{1}{\gamma}$, the solution can be approximated by $\langle x^2 \rangle \sim \frac{2kT}{a}t$.

Suppose we know the probability of the last position, $P(z-\Delta z, t-\Delta t)$, and we know the distribution of random walk steps, $q_{\Delta z}(\Delta z)$. Then we can simply multiply as shown in Equation 2.22 to find the probability of the next position of the particle, or molecular motor in our particular case.

$$P(z,t) = P(z - \Delta z, t - \Delta t)q_{\Delta z}(\Delta z) \tag{2.22}$$

In Equation 2.23, the integral is taken over all possible step sizes.

$$P(z,t) = \int_{-\infty}^{\infty} P(z - \Delta z, t - \Delta t)q_{\Delta z}(\Delta z)d\Delta z \tag{2.23}$$

In Equation 2.24, a Taylor series expansion is used to approximate the result.

$$P(z - \Delta z, t - \Delta t) = P(z,t) - \Delta t \partial_t P(z,t) - \Delta z \partial_x P(z,t) + \frac{1}{2}\Delta z^2 \partial_z^2 P(z,t) \tag{2.24}$$

Recall that a Taylor series expansion can be used to estimate a function as shown in Equation 2.25. For a function that depends on two variables, x and y, the Taylor series to second order about the point (a, b) is:

$$
\begin{aligned}
f(x,y) \approx \ & f(a,b) + (x - a)f_x(a,b) + (y - b)f_y(a,b) + \\
& \frac{1}{2!} \left[(x - a)^2 f_{xx}(a,b) + 2(x - a)(y - b)f_{xy}(a,b) + (y - b)^2 f_{yy}(a,b) \right]
\end{aligned} \tag{2.25}
$$

Now we can take the limits as was done previously for each of the terms in Equation 2.24, as shown in Equation 2.26.

$$
\begin{aligned}
P(z,t) = \ & \int P(z,t)q_{\Delta z}(\Delta z)d\Delta z - \\
& \int \Delta t \partial_t P(z,t)q_{\Delta z}(\Delta z)d\Delta z - \\
& \int \Delta z \partial_z P(z,t)q_{\Delta z}(\Delta z)d\Delta z + \\
& \int \frac{1}{2}\Delta z^2 \partial_z^2 P(z,t)q_{\Delta z}(\Delta z)d\Delta z
\end{aligned} \tag{2.26}
$$

In order to simplify the integrals in Equation 2.26, note that the integral is only over the step size Δz. Also, assume that the step distribution $q_{\Delta z}$ is normalized, $\int q_{\Delta x}d\Delta z = 1$. It is also assumed that $q_{\Delta z}$ is symmetric with zero mean, $\int \Delta z q_{\Delta z}(\Delta z)d\Delta z = 0$. Also, note that the variance, $\sigma_{\Delta z}^2$, is

$\int \Delta z^2 q_{\Delta z} d\Delta z = \sigma_{\Delta z}^2$. Applying these simplifications yields Equation 2.27.

$$P(z,t) = P(z,t) - \Delta t \partial_t P(z,t) + \frac{1}{2}\sigma_{\Delta x}^2 \partial_x^2 + \frac{1}{2}\sigma_{\Delta x}^2 \partial_x^2 P(z,t) \tag{2.27}$$

With some final manipulation, the result in Equation 2.28 is obtained.

$$\partial_t P(z,t) = \frac{\sigma_{\Delta z}^2}{2\Delta t}\partial_x^2 P(z,t) \tag{2.28}$$

Now we are in a position to understand the more general Fokker-Planck equation, also known as the Kolmogorov forward equation. All the assumptions we have made so far still apply; however, there are a few additional generalizations. One generalization is that previously we have assumed the mean of the random walk steps to be zero. It's possible, particularly for information transport with molecular motors, that the mean will be nonzero, either positive or negative, in the direction that transport is desired. Another generalization is that both the mean and the variance can be spatially dependent. This means that the distribution of the steps is dependent on the spatial location so the form of the distribution function must be $q_{\Delta z, z}$. Thus, we can rewrite Equation 2.26 to include the additional assumptions, as shown in Equation 2.29.

$$P(z,t) = \int_{-\infty}^{\infty} P(z - \Delta z, t - \Delta t) q_{\Delta z, z}(\Delta z, z - \Delta z) d\Delta z \tag{2.29}$$

Note that $q_{\Delta z, z}(\Delta z, z - \Delta z)$ is now the probability distribution for taking a step of size Δz at position z in time Δt. The Fokker-Planck equation can be derived in a manner similar to Equation 2.24, namely, by using a Taylor series expansion. It is important to note the Taylor expansion is in terms of z and not Δz.

Consider Equation 2.29 to be in the form of $P(z,t) = \int AB d\Delta z$. Then A is:

$$A = P(z,t) - \partial_t P(z,t)\Delta t - \partial_z P(z,t)\Delta z + \frac{1}{2}\partial_x^2 P(z,t)\Delta z^2 + \dots \tag{2.30}$$

$$B = q_{\Delta z, z}(\Delta z, z) - \partial_z q_{\Delta z, z}(\Delta z, z)\Delta z + \frac{1}{2}\partial_x^2 \partial_z^2 q_{\Delta z, z}(\Delta z, z)\Delta z^2 + \dots \tag{2.31}$$

In order to simplify the equation, we use all the previous techniques, such as the fact that the distribution should be normalized, $\int q_{\Delta z, z}(\Delta z, z) = 1$. Also, the mean value $\mu_{\Delta z}(z) = \int \Delta z q_{\Delta z, z}(\Delta z, z) d\Delta z$. The variance is shown in Equation 2.32.

$$\langle z^2 \rangle(z) = \int \Delta z^2 q_{\Delta z, z}(\Delta z, z) d\Delta z \tag{2.32}$$

Also, $\int \Delta z \partial_z q_{\Delta z, z}(\Delta z, z) d\Delta z$ is equal to $\partial_z \int \Delta z q_{\Delta z, z}(\Delta z, z) d\Delta z \equiv \partial_z \mu_{\Delta z}(z)$. After keeping all terms up to second order, the result is shown in Equation 2.33.

$$\partial_t P(z,t) = -\partial_z \left[V(z)P(z,t) \right] + \partial_z^2 \left[D(z)P(z,t) \right] \tag{2.33}$$

In the above equation, $V(z) \equiv \frac{\mu_\Delta(z)}{\Delta t}$ is the drift velocity and $D(z) \equiv \frac{\langle \Delta z^2 \rangle}{2\Delta t}$ is the diffusion term. Thus, the main difference from the earlier and simpler equations is the the new drift term, because the mean of the steps is no longer zero, and there is a dependency between drift and diffusion.

Another point that should be carefully noted about the derivation of the Fokker-Planck equation is the assumption required by the use of the Taylor expansion, namely, the system must be near equilibrium and the individual steps sizes must be small. It is possible to keep all the terms in the Taylor expansion, which yields a form known as the Kramers-Moyal expansion.

Let us go back to Equation 2.13, in which the Langevin equation was initially derived, and attempt to enhance the equation to describe more specifically the forces unique to the molecular motor analysis as shown in Equation 2.34. We assume that Brownian motion forces are at work, but also that the motor has some inherent chemical force that propels it in a given direction. Let $V_\mu(x)$ be a potential force operating within the motor at position $x(t)$ and time t and chemical state μ. $V_\mu(x)$ is complicated by the fact that it is not constant, but changes with time due to changes in the chemical states that cause the motor to operate. F_{ext} is an externally applied mechanical force, which is negative for load forces. $\xi(t)$ represents the random Brownian force.

$$-\gamma \frac{dx}{dy} - \frac{dV_\mu(x)}{dx} + F_{ext} + \xi(t) = 0 \qquad (2.34)$$

The goal is to combine the Brownian forces, due to the low Reynolds number as we discussed earlier, with the forces due to changes in chemical state that are driving the motor. The rate of change over time of the probability of being located at position x at time t due to the chemical forces in state μ is shown in Equation 2.35. The potential forces created by the chemical state are represented by $W_\mu(x)$, thus we can see the the chemical potential forces are dependent upon chemical state, position, and time.

$$\frac{\partial P_\mu(x,t)}{\partial t} = \sum_{\mu'} P_{\mu'}(x,t) W_{\mu' \to \mu}(x) -$$
$$\sum_{\mu'} P_\mu(x,t) W_{\mu \to \mu'}(x) \qquad (2.35)$$

Combining the Fokker-Planck equation with the chemical state changes yields Equation 2.36.

$$\frac{\partial P_\mu(x,t)}{\partial t} = \frac{1}{\eta} \frac{\partial}{\partial x} \left[\left(V'_\mu(x) - F_{ext} \right) P_\mu(x,t) \right] +$$
$$\left(\frac{k_B T}{\eta} \right) \frac{\partial^2 P_\mu(x,t)}{\partial x^2} +$$
$$\sum_{\mu'} P_{\mu'}(x,t) W_{\mu' \to \mu}(x) -$$
$$\sum_{\mu'} P_\mu(x,t) W_{\mu \to \mu'}(x) \qquad (2.36)$$

In statistical physics, a Langevin equation is a stochastic differential equation describing Brownian motion in a potential [63]. Consider a simple case in which a small particle is in a fluid environment moving in one dimension and subject to a potential, $\phi(x)$, where x is the location of the particle. In addition to the force caused by the potential $\phi(x)$, the particle experiences a viscous drag force and a Brownian force. Both the drag force and the Brownian force arise from collisions of the particle with the surrounding fluid molecules. The drag force on the particle, $-\zeta u$, is proportional to the velocity, u, where ζ is called the drag coefficient. The drag force always opposes the particle's motion. The Brownian force on the particle has zero mean. Using Newton's second law ($F = ma$) and incorporating the forces mentioned, the Langevin equation, Equation 2.37, can be derived.

$$m\frac{du}{dt} = -\zeta u - \phi'(x) + \sqrt{2k_B T \zeta} \frac{dW(t)}{dt} \qquad (2.37)$$

In the above equation, m is the mass, x is the particle's position, $k_B T$ is the Boltzmann constant, and T is the absolute temperature. $W(t)$ is a standard Wiener process satisfying $W(t+s) - W(t) \sim N(0, s)$, where N is a normal distribution of mean zero and variance s. The magnitude of the Brownian force is related to the drag coefficient and is given by $\sqrt{2 k_B T \zeta}$. This is a result of the fluctuation-dissipation theorem. The fluctuation-dissipation theorem relies on the assumption that the response of a system in thermodynamic equilibrium to a small applied force is the same as its response to a spontaneous fluctuation. Therefore, there is a direct relation between the fluctuation properties of the thermodynamic system and its linear response properties.

Regarding Equation 2.37, there are three time scales to consider: the time scale for the motor to forget about its initial velocity, the time scale of effects induced by the potential $\phi(x)$, and the time scale associated with the thermal diffusion. Because the molecular motor has so little mass, its momentum is negligible and its inertial time scale is the shortest time scale.

Let D be the diffusion coefficient of a particle, $D = k_B T / \zeta$. The diffusion coefficient generally is a dimensionless value that provides the proportion of a substance that diffuses through a given area divided by time. The diffusion coefficient as defined here increases with temperature and decreases with the drag coefficient, which makes intuitive sense. Also, let $t_0 = \frac{m}{\zeta}$, which purportedly has the dimension of time. Then, making the substitution yields into Equation 2.37 yields Equation 2.38.

$$\frac{du}{dt} = -\frac{1}{t_0} u + \frac{1}{t_0} \sqrt{2D} \frac{dW(t)}{dt} \tag{2.38}$$

In the limit as t_0 goes to zero yields Equation 2.39.

$$\frac{dx}{dt} = -D \frac{\phi'(x)}{k_B T} + \sqrt{2D} \frac{dW(t)}{dt} \tag{2.39}$$

A key feature of molecular motors is that the average velocity caused by the potential $\phi(x)$ is several orders of magnitude smaller than the instantaneous velocity. That is, the instantaneous velocity of the motor is approximately given by an assumption known as the equipartition of energy. The original idea of equipartition was that, in thermal equilibrium, energy is shared equally among all of its various forms; for example, the average kinetic energy in the translational motion of a molecule should equal the average kinetic energy in its rotational motion.

2.1.3 Molecular motor operation: The mechanics of walking

Now that we have discussed some background behind the analytical approach for modeling molecular motors, it's worth spending a little time to discuss what is known about the mechanics of the motility of molecular motors that ride on microtubules. In other words, to discuss what precisely is known about how a molecular motor actually moves. We have already suggested how the motor moves, but here we go into a little more detail.

The stepping motion of molecular motors on the microtubule rails is due to a small conformational change in a molecular complex powered by adenosine-triphosphate (ATP) hydrolysis. ATP plays an important role in cell biology as the molecular unit of currency of intracellular energy transfer. In this role, ATP transports chemical energy within cells for metabolism. It is produced as an energy source during the processes of photosynthesis and cellular respiration and consumed by many enzymes and a multitude of cellular processes. ATP is made from adenosine diphosphate (ADP) or adenosine monophosphate (AMP), and its use in metabolism converts it back into these precursors; thus ATP is continuously recycled in organisms, with the human body turning over its own weight in ATP each day. The ATP structure is comprised of three phosphate groups labeled α, β, and γ.

In simple terms, the motor has two heads (more like feet, but they are known as heads in the literature) that alternately attach and release from microtubule binding sites as shown in Figure 2.1. The binding sites are analogous to steps on a ladder. When a head releases from the microtubule binding site, it swings forward, landing on the next binding site on the microtubule. The process is then repeated with the other head releasing and swinging forward while the current head remains attached. A motion similar to legs walking is achieved. More specifically, a head is bound to the microtubule via ATP. The loss of the γ-phosphate group from ATP leaves a space of approximately 0.5 nm. This is thought to cause a rearrangement of conserved structural elements near the ATP-binding site [64]. Ultimately, this loosens the foot from its binding to the microtubule allowing it to swing forward on a lever comprised of an α-helix of variable length. An α-helix is a common structure in proteins, a right- or left-handed coiled conformation resembling a spring. The lever swings the head through an angle of up to 70 °. The lever swing is believed to be the ultimate cause for the working stroke; motors with longer necks take larger steps and move faster.

2.2 THERMODYNAMICS OF MOLECULAR MOTORS

Understanding the thermodynamics of the molecular motor and other nanoscale and cellular-scale networking systems is important because it enables precise knowledge of the amount of energy available, the efficiency, and ultimately the performance, of nanoscale networked communication systems. With knowledge of the thermodynamics of the system we can calculate the information transport capacity and the energy required for it to operate. We begin by relating energy and force to the biochemical operation via Gibbs free energy. From first principles in thermodynamics, the energy change of a system (e.g., a molecule being stretched, a motor moving along a track, etc.) can be separated into components related to the heat exchanged and the work done upon, or performed by, the system. When energy changes slowly enough that the system remains in quasi-static equilibrium, these quantities are the reversible exchanged heat and the reversible work (pressure-volume work and mechanical work), Equation 2.40.

$$
\begin{aligned}
dE &= dq_{rev} + dw_{rev} \\
&= (TdS) + (-PdV + \int F \cdot dx)
\end{aligned}
\tag{2.40}
$$

In the above, we see that the temperature T and the change in thermodynamic entropy S, which is the change in reversible energy, and the negative of the pressure P, and change in volume dV plus the forces times the change in distance gives the energy available for useful work. Gibbs free energy is the amount of energy available to do work. So it's the total energy minus heat energy plus the energy available through temperature and volume. Thus, energy that turns into heat or increases the entropy of the system does not count towards the Gibbs free energy. At this point we need to note that there are many forms and definitions of entropy, as discussed in Chapter 1, including information entropy, which we know plays a central role in information theory and thermodynamic entropy.

Temperature and pressure are usually controlled independent variables in an experiment, so we use Gibbs free energy, which is defined as $G = E - TS + PV$, Equation 2.41.

$$
dG = -SdT + VdP + Fdx
\tag{2.41}
$$

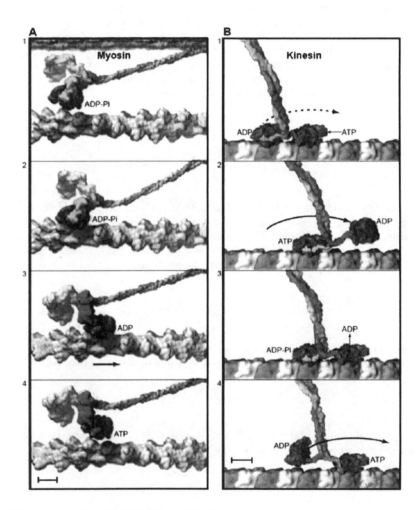

Figure 2.1 Molecular motor. (Copyright (2000) Science, U.S.A.)

One can think of stretching a molecule by an amount Δx. The work required is then shown in Equation 2.42.

$$W = \int_{x_o}^{x_0 + \Delta x} F dx \tag{2.42}$$

If done slowly (e.g., no heat loss), the work is reversible and equal to the free energy change in the system. Positive work means that the surroundings have done work on the system; the free energy of the molecule is increased. By measuring the free energy change as a function of temperature, it is possible to determine the change in entropy, $\Delta S_{stretch}$, and the change in enthalpy, $\Delta H_{stretch}$, upon stretching the molecule. Enthalpy relates the change in the internal energy of a system with the work done on its surroundings and is defined as $H = G - TS$. Using Equation 2.41, Equation 2.44 can be determined. Note that pressure P and length x are held constant in these equations.

$$\Delta S_{stretch}(x) \quad = \quad -\left(\frac{\partial \Delta G_{stretch}(x)}{\partial T}\right)_{P,x} \tag{2.43}$$

$$\Delta H_{stretch}(x) \quad = \quad -\left(\frac{\partial \Delta G_{stretch}(x)/T}{\partial 1/T}\right)_{P,x}$$

Let A and B be distinct observable states of the system; each occupies a local free energy minimum, at position x_A and x_B, respectively, separated by a distance Δx along the mechanical reaction coordinate. A and B could represent the states of a motor in sequential locations along its track, or folded and unfolded states of a protein. At zero force, the free energy difference between A and B is simply shown in Equation 2.44, where ΔG^0 is the free energy and $[A]$ and $[B]$ are the concentrations of the molecules in states A and B.

$$\Delta G(F = 0) = \Delta G^0 + k_B T \ln \frac{[B]}{[A]} \tag{2.44}$$

The energy required by a force applied over a given distance is $F(x_B - x_A)$; this energy simply linearly increases with the applied force and distance as shown in Equation 2.45.

$$\Delta G(F) = \Delta G^0 - F(x_B - x_A) + k_B T \ln \frac{[B]}{[A]} \tag{2.45}$$

When the system is at equilibrium, $\Delta G = 0$, so Equation 2.45 reduces to Equation 2.46, where $K_{eq}(F)$ is an equilibrium constant that depends upon the applied force.

$$\Delta G^0 - F(\Delta x) \quad = \quad -k_B T \ln \frac{[B]}{[A]}$$

$$= \quad -k_B T \ln K_{eq}(F) \tag{2.46}$$

It is interesting to note then, that by applying a force, either assisting or opposing the motor, the equilibrium state of the reaction changes, which changes the state or relative concentrations of A and B. In our particular case of a molecular motor information transport system, the energy comes from the hydrolysis of ATP, so the concentrations are $\frac{[B]}{[A]} \equiv \frac{[ADP][P_i]}{[ATP]}$.

From the standpoint of being able to transport information, whether the information is encoded as the concentration of molecular motors themselves, or a chemically encoded packet transported by a molecular motor, it is important to understand the relationship among the energy, efficiency, and force of the motor. Molecular motors are exergonic as opposed to endergonic. An exergonic process simply means that energy is released from the system in the form of work; if the pressure and temperature are held constant, then the Gibbs free energy, that is the energy available in the system to do work, decreases. In an endergonic process energy is absorbed by the system and, at constant temperature and pressure, the Gibbs free energy increases. Thus, a measure of the efficiency of the motor takes into account the reduction in Gibbs free energy and the actual energy that performs useful work. This is measured as $\eta_{TD} = \frac{F\delta}{\Delta G}$ where F is the applied force of the motor and δ is the step size of the motor. Since ΔG is the energy available to do work, then $F\delta \leq \Delta G$ and the efficiency must be less than or equal to one. As the motor applies more force, for example pulling a large payload or pushing against an object, the force F increases and thus the efficiency η_{TD} increases as well. Eventually, as the force against which the motor is working increases, the motor will reach a maximum value before stalling. This is known as the stall force, F_{stall}. Measured stall forces indicate efficiencies that vary widely from 15% up to 100%. Clearly, there may still be some uncertainty in the experimental measurement process. For those motors with efficiencies less than 100%, it is hypothesized that the energy not going into useful work is going into either heat or in generating forces that are orthogonal to the direction of motion. Another measure that is conceptually similar to the efficiency is the coupling constant, ξ. It is the probability that a motor takes a step per chemical reaction. If $\xi = 1$, the motor is tightly coupled. It should be noted that there is no direct relationship between coupling and efficiency; a tight coupling is not inherently more efficient. A motor step can require several chemical cycles and efficiently utilize all the energy from the chemical cycles.

As previously mentioned, the stall force is the maximum force the motor is capable of generating; the motor is exerting maximum force and its velocity has reduced to zero. Utilizing Equation 2.46, F_{stall} is shown in Equation 2.47.

$$F_{stall} = \frac{\Delta G^0}{\delta} + \frac{k_B T}{\delta} \ln \frac{[B]}{[A]} \qquad (2.47)$$

In Equation 2.47, ΔG^0 is the Gibbs free energy with no applied force and δ is the step size. Recall that δ here comes from the $x_B - x_A$ from Equation 2.45. Thus, with small step sizes, more force is required to stall the motor; this is analogous to low gear in a human-scale vehicle. It's interesting to note that, in some cases, the process can be reversed; pushing a motor backwards can reverse the chemical process thereby creating fuel. ATP synthesis via a reversible motor has been demonstrated; the motor, when a large force is applied that reverses the operation, causes ATP to be created from ATD and P_i.

Now that we have discussed the forces and motion involved in molecular motor movement, we discuss the highway system upon which the molecular motors travel, which are the network of microtubules.

2.3 MICROTUBULES

One fascinating aspect of nanonetworking is the greater potential to leverage self-assembly compared to human-scale networks. As discussed in the previous chapter, human-scale network connectivity has shown signs of being scale-free, rather than uniformly random. There are human-specific factors at work controlling how human-scale network structures form, for example, market dynamics and human interaction. With nanoscale networks. there are physical properties at work, which may ultimately be easier to control.

Another area of research in molecular motor-based communication is the method used to deploy the topology of microtubule rails needed before communication can begin. The two main focus areas are self-assembly and track deployment via lithographic methods [65]. The idea of microtubule tracks being developed via molecular self-assembly, a natural process [66], is the most exciting possibility and some exciting research is currently being conducted to investigate the feasibility of such network topology deployment. Self-assembly naturally occurs, usually in thermally activated self-assembly [67]. This naturally occurring biological method uses diffusion to bring the different parts of the microtubule in contact and allow for binding. This method for nanonetworks is slow and hard to control. The use of molecular motors to propel and control self-assembly is proposed in [68], which demonstrates the use of molecular motors for self-assembly and how they may speed up the process leading to strong bonds in the structure and more control over the structure, all required elements for nanonetworks. One of the issues in self-assembled networks is control over the process versus node complexity [52]. These issues will be discussed in Chapter 6, which discusses information theory.

The apparent self-organizing ability of microtubule networks has been an enigma to researchers. There have been suggestions regarding the possibility of collision-based computing using microtubules [69] to the possibility that quantum computation takes place within such networks giving rise to consciousness [70]. Microtubule networks form from within a "soup" of smaller tubulin components and there appears to be a form of communication and organization that takes place at this fundamental level. Living matter has evolved for over 3.5 billion years and has constructed nanoscale technologies that we still are discovering and trying to harness. Microtubules are dynamic network structures that are continuously forming and disassociating. This can be seen most clearly in the act of treadmilling, where a section of a microtubule grows at one end and disassociates at the other, but both processes occur at the same rate yielding an apparent consistent tubular structure. It has been suggested that there are tubulin trails, namely, locations of high tubulin concentration, perhaps from where a microtubule previously disassociated, that serves as a communication link for a likely new location for the formation of a new section of a microtubule. While shrinking, a microtubule releases a chemical trail of tubulin, which quickly become active building blocks for microtubules once they become associated with GTP again. New microtubules will tend to grow towards the location of such tubulin trails. The rate of growth when treadmilling occurs is on the order of 1 to 10 μm per minute. It is suggested that this behavior is similar to pheromones. Microtubule networks are also surprisingly sensitive to external fields such as gravity and magnet fields. Changes in such fields cause a change in the pattern formation that occurs in the microtubule networks. The microtubule networks contain self-similar structures at several scales and consume chemical energy in the form of guanosine triphosphate (GTP) in order to maintain their fight against entropy, that is, their organized state. Many phenomena have been observed indicating self-organized behavior under various conditions, for example, oscillations in pattern formation, concentration waves moving through the system, and more complex pattern formation when molecular motors are added to the system. Understanding and controlling microtubule pattern formation is an ongoing research effort; it will not only allow control of the nanoscale communication network, but it could have significant benefit in detecting and curing diseases caused by abnormalities in the cytoskeleton, including cancer.

2.3.1 Topology and persistence length

The topology of the microtubule or actin rail transport system plays an important role in molecular motor transport. One can imagine molecular motors as trains on a railroad or cars on roads. There are two main types of biological rail topologies: uniaxial and centered. Uniaxial systems are found in axons, tubelike bundles of filaments, while centered systems are the starlike, or aster, systems found extending from centrosomes in cells. Centrosomes are small organelles within cells. An organelle is any small portion of cell with a specialized function and is usually enclosed within its own membrane. The enclosure of organelles are often lipids, which are a broad class of biological molecules that includes fatty acids and their derivatives. The centrosome is a specific organelle that is the microtubule organizing center within animal cells. The centrosome regulates the cell division cycle.

Clearly, molecular motors take up space along the track system, whether the track system is composed of microtubules or actin filaments, and exclude other motors from occupying the same space. Thus, it is possible for "traffic jams" to occur. As one might expect, in uniaxial systems, or aligned, tubelike bundles, the area in which congestion occurs grows with motor density and spreads throughout the whole system. In centered systems, the traffic jams grow more slowly and dissipate in a more complex manner. The impact of such traffic jams is also influenced by the volume of the cargo being carried by the motors. A larger volume cargo will take up more space and a heavier cargo will require more force to move.

Keep in mind, based upon our previous discussion of Brownian motion and random walks, the the motor is subject to collision with surrounding fluid molecules; thus it is possible for the motor to be forced off its track. Experiments tracking the motion of motors detect a period of directed motion along the track followed by periods of random motion as the motor is loosened from its track and diffused through the surrounding fluid. If there are motor-motor or motor-filament interactions, there is the possibility for interesting self-organizing behaviors to occur. One simple interaction is spatial mutual exclusion: motors cannot occupy the same space. A related effect is that for a high density of motors on a filament, the ability for other motors to bind the filament is reduced. It is also known that mutual hindrance slows down the movement of motors in areas of high density.

The cytoskeleton plays a large part in the organization of the intracellular architecture. The speed of molecular motors, their processivity, the time they spend bound at microtubule ends, may have a strong influence on the patterns generated by the microtubules at steady state. An interesting motor-microtubule interaction occurs based upon experiments in which motors are added to randomly oriented microtubules. At low motor concentrations, microtubules remain randomly oriented. When the concentration of motors is increased, kinesin forms microtubule vortices at intermediate concentration and asters at higher concentration. As previously mentioned, an aster is a star-shaped structure. A vortex is a spinning formation of microtubules around a central mass. The speed and rate of rotation of the surrounding microtubules are greatest at the center, and decrease progressively with distance from the center. The motors appear to accumulate in the aster center. Motor complexes have been modeled as two identical motors, which could be bound to one or two microtubules or diffuse freely in the surrounding fluid. Motor complexes bound to two microtubules exert a force on the microtubules, causing them to move. This is the ultimate reason for self-organization in this system. Essentially, a motor complex, moving towards the end of two microtubules simultaneously, pulls them into a "V" shape; this could be the reason for the aster-shaped patterns that occur [71].

The microtubule network is comprised of chain segments that may not be straight, but rather have some curvature. In general, the tube chain can range from nearly straight line segments to spaghetti-like highly curved chains. Persistence length measures this curvature [72]; it is the distance over which the direction of the chain persists. Thus, a long persistence length indicates a chain that is very straight while a low persistence length indicates a highly curved spaghetti shape. More specifically, the persistence length is the rate at which the tangents taken along each segment of a microtubule become decorrelated from one another. If $R(s)$ is a point on a segment s, then let $u(s)$ be the unit tangent vector, $u(s) = \frac{\partial R}{\partial s}$. The orientations of the unit tangent vectors for segment 0 and segment s is quantified by the inner product, $\langle u(s) \cdot u(0) \rangle = e^{-s/\xi_p}$, where ξ_p is the persistence length. For longer persistence lengths, or for shorter tubes, the microtubules will be straighter. For longer tubes and shorter persistence lengths, the impact of decorrelation along the chain tangent becomes more significant. We can approximate the curved microtubules as many smaller random chains that happen to be connected end to end, but with decorrelated alignment. Thus, shorter persistence lengths will tend to decrease the percolation threshold, which is important in the explanation of network conductance that follows.

One way to estimate persistence length is via Equation 2.48. Here $\langle l \rangle$ is the mean bond length, $\bar{l}_1$ is the first bond in the chain and there are N bonds.

$$l_{ps} = \frac{\left\langle \sum_{i=1}^{N-1} \bar{l}_i \bar{l}_1 \right\rangle}{\langle l \rangle} \tag{2.48}$$

To review, microtubules have an outer diameter of $\approx$ 25 nm and lengths of several micrometers. Microtubules are important for defining morphology and providing mechanical stiffness to cells. The mechanical properties of microtubules have been studied intensively for over a decade. The consensus has been that microtubules can be characterized by a persistence length of 48 micrometers [73].

2.3.2   Towards routing and the ability to steer molecular motors

Molecular motors move unidirectionally along their actin or microtubule tracks. The direction is in reference to polarized ends of the tracks. Microtubule tracks are composed of proteins known as α-tubulin (minus-end) and β-tubulin (plus-end). The plus-end and minus-end refer to the polarity of the component proteins that are exposed, one pole on each end of the microtubule. Microtubules are dynamic network structures that are continuously growing from the plus-end. The minus-end may be capped or it may lose its cap and begin depolymerization, or disassembling. Kinesin motors move along microtubules predominantly towards the microtubule plus-end; dynein motors move towards microtubule minus-ends. Myosin motors move along actin filaments, which tend to have a more randomized network structure in vivo. Thus, long-distance transport in cells is predominantly via kinesin and dynein along microtubules. An interesting phenomenon is that kinesins do not all move with the same polarity on microtubules. While most kinesins move towards the microtubule plus-end, some move towards the minus-end. Thus, kinesin motors are able to carry out both plus- and minus-end transport in the cell. Research in [74] suggests that the directionality is caused by the neck, or "stalk," structure of the motor. The tilt of the stalk differs between the kinesin motors moving with different polarities. The tilt of the stalk may impact the motor heads to attach to the microtubule binding sites in a preferred orientation. Once the motors reach their polarized destination along the microtubule, they detach from the microtubule track and exhibit Brownian motion in the surrounding fluid until they happen to collide with, and attach to, another microtubule.

With regard to routing along a microtubule track, several possibilities occur when a motor reaches an intersection of microtubules. The motor can pass the intersection along the same microtubule, in other words, ignore the intersecting tube and continue along the same track. The motor can pause at the intersection. The motor can switch to the perpendicular microtubule, in other words, it can take a left or right turn following the new microtubule. Finally, the motor can simply dissociate, that is, it can detach from the track and simply float away. The majority of encounters detected experimentally indicate either passing or dissociation events occur at intersections [75].

Promising work related to steering such small structures has been demonstrated [76]. In experiments, kinesin motors are attached to a surface and a microtubule filament is transported by the operation of the stationary molecular motors. Thus, in this configuration, many molecular motor feet, on stationary bodies, are oriented up in the air and carry a segment of a microtubule by their walking motion. The microtubule segment is carried along like a person that falls into a crowded pit after a concert and the crowd carries them in the air, handing the person from one to another. Since the kinesin motors are stationary, it is the microtubule filament that moves. By applying an electric field, the section of microtubule being transported can change course by up to 90 degrees, running parallel to the electric field toward the positive electrode. Microtubule segments have been actively steered using this mechanism into the desired channel of a Y junction.

2.4 INTERFACING WITH MOLECULAR MOTORS

Now we come to the question regarding how to interface with the molecular motor, namely, how a transmitter emits the information and how the receiver picks it up. One approach is to use a form of ligand-receptor binding. A ligand, from the Latin "ligare," to bind, is a substance that is able to bind to a biomolecule and form a complex that can serve a biological purpose. It can be a signal triggering molecule binding to a site on a target protein, by intermolecular forces such as ionic bonds, hydrogen bonds and Van der Waals forces. The binding is usually reversible, which is ideal for cargo loading and unloading from a molecular motor. Ligand binding to receptors alters the chemical conformation, that is, changes the three-dimensional shape of the receptor protein. The conformational state of a receptor protein determines the functional state of a receptor. The tendency or strength of binding is called affinity. One particular form of this approach is mentioned in [77] and [78], by specifically using an avidin-biotin binding to load information molecules onto carrier molecules. Other suggested approaches utilize ultraviolet light to induce chemical reactions to control both the rate of motor transport by activating or inactivating the ATP energy source for the motor [77], or activating a coating that will reversibly bind with a cargo [79]. It should be noted that the tail of the molecular motor has a complex structure allowing it to bind to many different types of cargo. Different parts of the motor can bind to different surfaces.

The process of DNA hybridization, used in the transportation of the information molecule, is further explained in [80]. Hybridization is the process of combining complementary, single-stranded nucleic acids. Nucleotides will bind to their complement under normal conditions, so two perfectly complementary strands will bind to each other readily, which is also called annealing. However, a single inconsistency between the two strands will make binding between them more energetically unfavorable. Annealing may be reversed by heating the double-stranded molecule of DNA to break the hydrogen bonds between bases and separate the two strands, which is called denaturation. A similar communication network based on molecular motors is proposed in [80]. The key idea in this system is that the information molecule is moved over immobilized motor proteins, which are aligned along a microtubule. DNA hybridization takes advantage of the complementary nature of short- and long-stranded DNA. The key concept is that DNA hybridization exerts a stronger binding force with longer complementary pairs. The motor transporter has a short DNA complementary strand, with enough binding energy to pick up the longer complementary strand of the information cargo. When the motor transporter reaches its destination and thus passes over the longer complementary strands (complementary to the information cargo's long strand), the cargo will favor attachment to the longer complementary strand at the destination and thus detach from the motor. The motor, free of its cargo, can be reused to transport information cargo as necessary. This approach has been implemented in [81]. Further explanation of the microlithographic track system and methods for controlling it can be found in [78, 82].

2.5 MOTOR VELOCITY, RELIABILITY, AND BANDWIDTH-DELAY PRODUCT

Large numbers of molecular motors traversing a complex microtubule or actin network have the potential to carry a significant amount of information. Here we consider the impact of such a system upon protocols used to transmit information. While it's still an open research question as to how many biological cell signaling mechanisms work in detail within living organisms, our main focus in this chapter is the intersection of molecular motor information transport and communication networks. Molecular motors can be engineered to work in vitro, in which case one can imagine that standard communication protocols will be utilized. Recent communications research has looked to biological systems for biologically inspired approaches to help researchers innovate new improvements in networking. A more interesting question may be the reverse, to ask to what degree biological systems have evolved to incorporate communication protocols that are similar to currently engineered protocols. One of the main characteristics affecting a protocol is the bandwidth-delay product. Here we define and explain the bandwidth-delay product and then discuss its impact on protocols used to transfer information.

2.5.1 Automatic repeat request and reliable molecular motor communication channels

The bandwidth-delay product refers to a communication channel and refers to the bandwidth of the channel, for example in bits per second, and the propagation delay of the channel, in seconds. Thus, the bandwidth-delay product is in units of bits and, if one thinks of the channel as a pipe conducting fluid, it measures the amount of fluid in the pipe when the pipe is full.

Reliability in communications is critical both in human-scale and cellular communications. Any form of reliable communication requires some form of feedback in order order to ensure that proper transmission has taken place. In human-engineered networks, automatic repeat request (ARQ) protocols are some of the simplest and most effective protocols that combine error detection and retransmission to ensure that information is sent correctly despite errors. The simplest of the ARQ protocols is the Stop and Wait ARQ. In this reliable protocol, a message is sent and the sender waits for an acknowledgment before sending the next message. If no acknowledgment is received after a specified timeout, then the message is considered lost and the message is resent. If an acknowledgment is received, then the next message is sent. The acknowledgment acts as a simple feedback mechanism. Consider that in the molecular motor system, messages, in the form of molecules, are attached to molecular motors. As previously discussed, there is a probability that molecular motors may detach from their rails, or the motor may take the wrong route, and the transmitted message may not reach the destination. Without some type of feedback, in the form of an acknowledgment, the transmitter cannot know whether the message needs to be resent. Although the Stop and Wait protocol is simple, it can also be very inefficient, since one acknowledgment is required for each message and every message has to wait for an acknowledgment from the previously sent message. It is well-known that if the bandwidth delay product of a communication channel is high, then much of the channel is wasted when using such a protocol. In the case of a molecular motor system, the bandwidth, in terms of packets per second, where a packet is one motor cargo package, is dependent upon the velocity of the motors, the diameter of the microtubule and the size of the motor, and the maximum flow over all the branches in the microtubule network.

Note that, if the acknowledgment is transported by a molecular motor in the reverse direction, for example, then confusion can arise if the acknowledgment motor is lost due to detachment or taking an incorrect route. Sequence numbers are used as a solution in human-engineered communication systems. This allows a lost or delayed acknowledgment to be detected.

The efficiency is shown in Equation 2.49, where n_f is the number of bits in the information frame, n_a is the number if bits in the acknowledgment frame, and R is the bit rate of the transmission channel. t_{prop} is the time taken for the cargo to traverse the microtubule network and t_{proc} represents the processing time. t_f is the time to receive a frame, or cargo, starting from the beginning of the cargo, to the end of the cargo. In other words, the cargo itself may be very long and require time until the back end of the cargo is received after the front end has arrived. The processing time includes any time required to attach/detach the cargo from the motor as well as to either create or detect the biological signal within the cargo packets. This equation is stating the total delay from the time a message is transmitted until the corresponding acknowledgment is received and is comprised of the message propagation time from the sender to the receiver, the propagation delay time of the acknowledgment from the receiver back to the sender, the processing time at both the send and the receiver, the time to receive the entire cargo payload, and a similar time for the acknowledgment cargo.

$$t_0 = 2t_{prop} + 2t_{proc} + t_f + t_{ack} = 2t_{prop} + 2t_{proc} + \frac{n_f}{R} + \frac{n_a}{R} \tag{2.49}$$

Equation 2.50 shows the effective transmission rate for such a system. In this case, the effective transmission rate may be much less than the bandwidth of the channel. n_0 is the amount of any overhead in our cargo. This may be additional material in the cargo to help it attach/detach from the motor as well as any error coding information that may be part of the biological information payload. There is more about coding and biology in a section coming up shortly.

$$R^0_{eff} = \frac{\text{bits delivered}}{\text{total time}} = \frac{n_f - n_0}{t_0} \tag{2.50}$$

Equation 2.51 shows the efficiency of such a protocol. This is simply the ratio of the effective transmission rate from Equation 2.50 to the raw channel transmission rate R. For the efficiency to be large, the numerator should be large and the denominator should be small. The $\frac{n_0}{n_f}$ term reduces the size of the denominator and is the due to any cargo overhead. The $\frac{n_a}{n_f}$ term increases the size of the denominator which also reduces the efficiency. This represents the overhead due to the size of the acknowledgment cargo. The term $2(t_{prop} + t_{proc})R$ is the bandwidth-delay product.

$$\eta_0 = \frac{\frac{n_f - n_0}{t_0}}{R} = \frac{1 - \frac{n_0}{n_f}}{1 + \frac{n_0}{n_f} + \frac{2(t_{prop} + t_{proc})R}{n_f}} \tag{2.51}$$

More efficient reliable protocols include Go-Back-N ARQ and Selective ARQ, however, they also require more sophisticated processing. The trade-off between computation and communication is a fundamental problem in all forms of network communication at all scales. More will be said about this trade-off in Section 6.3 as it relates to self-organization and self-assembly. ARQ mechanisms are a form of reliability that is in contrast to forward error correction (FEC) techniques, which are clearly used in biological systems such as DNA which will be described in the next section.

The efficiency and effectiveness of communication protocols depends on being tuned to handle the bandwidth-delay capacity, for example, in terms of waiting for acknowledgments or handling congestion. For molecular motors, the channel is essentially mechanical and the capacity is related to the density of molecular motors that remain associated with the microtubule en route to a destination. ARQ protocols are most efficient when they allow the transmission pipe to remain fully utilized. However, there is evidence that greater molecular motor density increases the likelihood of motors disassociating from the microtubule.

2.5.2 Genetic communication and error correction

The previous section just discussed human-engineered ARQ protocols as a means of nanoscale communication. However, nature is far ahead of us in constructing nanoscale communication systems. It is extremely important for survival that information is conveyed without error in living systems in many ways and at the nanoscale. The genetic code is likely to be one of the first biological nanoscale communication mechanisms that the general public would consider if asked to consider how to implement such a concept. The word "code" in the term "genetic code" clearly indicates that we believe DNA contains information in encoded form. DNA plays a significant role in nanoscale communication for a variety of axillary reasons, including its use as an enabler of self-assembly, and its ability to act as a molecular computer, and finally it is itself a molecular scale transmission media. Let us a spend a moment to understand how much is known about DNA and the genetic code as an information channel, namely, what is known from an information theory point of view.

In 1957, Crick hypothesized that the genetic code was a comma-free code. A comma-free code is a code such that any partial reading of the code is an invalid code. Thus, a decoder will always know precisely when a full code has been read. In other words no separation between code words is required because the decoder cannot misinterpret reading part of a valid code. It would seem to make sense for DNA to use a comma-free code from the standpoint of making decoding easier. Unfortunately, nature is rarely that simple, and we always learn much from trying to understand it.

Let us take a step back and describe the mechanics of DNA. The information appears to be expressed in a linear string of letters represented by the following chemical bases: A (adenine), C (cytosine), G (guanine), and T (thymine). A and G are larger bases, known as purines, while C and T are smaller bases known as pyrimidines. This distinction will become important in understanding a promising proposal for the coding mechanism shortly [83]. The letters, represented by the bases, are held in place along a sugar backbone, which maintains their order. A wonderful feature of DNA is that certain bases fit snugly enough with one another that they form a weak bond. This process is known as hybridization. Understanding and controlling this feature allows engineering of a variety of applications, from self-assembly to DNA computing, and will likely be an aid to engineering any integrated nanoscale communication system. More specifically, A and T fit well together and C and G fit well together. These are known as base pairs. It is the hydrogen bonds between them that provide the weak, but stable and detachable, connection. As is well known, an entire sequence, or strand, of DNA that attaches with another strand using this mechanism provides the ability for DNA to replicate. A single strand can serve as a negative template for a complementary strand. When a new strand is created that fits this complementary strand, a duplicate of the original strand is produced.

In the above description of DNA replication, it was assumed that there were no errors. A true understanding of the DNA communication channel requires understanding the mechanism of coding and ultimately, the channel capacity. This requires understanding how much of the DNA communication channel is comprised of redundant error coding information. Even though work on understanding the DNA molecular channel has continued since Crick [84], it is only relatively recently that plausible coding mechanisms have been discovered. One such example is a relatively simple even-parity mechanism [83].

First, the notion of even parity is discussed, then the actual mechanism in DNA is explained. The goal of error coding is to enable easy and efficient detection of mistakes in the DNA sequence and at the same time, valid code words should be spread far enough apart in the entire set of possible code words that mistakes will not generate a new sequence that looks like a valid code. Even parity adds a single bit of information to a valid word that yields an even number of ones in a binary sequence. An odd number of ones would indicate an error. The "lock and key" nature of the base-pairing is enabled by hydrogen donors paired with hydrogen acceptors. Each of the DNA bases has a sequence of three connecting structures, either hydrogen donors or hydrogen acceptors. Thus, each nucleotide base can be considered a sequence of three binary values. The fact that the base is either a pyrimidine or a purine appears to serve as the final parity bit. Using this mapping, the DNA bases can conceptually be placed on nonadjacent corners of a cube such that any single bit error does not yield a valid code word. It has been shown that other chemical bases could also have served a role as DNA bases along with, or instead of, the current set of bases; however, they would not have formed a valid parity code.

2.6　INFORMATION THEORY AND MOLECULAR MOTOR PAYLOAD CAPACITY

A tacit assumption made thus far is that molecular motor cargo, in the form of molecules, particularly biological molecules, represents information. This brings up the question as to how much information may be contained within a biological molecule and how it relates to its physical properties, such as mass and shape. This is, in and of itself, a complex topic; information theoretic techniques have been applied to a wide variety of biological problems that may be leveraged for understanding the information capacity for engineered nanocommunications. Let us consider some of the prior applications of information theory to biological sequences in order to understand the information capacity.

2.6.1　Payload and information capacity

The most obvious application has been in attempting to characterize DNA and protein sequences; this work has attempted to determine how similar DNA is to a language, predicting likely sequences, and finding sequences that code for specific function both within and across species. In the case of nanonetworks, we are concerned with issues such as information carrying capacity, that is, what is the maximum payload and ultimately the bandwidth. But we are also concerned with any potential enablers for routing and addressing based upon the specificity of sequences and shapes to bind with receptors. This is has also been approached as an information theoretic problem.

DNA sequence analysis is an old and well-researched problem, so let us think about the more challenging problem of protein folding. This combines the dual problem of sequence analysis with conformational, or structural, changes. We start from a basic definition of entropy as discussed in Chapter 1 and shown in Equation 2.52 where p_i is the probability of each symbol that may occur in a sequence. Just like Boltzmann and Gibbs entropy in physics, entropy with regard to information is about measuring disorder; if the probabilities of the occurrence of the symbols are nearly equal, then the symbols are uniformly distributed and little can be inferred from the frequency of occurrence of any symbol. In this case, entropy is at its highest. If, on the other hand, some symbols occur more frequently that others, then it's likely that the frequently occurring symbol can be replaced with, or represented by, a shorter symbol. If one symbol significantly dominates another, then the entropy is significantly lower. A high information entropy in the molecular structure that is transferred implies a high transfer of information. A high information entropy also implies a potential to engineer a high information transfer rate, as the information transferred per molecule, or packet, should be higher.

$$H(x) = -\sum_{i=1}^{n} p_i \log_2 p_i \tag{2.52}$$

The basic definition of entropy in Equation 2.52 assumes symbols are independently generated; it is also known as zero order entropy, because it does not explicitly capture the correlation among adjacent symbols. An attempt to better capture the dependency between adjacent symbols is known as the kth order entropy and is defined in Equation 2.53. In this definition, w is a word of length k and represents every word of length k in the sequence.

$$H_k(x) = -\sum_{|w|=k} Pr[w \text{ occurs}] H_0(\text{successors}(w)) \tag{2.53}$$

Proteins consist of 20 different amino acids, or in our case, symbols. In Equation 2.54, p_i is the probability of finding the ith amino acid and $p(i|s)$ is the conditional probability of the ith amino acid occurring after the amino acid string s.

$$I_k(x) = - \sum_{i=1}^{20} \sum_{s}^{20^{k-1}} p_i p(i|s) \log_2 p(i|s) \qquad (2.54)$$

As k increases, more longer-range order is captured and the order entropy decreases. Ultimately, $I = \lim_{k \to \infty} I_k$. It has been found from analysis of protein sequences that I_1 is 4.18. This is known as the nonuniform entropy of a single amino acid in a protein. This provides one hint of the amount of information that could be transported by a molecular motor per molecule of cargo. However, there are other ways of representing the information payload, in particular, using the shape and conformational changes of a molecule to convey information, which are being explored.

2.6.2 DNA computing for nanoscale communication

There are fundamental and deep trade-offs between communication and computation that become more focused and of significant importance at the nanoscale. Much has been learned about this trade-off in the area of active networking, in which communication and computation become fundamentally integrated [1, 85]. In this section, we consider computational aspects that can aid nanoscale communication. In particular, the wide utility of DNA to perform critical computational tasks for nanoscale networking is discussed. DNA computing is a large topic in its own right; we only present a brief summary of the topic as it could relate to nanoscale networking. Benenson, in [86], describes a general mechanism for constructing a finite automata using DNA. Before describing the operation of this DNA finite automata, it is interesting to relate this DNA computational technique to discussions from Section 1.6.2 in which reversible computation and the energy required for computation were discussed. In particular, the cost of computation was related to the erasure, or loss of entropy of information, which required a corresponding increase in thermal entropy of $kT \ln 2$ per bit. In the Benenson automata, the input data, in the form of DNA, provides its own energy. The ordered sequence of input information encoded into the input DNA is cleaved and lost in the solution, resulting in an increase in entropy as the computation progresses. Simply put, the disordering of the input provides the energy required for the computation.

First, we begin with a quick review of the general Turing machine and then finite automata, which more closely resembles what the Benenson DNA computation resembles. A Turing machine is comprised of an input tape that has symbols encoded on it and a program in the form of states and transitions between states, depending upon the symbol that is read from the tape. In a Turing machine, an imaginary tape head reads a symbol from the tape and looks up its current state and the symbol that was just read from its set of rules, or program. The program indicates, based upon the symbol and current state, what the next state should be as well as which direction to move the tape, and whether, or what, symbol to write on the tape. This simple conceptual model is powerful enough to emulate any computational device.

A similar, but slightly more restricted device, is a read-only version. This is known as a finite automaton. It reads an input tape, but moves the tape only in the forward direction and does not write on the tape. The finite automaton's state may change depending upon its current state and what was read from the tape. Thus, the only way to observe a result from the finite automaton is to examine its state change, in particular its final state. In the general case, a finite automaton may have a set of states known as accepting states. If the final state ends in an accepting state, then we can recognize computationally significant information about the input as shown in Figure 2.2.

The Benenson DNA finite automaton operates upon a sequence of double-stranded DNA that acts as the finite automaton input. This double-stranded DNA has a special feature: one of the strands is longer than the other as shown in Figure 2.3. In other words, there is a single strand hanging off the end, similar to a tail. If a particular enzyme, known as a restriction enzyme, has a complementary strand to the tail that

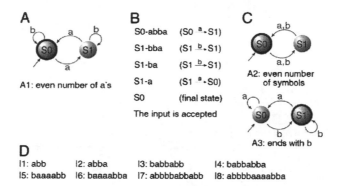

Figure 2.2 Example of a finite state machine that recognizes an even number of *a*s. Molecular motor. (Copyright (2003) National Academy of Sciences, U.S.A.)

is hanging off the data input sequence, it will hybridize, or connect, and then cleave, or chop off, a segment of DNA from the double-stranded DNA input sequence. This implements reading a symbol. The result of the cleaving by the enzyme leaves another tail at the end of the remaining input data. An important point to note is that the tail encodes both the current symbol and the last state. The exact point where the DNA input is severed determines the tail that remains, which encodes the current state. A key showing how states and symbols are recognized is shown in Figure 2.3. Thus, the restriction enzymes represent the state transition program. Only an enzyme that matches the current symbol and last state will hybridize and cleave the next symbol from the input sequence, leaving the next symbol and state exposed. When the final termination symbol remains at the end of the DNA input sequence, the resulting state will be encoded upon it as well; the result has been obtained. This final state, or nucleotide sequence, can be detected and used to trigger events as required depending upon the application. The transition rules are shown in Figure 2.3.

An example computation is shown in Figure 2.3. The set of enzymes shown in Figure 2.3 are available in the solution. The enzyme that forms a complementary match to the tail of the DNA sequence (or input tape) joins to the tail and breaks off a portion of the sequence corresponding to the appropriate transition rule. The next symbol and state remain exposed in the new tail on the DNA sequence. This process continues until a terminator symbol is reached. The final state once the terminator symbol is reached is the result of the computation.

This approach has been suggested for use in a variety of biologically programmed activities, including determining when to release therapeutic agents for drug delivery. It is not hard to imagine how each symbol might be a network address that is cleaved from a source-routed biological information packet or act as sequence number for more reliable communication as discussed previously related to molecular motor cargo delivery.

2.7 OTHER TYPES OF MOTORS

There are several types of cellular scale motors other than molecular motors that ride on rails. Here we introduce flagellar, synthetic DNA, and catalytic motors. All of these motors could provide mechanical motion for the transport of information. The molecular motors that ride on rails, that is, that ride along microtubules or actin, appear to be useful for short, very directed transport. Catalytic motors appear to be relatively medium range transport mechanisms, and flagellar motors are the longest range among the mechanisms discussed in this chapter. As always, there is a trade-off; the shorter range rail-riding molecular

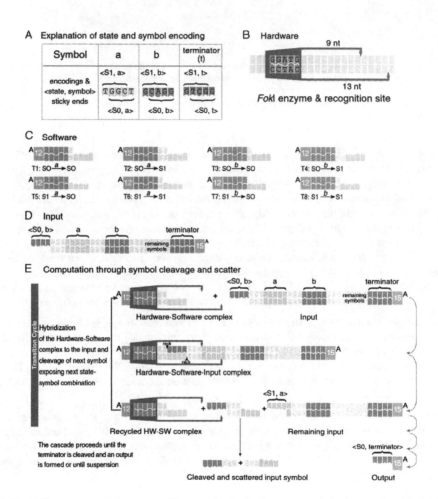

Figure 2.3   The DNA enzyme eats the input tape and leaves a tail indicating the current symbol and state. (Copyright (2003) National Academy of Sciences, U.S.A.)

motors can be highly directed and targeted given the rail system upon which they ride, while flagellar-driven motion appears to be less well-directed.

2.7.1 Flagellar motors

As an example of a flagellar motor, consider the bacteria known as *Serratia marcescens*, or *S. marcescens*. It is a single-cell organism and has an elliptical body that is 0.5 μm in diameter and 2 μm long. Its motion is caused by flagella, flexible helical structures that are 20 nm in diameter and 10 μm long. The *S. marcescens* bacteria has from four to ten such flagella. The flagella rotate at a surprisingly fast rate of 100 to 300 Hz. The propulsion force is approximately 0.45 pico-Newtons. Recall from our discussion of the Reynolds number and Brownian motion that this little organism is greatly effected by collision with the small particles of its fluid environment rather than its own inertia. It appears that the force of the flagella propels the bacteria in a random direction at approximately 47 μm/s for one second. Note that the flagella can switch the direction of their rotation. In the counterclockwise rotation, the flagella, being long and flexible, bundles up into a coil; this directed motion is called a run. When the bacteria switches to a clockwise rotation, the bundle is uncoiled and the bacteria goes into what is called a tumble. The tumble lasts for only a fraction of a second, much shorter than the directed motion that it undergoes in the counterclockwise motion, or run, phase. The result of alternating these two phases is our now familiar random walk. Just as with the rail walking molecular motors, cargo can be attached to organisms with molecular motors.

2.7.2 Synthetic DNA motors

It is possible not only to compute with DNA, but to construct synthetic molecular motors directly from DNA by creating the proper three-dimensional connections between DNA strands. A molecular switch component for a motor can be constructed such that it changes shape based upon external conditions, such as temperature or ionic concentration. The shape can change by the binding of a signaling molecular, which could be simply another DNA strand. Because of the ease with which we are now able to control DNA, it makes an ideal construction material for biomechanical devices. It is well known that complementary strands of DNA will unite, or hybridize. The specificity, which is the precise control over the location where DNA will join together, is very high. Thus, by generating the correct nucleotide sequences, one can confidently control what components will connect together, or hybridize. Application of DNA to other components to enable them to self-assemble will be seen in Section 4.6 when we discuss self-assembly of nanonetworks from nanotubes and nanowire.

It should be noted that other biological components, such as RNA and proteins in particular, offer a much greater variety of interactions that would allow for creating more complicated structures. The molecular motors that we have been discussing are an example of such a protein construction. However, DNA is simpler and better understood, so our first completely synthetic motors are being constructed from DNA [87]. DNA has a persistence length of 50 nm, thus for devices built from short sequences of several nanometers, the components can be considered to be stiff and straight. Single-stranded DNA has a much lower persistence length and is very flexible and can act as joints. Restriction enzymes can be used to cut DNA at specific points. A number of techniques can be used to dynamically change a DNA structure, for example, the spiral shape of the double strand can be flipped from a right-handed helix to a left-handed helix or a strand with a longer, better match can replace an existing strand causing a change in conformation. This strand-exchange technique has been used to implement a wide variety of different mechanical devices such as nanoscale tweezers. Autonomous DNA devices have been created that traverse along double-stranded RNA and can carry a payload [88].

2.7.3   Catalytic motors

A catalyst is a substance that either increases or decreases the rate of a chemical reaction by its presence. Catalysts generally react with input chemicals to form intermediate substances that subsequently give the final reaction product and, in the process, regenerate the catalyst. One of the most well-known examples of catalysis is the increase in the rate of decomposition of hydrogen peroxide into water and oxygen. Introduction of a catalyst greatly increases the rate at which the decomposition occurs. This interesting effect occurs if the catalyst is distributed nonuniformly; an unbalanced net force or net torque will be generated. If the catalyst is attached to one side of a cellular scale object, the object will move.

2.8   SUMMARY

This chapter focused on the molecular motor transport of information. The most important benefit of this chapter is the explanation of how the motion of molecular motors is modeled with the ultimate goal of determining the factors that influence how much information they may carry. To do this, we derived equations describing Brownian motion and the Langevin equation. These are widely used throughout physics and in other areas related to molecular communication; they are well worth the time and effort to understand. Introductory thermodynamics was presented in order to understand the energy and forces involved, thus yielding information about the payload weight that a molecular motor can carry and its impact on velocity. We also consider the amount of information that may be packed into a single molecule of a molecular motor's cargo. Finally, we reviewed a mechanism for computation using DNA; computation will be required to enable a complete network architecture to be implemented, which is discussed further in Chapter 7.

   Now that we have discussed molecular motors, what other nanoscale and molecular signaling mechanisms lie within the biological realm that we might harness for communication? What other mechanisms do cells use for communication, both within a single cell and among multiple cells? Are there useful nanoscale and molecular biological mechanisms for longer-range communication, such as neurons and olfactory-based mechanisms, that is, smell? We will explore these in the next chapter.

2.9   EXERCISES

Exercise 10   Molecular Motor Speed

Approximately one hundred times per second, kinesin hydrolyzes one adenosine triphosphate (ATP) molecule, generates a force of up to 7 pN and takes an 8-nm step towards the plus-end of a microtubule that is two microns in length. Assume the cargo is a molecule capable of representing an information entropy of four bits and weighs approximately 10 picograms.

   1. What is the bandwidth of this molecular motor channel?

Exercise 11   Molecular Motor Force

Approximately one hundred times per second, kinesin hydrolyzes one adenosine triphosphate (ATP) molecule, generates a force of up to 7 pN and takes an 8-nm step towards the plus-end of a microtubule that is two microns in length. Assume the cargo is a molecule capable of representing an information entropy of four bits per picogram.

1. What is the maximum bandwidth of the molecular motor channel?

Exercise 12   Molecular Motor Network Flow

Assume a section of a microtubule layout appears as shown in Figure 2.4. Assume information packets are attached as molecular motor cargo to the motor which travels at a rate of 1 nm/s on each segment of the microtubule. Assume motors with cargoes are released at node four and travel to node eight. The corresponding adjacency matrix for this graph is shown in Equation 2.55.

$$\begin{pmatrix} 0 & 1 & 0 & 1 & 0 & 1 & 0 & 1 & 1 & 0 \\ 1 & 0 & 1 & 1 & 0 & 0 & 0 & 1 & 0 & 1 \\ 0 & 1 & 0 & 0 & 1 & 0 & 1 & 0 & 1 & 1 \\ 1 & 1 & 0 & 0 & 0 & 0 & 0 & 0 & 0 & 0 \\ 0 & 0 & 1 & 0 & 0 & 0 & 0 & 0 & 0 & 1 \\ 1 & 0 & 0 & 0 & 0 & 0 & 0 & 0 & 0 & 1 \\ 0 & 0 & 1 & 0 & 0 & 0 & 0 & 1 & 1 & 1 \\ 1 & 1 & 0 & 0 & 0 & 0 & 1 & 0 & 0 & 1 \\ 1 & 0 & 1 & 0 & 0 & 0 & 1 & 0 & 0 & 0 \\ 0 & 1 & 1 & 0 & 1 & 1 & 1 & 1 & 0 & 0 \end{pmatrix} \qquad (2.55)$$

1. What is the total bandwidth?

2. What is the bandwidth-delay product?

3. How does the addition or removal of a link in the path from the source to the destination impact the performance?

Table 2.2

Microtubule Segments

Angle (Degrees)
10.721
-336.594
378.065
887.204
-478.892
143.909
747.586
-12.445
794.416
-184.214

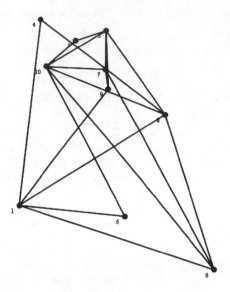

Figure 2.4   A portion of a random microtubule network.

Exercise 13   Microtubule Topology

The rate of decorrelation in tube angle with respect to distance is measured by persistence length. Assume the single microtubule has a length of 100 nm and is comprised of ten equal-length segments. The angles of each segment are as shown in Table 2.2.

1. What is the persistence length of this tube?

Exercise 14   Random Walk

Assume a cell is in the shape of a sphere with a radius of R μm, fluid density D, and a characteristic step of l μm.

1. How many steps does the particle take to move from the cell center to the cell wall?

2. What is the mean distance between collisions?

Exercise 15   Molecular Motor Thermodynamics

Assume we have a kinesin molecular motor. The average stall force has been experimentally reported to be approximately 7 pico-Newtons (10^{-12} Newtons). The step size is 8 nanometers. Assume the relative concentration of ADP to ATP is 2.5.

1. What is the standard free energy ΔG^0, that is, the free energy when no force is applied?

2. Assuming conditions remain the same, what is the stall force when the concentration of ADP to ATP is increased to 10?

3. What impact does the temperature have upon the stall force?

4. If the step size could be increased in order to increase the velocity and the corresponding bandwidth of the molecular motor, and information packets require a force of six pico-Newtons to transport, what is the maximum step size that could be taken?

Exercise 16   Molecular Motor Reliability

A one megabyte "file" of information is to be encoded as cargo attached to molecular motors. Each motor can carry four bits of information in this manner. Also assume that one motor out of ten thousand dissociates from the microtubule never to return. The average speed for a Myosin II motor is 8,000 nm/s. Assume a Myosin II motor is transporting the information over a distance of 100 nm. Also assume that the microtubule structure can hold a density of ten motors per nm of microtubule before dissociation becomes significant.

1. What is the bandwidth-delay product of this nanoscale network?

2. What is the error rate?

3. What is the probability that the entire file is transmitted without error?

4. Assume that it is possible to implement the simplest ARQ mechanism, namely, Stop-and-Wait ARQ. Break the file up into blocks of size N, where acknowledgments can be sent after every N blocks. How long does it take to deliver the information with blocks of size 80, 800, and 8,000?

Chapter 3

Gap Junction and Cell Signaling

Biology is complex, messy and richly various, like real life; it travels faster nowadays than physics or chemistry (which is just as well, since it has so much farther to go), and it travels nearer to the ground. It should therefore give us a specially direct and immediate insight into science in the making.

Sir Peter B. Medawar, Induction and Intuition in Scientific Thought (1969)

The previous chapter covered molecular motors as an information transport mechanism, particularly biologically inspired molecular motors that ride along the microtubule structure of the cell cytoskeleton. The nanonetwork mechanisms from the last chapter operate primarily as a nanoscale transport mechanism within cells. This chapter continues the exploration of biological nanoscale network mechanisms, but focuses upon nanoscale communication *between* cells, rather than within. There are a myriad of intercellular nanoscale communication mechanisms; cell signaling is a broad and complex subject. Our goal in this chapter is to focus upon a few potential approaches in depth and then briefly discuss broader mechanisms. We start with an explanation of gap junction signaling. Following that we cover biological signaling mechanisms over longer distances between source and destination cells. Cells, those of both single-celled organisms and within multicellular organisms, have their own individual and independent lives with a strong need to communicate with other cells. We will see that they use a variety of algorithms and coding techniques to communicate that are not so different from what we, as humans, have engineered into our communication systems. There are many examples, but a few of note are amplitude and frequency modulated calcium signaling, which is a nanoscale communication mechanism both within and between cells. Redundancy and error coding appear in chemical cell signaling techniques. In fact, cells, having evolved much longer than human technology, have innovated mechanisms that we are only now beginning to understand and may someday leverage for human-scale systems. At the nanoscale, biological systems are clearly masters.

Table 3.1 summarizes the concepts of molecular and nanoscale communication by comparing today's networking technology, termed "traditional" networking, with nanoscale communication [89]. This contrast and comparison helps to differentiate and explain nanoscale networking and its fundamental technology, advantages, and challenges. The physical medium of today's traditional networking technology is largely electromagnetic; while nanoscale networking also includes electromagnetic mechanisms, in the form of a single carbon nanotube radio for example, it is primarily focused upon single or small sets of molecules as the physical medium. The types of signals in today's networks are primarily electrical and optical, while in nanoscale systems there is a tendency towards chemical and mechanical signals, as well as electrical and optical. This has implications on the speed of propagation, namely, instead of the speed of light, nanoscale systems include the relatively slower propagation of molecules, which, for the nanoscale distances involved, can be a more effective form of information transport. While traditional networks are primarily wireless or fiber-optic, those at the nanoscale tend to include more liquid environments. Traditional human-scale networks have receivers designed to decode information into forms readily interpreted by humans, such as voice, video, or text, while nanoscale receivers are more likely to decode molecular-scale information for use by other molecular and nanoscale components. Thus, coding and decoding are likely to involve chemical reactions. Finally, traditional human-scale networks are designed for human consumption of the resulting information; much emphasis is placed upon the reliable and timely transport of the information perceptible by the human eye or ear. This requires a significant amount of bandwidth and energy. However, these signals are ultimately perceived within the human body using nanoscale and cellular scale mechanisms. Nanoscale networks, on the other hand, are likely to be more tolerant of error, operate in a more stochastic manner, and require much less energy.

Table 3.1
Comparison of Communication Features

Property	Traditional	Nanoscale
Carrier	Electromagnetic field	Molecule
Type of signal	Electrical-optical	Mechanical-chemical
Propagation speed	Speed of light (3×10^5 km/s)	Relatively slower
Environment	Airborne or fiber	Aqueous
Information transmitted	Voice, text, video (human interaction)	Physical/chemical states
Receiver behavior	Receiver interprets encoded data	Chemical reactions
Other features	Accuracy and high energy required	Stochastic and low energy

3.1 INTRODUCTION

Our focus in this chapter is upon the communication and networking aspects of gap junction and cell signaling; however, it is important to have an understanding of a few topics related to basic biochemistry and cell signaling. This chapter reviews another method for realizing nanonetworks via molecular communication, namely, that of signaling molecules. Rather than using a system of motor proteins as explained in the previous chapter, this chapter considers using signaling molecules that are not necessarily moved along a microtubule rail topology. The basic method is to either use an aqueous medium that separates cells or nanomachines, or have the cells connect through gap junctions. In both cases, an information molecule is released, either into the medium to propagate through Brownian motion or through a gap junction into an adjacent cell.

Chemical signaling among cells is the basis for this approach to nanoscale communication networks. Largely biological in nature, this approach can take advantage of previous work from biology to understand the best way to approach this challenge. As outlined in [90], cells can use three methods of signaling: (1) they can secrete chemicals that signal another cell, (2) they can contain plasma-membrane-bound signaling molecules that influence other cells that make direct physical contact, and (3) they can form gap junctions that directly join the cytoplasm of the interacting cells.

There are gaps between cells that mediate the passage of small molecules from one cell to the other. Communicating junctions can be divided into two general types: (1) gap junctions that allow molecular passage, and (2) chemical synapses that communicate via indirect communication between two neurons. Of these two mechanisms, the gap junction has been the method most explored so far in proposed molecular-based communication networks. Gap junctions are found in most animal tissue cells [90], and allow small water-soluble molecules to pass directly from one cell to another. The estimated sizes of these gaps are about 1.5 nm [90]. This imposes a limit on what could be used as a signaling molecule. Small molecules, like amino acids and nucleotides, may be used, but larger nucleic acids and polysaccharides would not pass through. It also has been shown that the permeability of gap junctions can be controlled, by using pH or calcium ions, Ca^{2+}. This is a useful ability in constructing nanoscale network applications.

Chemical signaling is an important molecular communication method that is actively being researched for nanonetwork applications. Several methods have been found by which chemical signaling is achieved. Currently the most actively researched method for nanonetwork applications involves local chemical mediators that most cells in the body can secrete to influence the local environment. There are two other methods: (1) hormone secretion by cells within glands, and (2) the use of neurotransmitters. Multiple methods can be used to stimulate a cell to generate signals. Two of the most common methods are: (1) control of the intracellular concentration of cyclic adenosine monophosphate (also known as cyclic AMP or cAMP), and (2) control of the membrane-bound gated Ca^{2+} channel. It should be reiterated at this point that this book is oriented towards engineers, many of whom are not expected to be familiar with biochemistry. Thus, it it is important not be intimidated by the introduction of biochemicals, such as cAMP; it is not expected that the reader be familiar with the properties of such biochemicals. The nature and properties of such chemicals will be explicitly mentioned as they relate to information transport and networking. cAMP is synthesized from ATP (ATP was discussed in the last chapter) near the cell membrane. Ca^{2+} is an important molecule with many uses in natural molecular communication. Its use in molecular-based nanonetworks is an active area of research [91], which will be discussed in more detail in this chapter.

Cells use special receptors, which are proteins that bind to specific signaling molecules, giving cells the ability to respond to signaling molecules. This high affinity for signaling molecules is important in molecular-based nanonetworks, allowing specific molecules and concentrations of molecules to be used for information encoding. Another important feature of molecular communication useful to nanonetworks is the ability for cells of the same type to be engineered to give differing responses to a specific signal. Much variety exists in molecular signaling, including the ability to use signaling molecules that can cross the plasma membrane of cells, those that can only bind to the surface, and a combination of the two [90]. Signaling molecules move from one cell to another via a diffusive process through extracellular media, or in the case of the gap junction, diffuse through connecting cells. This process can be modeled using Brownian motion, discussed in the previous chapter.

The basic communication method using cell signaling takes place as follows. Information is encoded using an information molecule; the concentration of that molecule encodes the information. Calcium is currently one of the most widely used information molecules. A method for encoding information using concentrations of calcium is given in [92] and explained later in this chapter. Transmission begins when a nanomachine is stimulated to begin the propagation of the information molecule [91, 93]. The information molecule is propagated either through Brownian motion in an aqueous medium or through gap junctions in cells that are in direct contact. Information molecules are received by the target cells, either through gap junctions or through receptors.

3.1.1 Calcium signaling in nature

Long ago, cells evolved into multicellular organisms; thus, there was a need for coordinated action among the constituent cells. Messenger molecules were secreted to form a channel to carry the information required for multicellular coordination. Calcium (Ca^{2+}) and phosphate ions have become the ubiquitous channels for such coordination messages. The binding of both calcium and phosphate (phosphorylation) changes both the shape and charge of proteins and plays a significant role in cell signaling. More specifically, phosphorylation is the binding of a phosphate group (PO_4) to a protein (or other organic molecule). Cell signaling is also know as signal transduction; many enzymes and receptors are switched "on" or "off" by phosphorylation and dephosphorylation. With regard to Ca^{2+}, cells use a considerable amount of energy to control the Ca^{2+} concentration. Cells must constantly pump it outside the cell, in order to keep the concentration within the cell low. Many proteins have evolved within the cell to bind with Ca^{2+} in order to buffer it or to trigger responses. ATPase pumps continually push Ca^{2+} into the endoplasmic reticulum or out of the cell. ATPase was explained in Chapter 2; it is a class of enzymes that catalyze the decomposition of adenosine triphosphate (ATP) into adenosine diphosphate (ADP) and a free phosphate ion, which releases energy to drive chemical reactions. There exist voltage-gated Ca-selective channels that are able to conduct approximately a million Ca^{2+} ions per second yielding a greater than 10-fold change in intracellular Ca^{2+} levels within milliseconds. The channels appear to be highly selective, allowing only Ca^{2+} to exit the cell through cell gates. The selectivity appears to be controlled by the presence of external Ca^{2+}. When the external Ca^{2+} is removed, the cell gate becomes less selective and allows K^+ and Na^+ to exit as well. A mathematical model of calcium signaling that incorporates the frequency and amplitude modulation of the signal will be explained later in this chapter.

3.1.2   Gap junction signaling

Gap junctions are physical channels between adjacent cells. The cytoplasm within each cell is directly connected by the gap junction. The channel itself is comprised of a pore surrounded by hexamers of a protein called connexin. Connexin forms around the pores of both adjacent cells; the pores in both cells are adjacent to one another such that if they were both open, a direct flow could occur from one cell to the other. There are more than 20 different types of connexin proteins found so far and each imparts different physical properties upon the intercellular channel in terms of permeability and selectivity. The channels are dynamic in terms of controlling their permeability and selectivity and can respond to many internal and external factors.

A primary purpose of gap junctions appears to be the propagation of calcium as a signaling mechanism. Calcium waves are induced and appear to signal information relevant to cell processes such as cell growth and death. The wave nature of the calcium concentration fluctuations is intriguingly like the carrier wave of amplitude and frequency modulated signals in human-engineered radio transmission. In terms of signal transduction, calcium signaling appears to be a widely used second-level messaging system, that is, a signal originating from one signaling technique that is translated into a calcium signal for further propagation. The fact that the calcium signal appears in the form of waves indicates that a sophisticated encoding mechanism may be used to communicate a significant amount of information, not simply a binary on/off signal.

3.1.3   Intercell signaling

The cells within a multicellular organism form a complex network of communicating agents. The intra- and intercellular mechanisms of communication must work together in a coordinated manner in order to correctly transmit and receive the information necessary for the survival of an organism as a whole. This communication follows the same fundamental principles as human-scale communication although, having evolved over many billions of years, more advanced techniques are used that continue to inspire communication researchers. Multicellular organisms, with their need for complex intercellular communication mechanisms, did not appear for 2.5 billion years after single-cell organisms appeared. This time delay is indicative of the communication advances that were required.

Cell signaling has evolved primarily in the form of signaling molecules, some of which may convey information over long distances through an organism and others that only impact adjacent cells. Most cells within an organism send and receive information. Cells emit signals in the form of molecules that are received at destination cells by receptor proteins. Receptor proteins are usually on the surface of the cell in order to easily bind to the signal molecule. The binding of the signaling molecule with the receptor activates intracellular signaling pathways, which interpret the signal causing the cell to generate the appropriate response to that particular signal. In other words, information has efficiently been transmitted from a source to a target at the cellular- and nanoscale. Cell signaling is part of a complex system of communication that controls fundamental cellular activities and coordinates cell behavior. The ability of cells to perceive and correctly respond to their situation is the basis of organ and organism development, tissue repair, and defense. Errors in cellular information processing and communication are responsible for diseases such as cancer, autoimmunity, and diabetes.

Early signs of intercellular communication originated among single-celled organisms such as bacteria and yeast. While these are single-celled organisms, leading independent lives, they are able to sense their environment and their neighbors. They emit molecules that indicate their presence; the concentration of molecules increases with the density of the population of cells. This is known as quorum sensing and is used to coordinate group behavior among these independent single-celled organisms. Consider the complexity of the information that must be transfered, even in this early and primitive mechanism, given that communication is required to control their movement, antibiotic production, spore formation, and rate of reproduction.

3.1.4 Long-distance nanoscale biological signaling

There are many other forms of cellular and nanoscale signaling within biological organisms. Neural, endocrine, and pheromone signals communicate over relatively long distances. An introduction to these mechanisms is mentioned here; more will be said regarding neural and pheromone communication later in this chapter.

3.1.4.1 Neural and endocrine signaling

Longer-range communication can occur via neurons and endocrine signaling. Neurons are relatively long, specialized communication cells, which connect with other neurons at synapses, or neuron junctions. Electrical signals, or action potentials, travel the length of the axon, or long portion of the neuron. When the electrical signal reaches the synapse, a chemical neurotransmitter is released. The neurotransmitter carries the signal across the synapse to receptors in the post-synaptic portion of the target. The action potential can travel at speeds of up to 100 meters per second. The neurotransmitter communicates approximately 100 nm across the synapse in less than a millisecond. We discuss the potential for human-engineered neuron-based communication later in this chapter.

Another long-range mechanism is endocrine signaling, which relies on the release of signaling molecules, known as hormones, into the blood stream, where they may travel long distances through the body to reach their targets. While the neurotransmitter concentration is relatively high, because it is well contained, the endocrine signaling molecule concentration is generally very low. However, the neurotransmitter has relatively lower affinity for its target, which allows it to disassociate rapidly. This is necessary in order for the synapse to rapidly reset for the next signal. Thus, we can see some of the physical trade-offs that evolved in biological communication.

3.1.4.2   Odor and pheromones

A pheromone is defined as a chemical signal that triggers a natural response in other organisms. Examples of natural responses that may be triggered are alarm pheromones, food trail pheromones, sex pheromones, and many others that affect behavior or physiology of organisms within a species. The responses may give advantage to the receiver (kairomones), or benefit the sender (allomones). For example, production of allomones is a form of defense, particularly by plant species against insect herbivores. It is interesting to note that some insects have developed ways to defend against plant allomones. One example is to develop a positive reaction to them; the allomone becomes a kairomone.

Pheromone signals are also known as semiochemicals and are produced in very minute amounts. They are diluted in the environment with a complex mixture of chemical compounds derived from a myriad of sources. Because of this, the olfactory system in insects has evolved into a remarkably selective and sensitive system, which approaches the theoretical limit for a detector. Clearly, a pheromone is a molecular-scale chemical signal and encodes information. A semiochemical (semeon means a signal in Greek) is a generic term used for a chemical substance that carries a message. The notion of using algorithms based upon a species interacting through pheromones for optimization and to improve tasks within communication networks goes back many decades. Chapter 8 goes into greater detail on engineering an odor-based communication network.

3.2   GAP JUNCTIONS

A gap junction is one of the most direct connections that can exist between cells. It is a narrow aqueous channel connecting the cytoplasm of two cells. The channel is large enough to allow ions and other small water-soluble molecules to pass through, but does not allow larger, more complex molecules such as proteins or nucleic acids, to pass through. The junction itself is a nexus or a specialized intercellular connection generally found in animal cells. One gap junction is composed of two connexons (or hemichannels) that connect across the intercellular space. Communication is generally bidirectional and the tendency is to enable a more uniform concentration of molecules between the communicating cells. Intracellular Ca^{2+} and cyclic adenosine monophosphate (cAMP) are typical signaling molecules that pass through a gap junction channel.

A specific example of the operation of the gap junction channel occurs when blood glucose levels fall. Noradrenoline is released from cells that recognize the need for more glucose, which stimulates the production of more glucose. Not all cells in an area may recognize the need for more glucose, so cAMP is propagated via gap junctions to encourage other cells to increase production of glucose.

3.2.1   Liposomes: artificial containers

Molecular information within a nanoscale network is subject to problems similar to those encountered in traditional networks. Signals in a traditional network are subject to noise and attenuation and they also collide with other transmitters that are trying to transmit simultaneously. In a traditional network, physical techniques can be utilized to maintain the integrity of the signal and limit its propagation such that it travels only toward the intended target. For example, wires are shielded and beam-forming antennae shape the beam such that the signal follows as narrow a path as possible towards its intended target. In a similar manner, signaling molecules can be encapsulated within a material both to maintain their integrity and to keep the molecules from propagating where they are not needed.

Natural encapsulating materials in the biological environment at the cellular level are known as vesicles and liposomes. A vesicle is essentially a small bubble of liquid within a cell. It is a small, intracellular, membrane-enclosed sac that stores or transports substances within a cell. Vesicles form naturally due to the properties of lipid membranes. Many vesicles can have specialized functions depending on the materials that they contain. Vesicles are separated from the intracellular fluid by at least one phospholipid bilayer. If there is only one phospholipid bilayer, they are called unilamellar vesicles; otherwise they are called multilamellar, where a "lamellar" refers to membrane. Vesicles store, transport, or digest cellular products and waste. The membrane enclosing the vesicle is similar to that of the plasma membrane, and vesicles can fuse with the plasma membrane to release their contents outside of the cell. Vesicles can also fuse with other organelles within the cell. Because it is separated from the intracellular fluid, the inside of the vesicle can be made of a different chemical composition from the intracellular fluid. For this reason, vesicles are a basic tool used by the cell for organizing cellular substances. Vesicles are involved in metabolism, transport, buoyancy control, and enzyme storage. They can also act as chemical reaction chambers.

A molecular communication system has been proposed that uses liposomes, an artificially created vesicle that has the lipid bilayer membrane structure similar to cell vesicles [94]. This liposome is used to encapsulate information molecules. This is then passed through the gap junction. The advantage of the liposome is the same as the vesicle in the motor-based molecular communication network; it can store multiple molecules and can protect the information molecules from chemical reactions as it propagates to the receiving nanomachine. This is important as some molecules will not last long in the intercellular medium [90]. Again the motion of this molecule may be analyzed based on Brownian motion [95].

A liposome is a tiny vesicle, made out of the same material as a cell membrane. Note that liposomes can be filled with drugs and used to deliver drugs for cancer and other diseases. As a channel in an information transport network, the liposome would contain gap junctions on its membrane, similar to a normal cell. Information molecules inside the transmitting cell propagate to the liposome container through the gap junctions connecting the adjacent liposome to the transmitting cell. The liposome is transported to a receiver cell, where the information molecules propagate from the liposome to the receiver, again through gap junctions when they are adjacent [96, 97]. The liposome acts like a shuttle, safely transporting the information from one cell to another.

3.3 CELL SIGNALING

Cell signaling was introduced earlier in this chapter; however, in this section we go a little deeper and discuss human-engineered attempts at cell signaling and protocols. There are many types of cell signaling mechanisms, including proteins, small peptides, amino acids, nucleotides, steroids, retinoids, fatty acids, and dissolved gases such as nitric acid and carbon monoxide. These molecules may be released by numerous means. Cells may release signaling molecules that are bound within vesicles, or membranes, through a process known as exocytosis. Some signals may be released via simple diffusion and travel via Brownian motion as discussed in Chapter 2. Some signaling molecules remain attached to the cell membrane and are only activated when another cell makes contact. It should be noted that signal molecule concentrations are very small $\leq 10^{-8}$ mole and bind with a very high affinity constant $K_a \geq 10^8$ liters/mole. The affinity constant is simply the ratio of the bound concentration to the product of the unbound concentrations at equilibrium. A high affinity constant indicates that the molecules are tightly bound as the unbound concentration is very low relative to the bound concentration. Also, this binding between the signaling molecule and the target receptor occurs with high specificity. Specificity is a measure of the degree to which an exact match to a particular target is required before binding will occur.

Signaling mechanisms can be classified by the distance over which they operate as well as the manner in which they operate. Contact-dependent signaling requires direct membrane contact. Signaling molecules may be emitted into the local area surrounding the cell. If these signals affect cells of a different type, they are known as paracrine signals. If the signals affect cells of the same type, they are known as autocrine signals.

An important mechanism in the operation of signaling is control of the signaling molecules. In order for the molecular concentration to work properly and remain local, it must be removed rapidly. There are several mechanisms by which this occurs: the molecules can be taken up by neighboring cells, destroyed by enzymes, or caught within the extracellular matrix. Also, proteins known as antagonists can bind to signaling molecules and render them ineffective.

Traditional work in biology has focused on studying individual parts of cell signaling pathways. Systems biology research helps us to understand the underlying structure of cell signaling networks and how changes in these networks may affect the transmission and flow of information. Such networks are complex systems and may exhibit a number of emergent properties.

In summary, cells communicate with each other via direct contact (juxtacrine signaling), over short distances (paracrine signaling), or over large distances and/or scales (endocrine). Some cell-to-cell communication requires direct cell-cell contact. Some cells can form gap junctions that connect their cytoplasm to the cytoplasm of adjacent cells. In cardiac muscle, gap junctions between adjacent cells allow for action potential propagation from the cardiac pacemaker region of the heart to spread and cause coordinated contraction of the heart.

3.3.1 Network coding

Just as we will see that calcium signaling appears to use a biological analog of amplitude and frequency modulation later in this chapter, it has been suggested that biological systems have also evolved the mechanism of network coding [98]. What follows is a brief introduction to network coding so that we can discuss its potential analog in biological systems.

In a nutshell, network coding [99] maximizes the flow of information through a communication network by placing common information on common links such that the information is meaningful to all the nodes that share the common links. Without the concept of network coding, human-engineered communication systems would establish a single path for each source-destination pair through a network. This requires each source-destination pair to utilize the bandwidth through the network equivalent to the amount of information that is transmitted; each path is treated as an independent channel and there is, in essence, only one sequential path for each source-destination pair. However, if the communication network is highly interconnected, just as biological signaling systems are, it should be possible for transmitters of information to share common links such that a single packet of information on the common link conveys meaningful information to all receivers listening on the common link. In other words, one packet of information is simultaneously received by two different receivers, but is interpreted differently depending upon the receiver. Thus, this same packet of information simultaneously conveys meaningful information to both receivers.

One of the simplest examples of network coding is shown in Figure 3.1. In this example, there are assumed to be a transmitter on node 1, transmitting to node 5 and node 6. The link from node 2 to node 7 is the common link. Information flows from the transmitter through nodes 1, 3, and 5 as well as the common path 1, 3, 2, 7, and 5. The transmitter also sends information through nodes 1, 4, and 6 as well as through the common link 1, 4, 2, 7, and 6. To make the illustration even more clear, the example of transmitting only two bits is given. In this case, the transmitter at node 1 transmits the first bit b_1 of information as it normally would without network coding through the paths that are not common, namely 1, 3, 5, and it sends bit b_2 normally through the path 1, 4, 6. The difference with network coding is that even though each receiver has received only one bit, only a small amount of additional common information allows both receivers to obtain both bits. Namely, sending $b_1 \oplus b_2$, or bit one exclusive-or bit two, allows both receivers to easily decode the missing bit. Specifically, the receiver at node 5 will have b_1 and $b_1 \oplus b_2$ from which it can decode the second bit by applying $b_1 \oplus (b_1 \oplus b_2)$. The second receiver can do the analogous procedure with b_2 to obtain b_1. In cell signaling pathways, or networks, there exist configurations in which a common signal is interpreted differently by two different receptors, resulting in a more efficient flow of information.

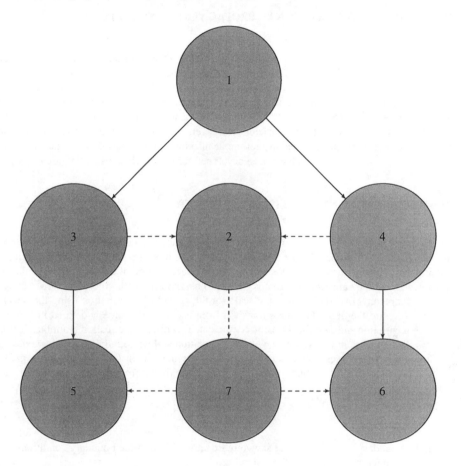

Figure 3.1 The canonical network coding example.

Much has been written about network coding for human-engineered communication systems; going into detail on this particular topic would take us away from the main topic of nanoscale communication. However, the example just illustrated for two bits can be extended to more bits d by considering the transmission of a d-dimensional vector of bits. Instead of using exclusive-or, $\oplus$, a d dimensional vector $v(XY)$ is assigned to every edge XY where X and Y are nodes connected by edge XY. One can think of $v(XY)$ as a column vector and the information that flows along XY as a matrix product of the incoming information vector and $v(XY)$. The result is that the output of a node is a linear combination of the input vectors. A key point is that the amount of information reaching the receiver R is then considered to be the dimension of the vector *space* for $v(R)$, where the input to R is a linear combination of all the input vectors.

   In the next section, we examine an intriguing concept: biological forms of amplitude and frequency modulated signaling. Calcium ions appear to be the first widely studied channel using these common human-engineered forms of communication.

3.4   MODELING BIOLOGICAL SIGNAL PROPAGATION AND DIFFUSION

This section delves into the analysis of the communication mechanisms introduced in this chapter. First we consider the general case of diffusion, how it works and how efficient it may be for information transport. Next, we spend a significant amount of time covering the more specific case of calcium signaling. Although some effort is required to understand the analysis, the results are rewarding for several reasons. First, the intriguing analogy between radio waves and calcium waves as a signaling mechanism becomes apparent. Second, the analysis used in understanding calcium waves relates to other biological signaling mechanisms, including a similar oscillatory pattern used in pheromone information transport. Finally, hardware platforms utilizing calcium wave communication have been developed for nanoscale network research [100].

3.4.1   Information concentration and propagation distance

Signal propagation and diffusion are key components of the nanoscale communication mechanisms discussed in this chapter. Let us consider the fundamental nature of how signaling molecules spread to neighboring cells. Recall that the transmitter cell secretes a given concentration of a signaling molecule and these molecules spread until they are received, that is, bound to neighboring cells' receptors. We want to define and derive the characteristic distance and time scale for the spread of information using this mechanism. Diffusion, based upon time-dependent mass transport from a spherical source is defined in Equation 3.1. Here c is the concentration and D is the diffusion coefficient. This diffusion equation is applicable because it describes particles moving randomly, that is, without correlation and without the use of Newtonian physics to predict individual trajectories. We described the Reynolds number in Chapter 2. With regard to this equation, it implies particles in a fluid with a very low Reynolds number, that is, nonlaminar, turbulent flow.

$$\frac{\partial c}{\partial t} = D \frac{1}{r^2} \frac{\partial}{\partial r} \left[r^2 \frac{\partial c}{\partial r} \right] \tag{3.1}$$

Assume that the boundary conditions are as shown in Equation 3.2. In these boundary conditions, F_0 is the signal molecule production rate and ρ is the cell radius.

$$-D \frac{\partial c}{\partial r} = F_0 \text{ at } r = \rho \text{ and } c \to 0 \text{ as } r \to \infty \tag{3.2}$$

   The solution to Equation 3.1 using the boundary conditions in Equation 3.2 is shown in Equation 3.3. erfc is the complementary error function. Recall that if the error, when the result of a series of measurements is taken, is described by a normal distribution with a standard deviation of σ and an expected value of zero, then $\mathrm{erf}(\frac{a}{\sigma\sqrt{2}})$ is the probability that the error of a single measurement lies between $-a$ and $+a$, for positive a. This can be useful, among other applications, in determining the bit error rate of a digital communication system. The complementary error function is $\mathrm{erfc}(x) = 1 - \mathrm{erf}(x)$.

$$c(r,t) = \frac{F_0 \rho}{2Dr}\sqrt{\frac{4Dt}{\pi}}$$
$$\left[e^{\frac{-(r-\rho)^2}{4Dt}} - e^{\frac{-(r+\rho)^2}{4Dt}} - \frac{|r-\rho|}{\sqrt{\frac{4Dt}{\pi}}} \times \mathrm{erfc}(\frac{|r-\rho|}{\sqrt{4Dt}}) + \frac{(r+\rho)^2}{\sqrt{\frac{4Dt}{\pi}}} \mathrm{erfc}(\frac{|r+\rho|}{\sqrt{4Dt}}) \right] \quad (3.3)$$

The reference length scale is defined as the distance from the center of the cell to its idealized circumference, ρ. The reference time scale is the time taken for the diffusion to occur over the reference length distance. The reference length scale is shown in Equation 3.4. The reference time scale is shown in Equation 3.5.

$$\zeta = \frac{r}{\rho} \quad (3.4)$$

$$\tau = \frac{4Dt}{\rho^2} = \frac{t}{t_{diff}} \quad (3.5)$$

The concentration as a function of distance and time can be expressed in terms of the reference time and length scale as shown in Equation 3.6.

$$c(\zeta, \tau) = \frac{F_0 \rho}{D} \psi(\zeta, \tau) \quad (3.6)$$

The expression for $\psi(\zeta, \tau)$ from Equation 3.6 is shown in Equation 3.7 and the values of A and B are shown in Equations 3.8 and 3.9.

$$\psi(\zeta, \tau) = \sqrt{\frac{\tau}{4\zeta^2}} \left[\frac{e^{-A^2} - e^{-B^2}}{\sqrt{\pi}} - A\,\mathrm{erfc}(A) + B\,\mathrm{erfc}(B) \right] \quad (3.7)$$

$$A = \frac{(\zeta - 1)}{\sqrt{\tau}} \quad (3.8)$$

$$B = \frac{(\zeta + 1)}{\sqrt{\tau}} \quad (3.9)$$

The diffusion reaches a steady state at any given distance from a cell after an infinite amount of time. Thus, it is useful to define the time at which only a fraction of the steady state value is reached, τ_f as shown in Equation 3.10.

$$\psi(\zeta, \tau_f) = f\frac{1}{\zeta} \quad (3.10)$$

It turns out that the above equation can be evaluated numerically as explained in [101] and approximated in the form shown in Equation 3.11.

$$\tau_{0.5} = (2.097\zeta)^2 \quad (3.11)$$

An interesting result from this analysis is that one might expect faster diffusion to lead to longer communication distances. However, the opposite is true, since a slower diffusion rate allows a higher concentration to build up before random movement dilutes the density of molecules. It is important to note that this analysis assumes the ideal case of a cell secreting molecules from a central open location. In the next section, let us look more closely at how a specific form of molecular diffusion, namely, calcium signaling, operates. Remember that our goal is to gradually work towards understanding the amplitude and frequency modulation aspects of calcium signaling.

3.4.2   Calcium waves

Calcium signaling is a common intercellular communication method and is a focus area in nanonetworks based on molecular communication. Cells are able to release and detect levels of calcium through gap junction communication or diffusion using several methods. Special binding proteins can be used to bind to and possibly alter functionality of the cell. Also storage proteins can allow for calcium stores in cells. These stores can be released when the cell is stimulated. Typical Ca^{2+} signaling contains a sudden increase in calcium, followed by a sustained oscillation of calcium. This ability to oscillate is one of the key features that makes calcium signaling a prime candidate for implementing molecular-based nanonetworks.

Calcium signaling has been found to have two types of oscillations, baseline and sinusoidal [95]. The sinusoidal oscillations are symmetrical fluctuations on an elevated calcium level and can be produced via a protein kinase C feedback that inhibits the G-protein or the receptor. The sinusoidal calcium oscillations have been found to have a constant frequency, another quality aspect for networks. Baseline oscillations are produced via repetitive spikes in the calcium levels. While the rising phase is often fast and similar, the declining phase of these spikes can be slower and differ between cells, making this form of oscillation harder to use for encoding information. Calcium dispersion and speed has been measured experimentally in cells; two osteosarcoma cell lines giving results for the speed of dispersion, control, depletion of calcium stores and other factors have been studied. Identifying details like these may be key to selecting ideal cells for networks. For example, calcium stores can be depleted, after which a delay is required to restore the calcium before the cell can again be used for sending information via calcium waves [102].

A mathematical model of intercellular calcium oscillations based on the gap junction signaling has been derived [103], which models the calcium oscillation based on experimental data from calcium oscillations in liver tissue. Models such as these will be important to future models for larger molecular networks and simulation platforms. Mathematical analysis of the information theoretic development of molecular-based communications and a method for using concentrations of calcium molecules to encode information have been explored [92], including hints at mechanisms for AM and FM concentration manipulation as we will discuss. This is a further development of the special properties of calcium signaling and another key step to realizing molecular-based nanonetworks. With these methods, signaling can encode more information than could be encoded if only a single information molecule were used, improving the ability to encode more information. A design for a nanoscale molecular communication system using calcium signaling including how nanomachines could be engineered to have calcium sensitive molecules, like calmodulin, allowing the nanomachines to detect the presence and concentration of calcium has been suggested [91]. The basic outline for communication as proposed is:

- *Encoding*: Sender nanomachine initiates signaling via stimulation to its neighboring cells, this leads to the generating of calcium waves that encode the information. The waves may propagate through a gap junction in this case each cell moves the encoded information along, or through, some medium.

- *Propagation*: Calcium signals may be broadcast or directed depending on the network. The network could consist of similar or differing interconnected cells via gap junctions or dispersed cells that can also be mixed. The mix of cells can dictate the way the wave is propagated. For example, waves could propagate through similar gap junctions, allowing the same wave encoding, or through differing junctions, requiring a recoding of the information.

- *Decoding*: The calcium is either detected as it passes through the gap junction, to the receiving nanomachine or through special protein receptors on the receiving cell. The reception of the calcium signal results in some reaction to the influx of calcium [28, 29].

3.4.3 Calcium stores and relays

As mentioned in the introduction, cytoplasmic Ca^{2+} concentration is widely utilized as a key signaling messenger in nearly all organisms to regulate a broad spectrum of distinct processes [104]. Cytoplasmic Ca^{2+} concentration can be increased either via release from intracellular stores or via influx into the cell. While Ca^{2+} is a signaling mechanism, it is also influenced and controlled via other signaling mechanisms, known as secondary messengers. A relatively small number of second messengers are thought to release Ca^{2+} from internal stores. Second messengers include inositol 1,4,5-trisphosphate ($InsP_3$) among others.

The release and intake of Ca^{2+} is described in more detail here. Ca^{2+} release from the cell is relatively quick, fully deactivating within a few tens of seconds. The release is due to clearance of Ca^{2+} from the cytosol, or liquid filling the volume of the cell, by resequestration into other organelles (notably mictochondria and endoplasmic reticulum) as well as extrusion from the cell by Na^+/Ca^{2+} exchangers and Ca^{2+}-ATPases in the plasma membrane.

Regarding calcium ion intake, the concentration gradient for Ca^{2+} across the plasma membrane of resting cells differs by $\approx$10,000-times. There is a much larger concentration ready to diffuse into the cell. In addition, the cell membrane has a hyperpolarized resting membrane potential. Hyperpolarization in biology means that the cell's membrane has a negative charge, which would attract the Ca^{2+} ions. The combination of the concentration difference and charge difference both serve to attract calcium ions into the cell. Thus, even with a low membrane permeability for Ca^{2+}, small increases in permeability result in a large Ca^{2+} influx. The permeability increase occurs by opening ion channels in the plasma membrane. Calcium can enter cells via several classes of channels. These include voltage-operated channels, second messenger-operated channels, store-operated channels, and receptor-operated channels. Voltage-operated channels are activated by membrane depolarization, and second messenger-operated channels are activated by any of a number of small messenger molecules, the most common being inositol phosphates, cyclic nucleotides, and messengers derived from lipids (recall the earlier discussion of lipids and vesicles). Store-operated channels are activated by depletion of intracellular Ca^{2+} stores, and receptor-operated channels are activated by direct binding of a neurotransmitter or hormone agonist. An agonist is a general term used to describe a type of ligand that binds and alters a receptor. The agonist's efficacy is a property that distinguishes it from antagonists, a type of receptor-ligand that also binds to a receptor but does *not* alter the activity of the receptor. An agonist may have a positive efficacy, causing an increase in the receptor's activity, or negative efficacy, causing a decrease in the receptor's activity. Under some conditions, Ca^{2+} can enter cells via the Na^+-Ca^{2+} exchanger operating in reverse mode.

Excitable cells, such as nerve and muscle, contain voltage-gated Ca^{2+} channels, while these channels do not appear in nonexcitable cells. Receptor-operated channels open rapidly upon binding to an external ligand (usually a neurotransmitter) and are also often found in excitable cells. Second messenger-operated channels are found less often in general and seem evenly distributed between excitable and nonexcitable cells. Store-operated Ca^{2+}-permeable channels, the main focus of this section, appear to be widespread, present in all eukaryotes from yeast to humans. Hence, it would appear that store-operated Ca^{2+} channels represent the primordial Ca^{2+} intake pathway.

The concept of store-operated Ca^{2+} was proposed in 1986. This idea originated from a series of experiments in cells investigating the relationship between Ca^{2+} release from internal stores, Ca^{2+} entry into the cell, and refilling of the Ca^{2+} store. It appeared that the amount of Ca^{2+} in the stores controlled the extent of Ca^{2+} influx in nonexcitable cells, a process that was originally called capacitative calcium entry. When the stores were full, Ca^{2+} influx did not occur but, as the stores emptied, Ca^{2+} entry into the cell occurred.

A variety of cells such as hepatocytes (liver cells), pituitary cells, endothelial cells (within blood vessels), fibroblasts (connective tissue cells), as well as newly fertilized eggs have demonstrated repetitive

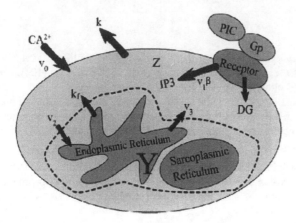

Figure 3.2   Schematic for the calcium signaling analysis.

spikes in the concentration of their intracellular calcium caused by either hormone or neurotransmitter signaling mechanisms. The period of these oscillatory spikes in concentration range from less than one second to up to 30 minutes. It has been clear ever since the discovery of these oscillations that they encode and transmit information triggered by cell signals. As mentioned earlier, $InsP_3$ plays a significant role in controlling the calcium signal frequency. Many theories for the cause of calcium signaling frequency have been proposed; the simplest, or minimal, model that explains it is the most likely, based upon Occam's razor.

We follow the model in [105], which begins by assuming that an internal agent triggers the production of $InsP_3$. This causes the cell to release previously stored Ca^{2+}, which raises the internal concentration level. Recall that the cell is able to rapidly replenish Ca^{2+} into its internal stores by pumping more Ca^{2+} into the cell as necessary. The mechanism operates as follows. Recall that an agonist is a ligand that binds to a receptor and influences its behavior. Agonists act through a transducing mechanism comprising a receptor (R), a G-protein (Gp), and phosphatidylinositol-specific phospholipase C (PIC) to hydrolyze, where hydrolysis occurs when one or more water molecules are split into hydrogen and hydroxide ions and is used to break down polymers, the inositol lipid phosphatidylinositol 1,4-bisphosphate into diacylglycerol (DG) and $InsP_3$ (IP_3). This generates the IP_3 signal which then controls, or modulates, the release of Ca^{2+} from an $InsP_3$-sensitive store into the cytosol, or the cell's intracellular fluid. The rate at which this occurs is represented by v_1. It also indirectly controls the influx of external Ca^{2+} into this store through a capacitative mechanism.

Let us define the variables of the model as illustrated in Figure 3.2. Each level of $InsP_3$ controls a constant flow of Ca^{2+} into the cytosol, determined by $v_1\beta$. Let Y be the concentration of the $InsP_3$-insensitive pool of Ca^{2+} within the cell. Z represents the concentration of Ca^{2+} within the intracellular fluid. The concentration level of Ca^{2+} is initially low, during which primer Ca^{2+} is transferred at a rate v_2 into the $InsP_3$-insensitive pool. Then it is interspersed with brief Ca^{2+} spikes when Ca^{2+} stored in that pool Y is released to the cytosol at a rate of v_3 in a process activated by the concentration of Ca^{2+} within the intracellular fluid.

Now, let v_o be the rate of influx of extracellular Ca^{2+} into the intracellular fluid. k is the rate of efflux of cytosolic Ca^{2+} from the cell. k_f is the rate of the passive leak of $InsP_3$-insensitive pool of Ca^{2+}, represented by Y, into intracellular fluid Z. $v_1\beta$ is used to describe the rate of influx of Ca^{2+} from the $InsP_3$-insensitive pool and is assumed to be proportional to the saturation function β of the receptor for $InsP_3$.

The main focus of this nanoscale signaling mechanism is the carrier signal Z. The rate of change of Z is shown in Equation 3.12. We can see that it is the sum of the Ca^{2+} input rates minus the Ca^{2+} output rates. v_0 is the input rate into the cell and is assumed to be constant regardless of cell signaling, while kZ is the output flow rate that is also independent of signaling and proportional to the input concentration.

Recall that ATP-driven pumps continuously move Ca^{2+} either out of the cell or into stores within the cell. Here v_2 is the rate of pump-driven Ca^{2+} into the $InsP_3$-insensitive pool. v_3 represents the rate of transport from this pool into the intracellular fluid Z. $k_f Y$ refers to a passive, leaky transport of Y into Z.

$$\frac{dZ}{dt} = v_0 + v_1 \beta - v_2 + v_3 + k_f Y - kZ \tag{3.12}$$

In Equation 3.13, the rate of change of the $InsP_3$-insensitive pool of Ca^{2+} is shown. Essentially, the rates of Ca^{2+} leaving the previous equation (negative values) are entering this equation (positive values), with the exception of v_0 and kZ which are the rates entering and leaving the entire cell and $v_1\beta$, which is the rate of $InsP_3$-sensitive Ca^{2+} input.

$$\frac{dY}{dt} = v_2 - v_3 - k_f Y \tag{3.13}$$

To summarize, the signal begins when an external signal triggers an increase in $InsP_3$. This leads to an increase in the intracellular concentration of Ca^{2+}. The increase in the intracellular concentration is a positive feedback mechanism that increases v_3.

Next the cooperative nature of the processes is included in the analysis. In general, the binding of a ligand to a macromolecule can be enhanced if there are already other ligands present on the same macromolecule (known as cooperative binding). The Hill coefficient provides a way to quantify this effect. There are two Hill functions used here.

First, there is a cooperative process occurring between the Ca^{2+} pumping process and the Ca^{2+} release from the intracellular store. Both serve to increase the intracellular Ca^{2+} concentration. This is represented in Equation 3.14. n is the cooperativity coefficient and K_2 is a threshold constant for pumping Ca^{2+} into the cell. V_{M2} is the maximum rate of Ca^{2+} pumping into the intracellular store.

Recall that v_3 represents the rate of transport from the $InsP_3$-insensitive pool into the intracellular fluid Z. Thus, in a similar fashion to the previous equation, Equation 3.15 represents the positive feedback of Z and Y on v_3. Here m and p are the cooperativity coefficients, K_R is the release threshold constant from the $InsP_3$-insensitive pool, and K_A is the activation threshold constant for the activation of v_3 by Z. V_{M3} is the maximum rate of release from the intracellular store.

$$v_2 = V_{M2} \frac{Z^n}{K_2^n + Z^n} \tag{3.14}$$

$$v_3 = V_{M3} \frac{Y^m}{K_R^m + Y^m} \frac{Z^p}{K_A^p + Z^p} \tag{3.15}$$

From Equations 3.14 and 3.15 we can see that if the cooperativity coefficients are one, then there is no gain or loss from cooperation and as the coefficients increase, there is more cooperative gain. In general, it is also possible for the coefficients to be negative, in which there is a negative impact of one process upon the other. In this particular situation, we expect the coefficients to be positive. Recall that β is the saturation coefficient for the $InsP_3$ receptor and is thus one of the initial control points for the whole process. The solutions of Equations 3.12–3.15 for the intracellular Ca^{2+} and the Ca^{2+} store generate the characteristic oscillatory Ca^{2+} concentration spikes over time. This is the fascinating signal that looks like an AM or FM modulated communication signal; let us now find out why this is the case.

It is immediately apparent that information is encoded within the frequency of calcium concentration changes. This is a form of biological frequency encoding and is analogous to frequency coding in communication networks, except that it occurs at the cellular scale. This direct analog of human-engineered frequency coding has existed for billions of years before humans existed. Thus, it is always fascinating to see how nature has evolved systems equivalent to, or in most cases better than, human-engineered systems. Let us understand more about what is known regarding calcium signaling, biological frequency coding, and how biological receivers may decode the information, all at subcellular scales.

It has been determined that calcium spikes increase their frequency and amplitude with the increase in the strength of a given stimulus. But exactly how is the information encoded by nature? It is also curious how the same calcium signal, which diffuses everywhere, can selectively address different receivers, apparently based upon frequency and amplitude of oscillation. The analysis that examines how this might be accomplished in nature comes from [106]. It is known that Ca^{2+} oscillations of a given frequency are more effective at promoting protein phosphorylation than a constant Ca^{2+} signal of equal average concentration. Recall that protein phosphorylation is a downstream signal induced by Ca^{2+} signaling. The Ca^{2+} concentration is assumed in this analysis to be controlled by a kinase signal. When the kinase is phosphorylated, it induces an increase in the concentration of Ca^{2+}. When the kinase is dephosphorylated, it reduces the concentration of Ca^{2+}. Note that there is a cooperative increase in the effectiveness of kinase-induced Ca^{2+} concentration increase.

Let $S(t)$ be the Ca^{2+} concentration, Y be the fraction of active kinase, and X be the fraction of phosphorylated target protein. Then we let Y_T be the total concentration of kinase. α_Y and β_Y represent the Ca^{2+} binding and release constants. The rate constants for phosphorolaytion and dephosphorolation are α_X and β_X. Then the rate of change of Y and X are shown in Equations 3.16 and 3.17.

$$\frac{dY}{dt} = \alpha_Y S(t)^n (1 - Y) - \beta_Y Y_T \tag{3.16}$$

$$\frac{dX}{dt} = \alpha_X Y_T Y (1 - X) - \beta_X \beta_X X_T \tag{3.17}$$

It can be assumed that the Ca^{2+} binding is very fast relative to dephosphorylation, so $\frac{dY}{dt} \approx 0$ and the previous equation can be simplified as shown in Equation 3.18. The term $\alpha(S)$ will be explained next.

$$\frac{dX}{dt} = \alpha(S) - (\alpha(S) + \beta_X) X_T \tag{3.18}$$

The phosphorylation rate and Ca^{2+} concentration are cooperative and take the form of a Hill equation as explained earlier, and is shown in Equation 3.19.

$$\alpha(S) = \hat{\alpha} \frac{(S/K_S)^n}{(S/K_S)^n + 1} \tag{3.19}$$

$\hat{\alpha}$ is the maximum value $\alpha_X Y_T$, where $K_S = \sqrt[n]{\beta_Y/\alpha_Y}$, and n is the Hill coefficient. The Ca^{2+} oscillations themselves are described by Equation 3.20 with period T and amplitude S_0, and the width of the spike is Δ.

$$S(t) = \begin{cases} S_0, & iT \le t < iT + \Delta \\ 0, & iT + \Delta < t < (i+1)T \end{cases} \tag{3.20}$$

Equation 3.18 can be solved for the time during the calcium spike and for the time between spikes, which are given in Equation 3.20. $\zeta = t - iT$ is introduced to simplify the time variable. The solution for each part is shown in Equation 3.21. Note carefully the subscript i in X_i in the equation. This refers to the ith cycle.

$$X(t) = X_i(\zeta) = \begin{cases} e^{-\alpha_0 + \beta} A_i + X_{max}, & 0 \le \zeta < \Delta \\ e^{-\beta\zeta} B_i, & \Delta < \zeta < T \end{cases} \tag{3.21}$$

The new phosphorylation rate for cycle zero is α_0 in Equation 3.22.

$$\alpha_0 = \alpha(S_0) = \alpha \frac{(S_0/K_S)^n}{(S_0/K_S)^n + 1} \tag{3.22}$$

The concentration of Ca^{2+} at the beginning of one cycle is equal to the concentration at the end of the previous cycle, so the conditions in Equation 3.23 can be used to obtain the difference equation in Equation 3.24. The maximum activation of the target is $X_{max} = \frac{\alpha_0}{\alpha_0 + \beta}$.

$$\begin{aligned} X_i(T) &= X_{i+1}(0) \\ X_i(\Delta^-) &= X_i(\Delta^+) \end{aligned} \tag{3.23}$$

$$\begin{aligned} A_{i+1} &= e^{-\alpha_0\Delta + \beta T} A_i - X_{max}(1 - e^{-\beta(T-\Delta)}) \\ B_i &= e^{-\alpha_0\Delta} + X_{max} e^{\beta(T-\Delta)} \end{aligned} \tag{3.24}$$

At initial conditions, $X_i(0) = 0$ yields $A_n = -X_{max}$. An ansatz can used [106], which is to suppose the solutions of a homogeneous linear differential equation and difference equation have, respectively, an exponential and power form. So assuming $A_i = a + be^{-i(\alpha_0\Delta + \beta T)}$ and $e^{-i(\alpha_0\Delta + \beta T)}$ is a solution of the homogeneous difference equation yields Equation 3.25.

$$A_i = A_\infty \left(1 + e^{-(i+1)(\beta T + \alpha_0\Delta)} \frac{e^{(\alpha_0+\beta)\Delta} - 1}{1 - e^{-\beta(T-\Delta)}} \right) \tag{3.25}$$

Note that as the number of cycles goes to infinity, the coefficient $A_\infty = -X_{max}(1 - e^{-\beta(T-\Delta)})/(1 - e^{-\beta T + \alpha_0\Delta})$. Equations 3.24 and 3.25 give Equation 3.26.

$$B_i = B_\infty \left(1 - e^{-(i+1)(\beta T + \alpha_0\Delta)} \right) \tag{3.26}$$

$B_\infty = -X_{max}(e^{-\beta\Delta} - e^{\alpha_0\Delta})/(1 - e^{-\beta\Delta + \alpha_0\Delta})$. Filling in all the values for Equation 3.21 yields Equation 3.27, the dynamics of the target protein as a function of the calcium ion oscillation.

$$\begin{aligned} X(t) &= \\ X_i(\zeta) &= \begin{cases} X^+(\zeta)(1 - e^{-i\beta T + \alpha_0\Delta}) + X_{max}(1 - e^{-(\alpha_0+\beta)\zeta})e^{-i(+\beta T + \alpha_0\Delta)}, & 0 \le \zeta < \Delta \\ X^-(\zeta)(1 - e^{(i+1)\beta T + \alpha_0\Delta}), & \Delta < \zeta < T \end{cases} \end{aligned}$$

$X^+(\zeta)$ is the stationary target activation during a Ca^{2+} spike which takes place from $0 \le \zeta < \Delta$ and X^- is the similar value for the Ca^{2+} inter-spike time which occurs during $\Delta \le \zeta \le T$. These are shown in

Equation 3.27.

$$X^+(\zeta) = X_{max}\left(1 - \frac{1 - e^{-\beta(T-\Delta)}}{1 - e^{-(\beta T + \alpha_0\Delta)}}e^{-(\alpha_0+\beta)\zeta}\right)$$

$$X^-(\zeta) = X_{max}\left(\frac{e^{-\beta\Delta} - e^{-(\alpha_0\Delta)}}{1 - e^{-(\beta T + \alpha_0\Delta)}}\right)e^{-\beta\zeta} \tag{3.27}$$

The mean target protein activity has already been discussed and is shown as the integral in Equation 3.28.

$$X_i = \frac{1}{T}\int_{iT}^{(i+1)T} X_i(\zeta)d\zeta \tag{3.28}$$

Thus, integrating Equation 3.27 yields Equation 3.29.

$$\begin{aligned}\bar{X}_i &= X_{max}\left[\frac{\Delta}{T} + \frac{\alpha_0}{\beta T(\alpha_0 + \beta)}\frac{(1 - e^{-\Delta(\alpha_0+\beta)}(1 - e^{-\beta(T-\Delta)}))}{1 - e^{-(\beta T + \alpha_0\Delta)}}\right](1 - e^{-(\beta T + \alpha_0\Delta)}) \\ &\quad + e^{-i(\beta T + \alpha_0\Delta)}X_{max}\left(\frac{\Delta}{T}\frac{(1 - e^{-(\alpha_0+\beta)\Delta})(\frac{\alpha_0}{\alpha_0+\beta} - e^{-\beta(T-\Delta)})}{\beta T}\right)\end{aligned} \tag{3.29}$$

Finally, as the number of cycles goes to infinity, the terms with negative i go to zero, leaving the final result as shown in Equation 3.30.

$$\bar{X}_i = X_{max}\left[\frac{\Delta}{T} + \frac{\alpha_0}{\beta T(\alpha_0 + \beta)}\frac{(1 - e^{-\Delta(\alpha_0+\beta)})(1 - e^{-\beta(T-\Delta)})}{1 - e^{-(\beta T + \alpha_0\Delta)}}\right] \tag{3.30}$$

The stationary value for X, signified by $\bar{X}$ is the limit as the number of cycles goes to infinity, $\bar{X} = \lim_{i\to\infty} \bar{X}_i$. From Equation 3.30 and some algebraic manipulation, the result in Equation 3.31 can be obtained. The values of σ, ω, and γ will be explained next.

$$\bar{X} = x_{max}\left[\gamma\frac{\omega\sigma}{1+\sigma}\frac{(1 - e^{-\gamma\frac{(1+\sigma)}{\omega}})(1 - e^{-(\frac{-(1-\gamma)}{\omega})})}{1 - e^{\frac{(1+\gamma\sigma)}{\omega}}}\right] \tag{3.31}$$

Let us take a minute to recall the goal of this analysis. Ultimately, we would like to understand how the calcium concentration waves encode information that controls the target activity. The target, in this case, is the next downstream signal, namely, activation of kinase. σ in Equation 3.32 indicates the degree to which kinase senses the amplitude of calcium concentration. This is done with reference to the kinase binding affinity $\frac{1}{K_s}$ and the cooperativity of the kinase activation by means of the Hill coefficient n. It is also based upon the rate of phosphorylation relative to dephosphorylation ($\frac{\hat{\alpha}}{\beta}$). Thus, if the calcium concentration is effective with a low binding affinity and the rate of phosphorylation is high relative to dephosphorylation, then σ will be large and the activation rate is very effective.

$$\sigma = \frac{\hat{\alpha}}{\beta}\frac{(\frac{S_0}{K_S})^n}{1 + (\frac{S_0}{K_S})^n} \tag{3.32}$$

Recall that β is the rate of Ca^{2+} release and dephosphorylation. The time period of a Ca^{2+} spike is T. Thus, $\frac{1}{\beta}$ is the time period required for release to occur, known as the basal response time. The relative oscillation frequency in Equation 3.33 indicates how fast the calcium oscillations are relative to the basal response time. If σ is small, the target kinase is released between calcium spikes and the target activation tends to follow the calcium spikes. If ω is large, then very little kinase is dephosphorylated, or released, between calcium spikes and phosphorylation builds up, or integrates over calcium cycles.

$$\omega = \frac{1}{\beta T} \tag{3.33}$$

The duty ratio is shown in Equation 3.34. Recall that Δ is the time of the cycle during which the calcium level is high and T is the full cycle time. Thus, this ratio is a simple duty cycle. Let S_0 be the amplitude of the first calcium cycle. Then γS_0 is the average calcium concentration.

$$\gamma = \frac{\Delta}{T} \tag{3.34}$$

Given that σ is the effective rate of target activations, the maximum activation rate is as shown in Equation 3.35.

$$X_{max} = \frac{\sigma}{1 + \sigma} \tag{3.35}$$

For relatively fast calcium oscillations, as compared to target activity, the resulting impact on target activity is shown in Equation 3.36.

$$\bar{X}(\omega \to \infty) = \frac{(1 + \sigma)\gamma}{1 + \sigma\gamma} X_{max} = \frac{\sigma\gamma}{1 + \sigma\gamma} \tag{3.36}$$

The optimal signal shape for activating a target can be found by solving Equation 3.37.

$$\max_{\gamma} \bar{X}$$
$$\bar{S} = \gamma S_0 = constant \tag{3.37}$$

From Equation 3.36, which defined target activation for relatively fast calcium oscillations, the maximum value for the duty ratio γ is shown in Equation 3.38. If the Hill cooperativity coefficient is less than one, then $\bar{X}$ simply increases monotonically with γ. This is a somewhat obvious result, namely, target activity increases with the duty ratio of calcium. However, a more interesting result follows next.

$$\gamma_{max} = \frac{\bar{S}/K_S}{\sqrt[n]{n - 1}} \tag{3.38}$$

When the cooperativity coefficient n is greater than one, the maximum target activity can be obtained with a duty ratio of less than one. From Equation 3.38, the condition in which maximum activity is attained when γ is less than one, namely, when oscillations occur, is shown in Equation 3.39.

$$K_S > \frac{S_0}{\sqrt[n]{n - 1}} \tag{3.39}$$

Thus, the key point is that maximum target activation is greater during oscillations meeting the condition in Equation 3.39 than with a constant calcium concentration of the same value. A waveform is more potent than a constant signal. Further analysis with the same model shows that oscillations of various frequencies and amplitudes can selectively activate target responses. In other words, biology has implemented a beautiful form of frequency and amplitude encoding that takes place on the nanoscale within our body every day.

3.4.4   Gene and metabolic communication networks

This section is about the role genetic programming plays in creating new communication mechanisms among cells. Genes are the code that programs a cellular machine. A gene regulatory network is a collection of DNA segments in a cell that interact with each other and with other substances in the cell to govern the transcription of genes into messenger RNA. Thus, the genetic code is, in this sense, self-regulating. The genetic regulatory network is analogous to an active network [1, 56], in which code that programs a network is carried by the network itself. In this section we discuss how programming and engineering with genes is advancing via standardized approaches taken from the Internet Engineering Task Force (IETF). The technical implementation of genetic engineering is using network communication protocols and the process of standardizing genetic engineering protocols is being patterned after IETF procedures complete with similar documentation guidelines because of the strong analogy with networking.

An example of this is the BioBrick project whose goal is to provide standardization for bio-nanotechnology by employing well-defined biological parts. Each BioBrick part is a DNA sequence held in a circular plasmid. The "payload" of the BioBrick part has universal and precisely defined sequences at each end that are not considered to be within the usable portion of the BioBrick part. The goal is to enable ease in assembling larger BioBrick parts by chaining together smaller parts in any order required. The particular sequence where DNA may be cut by an enzyme is known as a restriction site. The restriction sites between two parts are removed, allowing the use of those restriction enzymes without breaking the new, larger resulting BioBrick part. Note that to facilitate the assembly process, it is invalid to include any of the restriction sites within a BioBrick part [107].

Like any well-engineered system, the BioBrick parts are organized into a hierarchy that includes "parts," "devices," and "systems." Parts are the fundamental building blocks and include basic functions. Examples include encoding a given protein, or providing a promoter to let RNA polymerase bind and initiate transcription of downstream sequences. Devices are a set of parts that implement a human-defined function. An example is the production of a fluorescent protein whenever the environment contains a certain chemical. Finally, systems perform the highest-level, most complex tasks. An example is the oscillation between two colors at a predefined frequency.

The simple fact that the BioBrick assembly is standardized allows a biological engineer to assemble any two BioBrick parts such that the resulting part is a valid BioBrick part so that it can be combined with any other BioBrick part. BioBrick parts take advantage of idempotency, namely, that multiple applications of an operation do not change the result. This has two fundamental advantages. First, the BioBrick standard enables distributed design and development of compatible biological parts. Engineers at different locations can design a part that conforms to the BioBrick assembly standard, and those two parts will be physically composable via the standard. Second, since the process is standardized, it is much easier to optimize and automate than more traditional ad hoc approaches, not unlike network protocol standards.

BioBrick parts are maintained within *E. coli* plasmid vectors. A plasmid is a DNA molecule separated from the chromosome and is capable of replicating independently of the chromosomal DNA. It is often circular and double-stranded. Plasmids usually occur naturally in bacteria and are sometimes found in eukaryotic organisms; their size varies from 1 to over 1,000 kilobase pairs (kbp).

A BioBrick part has the look and feel of a standardized active network packet with well-defined fields that aid in the reliable transfer of the information within the packet. Each BioBrick has a well-defined cloning site to aid in propagation. Within the cloning site, the BioBrick part contains a positive selection marker, which is analogous to an error detection code. Just as an error detection code will flag a packet as corrupt and cause the packet to be dropped, the positive selection marker will cause any contaminated parts to grow with a toxic gene that will kill the cell. BioBrick parts contain terminators and translational stop codons on each side of the cloning site to ensure that no accidental encoding takes place at the wrong location. Finally, just as an information packet usually contains a length or size field, primer annealing sites are included in the standard BioBrick part allowing the length and sequence to be checked.

3.5 OLFACTORY AND OTHER BIOLOGICAL COMMUNICATION

This section discusses the communication paradigm inherent in pheromones, which appears to be a good example of long-distance communication at the nanoscale. First, the nature of pheromones and their modeling is discussed, then quorum sensing, usually exhibited by bacteria, is discussed. Next is a topic that at first glance may seem odd in this section, namely, memristors. While the physics of memristors are more akin to topics in the next chapter involving nanoscale structures, memristors also have interesting properties that are analogous to biological phenomena and thus memristors are introduced here. Finally, this section ends with an analysis of pheromones as a communication channel.

Pheromone communication is already widely used in the animal world and models have been developed to describe this type of information channel [108]. Pheromone communication operates similar to molecular communication using signaling molecules. The difference is that the medium through which the molecules would propagate. Unlike cell-based systems that are very dependent on pH level and temperature, pheromones can operate outside a protective host system. Pheromones can still be affected by temperature and effects in the medium, for example, wind. The communication system would work much like the above systems: Encoding could be done using molecules. The signaling molecule is released into the medium, for example, in a gas form. It is important that the release is able to be controlled, or that it is voluntary. If the signaling and passing of information cannot be controlled, then a communication system cannot be based on it. Brownian motion again plays a role in modeling the propagation of the molecules in this medium, for example propagation of particles through a gas, Brownian motion models the effects well. Models discussed in more detail in Chapter 8 help in selecting the level of molecules needed for release and the sensitivity of the receptors for the receiver. The receiving nanomachine can again use special receptor proteins that bind to the signaling molecule. This allows the receiver to be selective and detect multiple molecules. Once the molecule is detected it can cause a reaction in the receiver that can trigger some engineered response to the information.

One example of a pheromone is a scent picked up by olfactory receptor cells. The sequence of events is compromised of cell signaling processing as described earlier in this chapter. The ligand molecules, which carry the scent information, attach to receptor proteins on the receptor cell membrane. Once the binding occurs and the receptor is activated, a guanine nucleotide-binding protein (GTP-binding) G-protein is activated which then acts on enzymes that trigger second messengers, for example, cyclic AMP. In general, G proteins function as "molecular switches" because they alternate between an inactive guanosine diphosphate (GDP) and active guanosine triphosphate (GTP) bound state. These states continue a signaling process that regulates downstream cell processes. The second messengers open ion channels in the cell membrane which change the cell's membrane potential. Odorant molecules cannot leave the system and must be broken down by enzymes.

The next section discusses memory, via memristance, in a biological context. Memristance will play an increasing role in memory and adaptive behavior of nanoscale systems, including nanoscale networks.

3.5.1 Memristors

Memristors are a nanoscale phenomenon that appear to have an analog in biological systems, namely, the notion of memory. A memristor is similar to a resistor, except that the resistance is not fixed, it depends upon the past history of voltage applied to the memristor. In its most general form, the memristor follows Ohm's law and is given by Equation 3.40 [109].

$$v = R(w)i$$
$$\frac{dw}{dt} = i \tag{3.40}$$

The important point here is that w is an internal state variable for the device and R varies as a function of w. Furthermore, the rate of change of the state variable with respect to time is given by the charge i. A simplified implementation of such a device has a low resistance R_{ON}, doped volume of width w, high resistance R_{OFF}, and undoped volume within the device structure. The total distance of the doped and undoped regions is D. The main feature of this device is that the width of these regions can change dynamically.

Equation 3.42 shows the total resistance, which is simply the sum of the individual resistances in series over the total width D. If we assume that the charged dopants drift due the applied charge, then we let their mobility be μ_V and the rate of change of the doped region $dw(t)/dt$.

$$v(t) = \left(R_{on}\frac{w(t)}{D} + R_{OFF}\left(1 - \frac{w(t)}{D}\right) \right) i(t)$$
$$\frac{dw(t)}{dt} = \mu_V \frac{R_{ON}}{D} i(t) \tag{3.41}$$

Integrating over time yields Equation 3.42.

$$w(t) = \mu_V \frac{R_{ON}}{D} q(t) \tag{3.42}$$

Placing Equation 3.42 into Equation 3.41 yields Equation 3.43. Note that this assumes that $R_{ON} \ll R_{OFF}$.

$$M(q) = R_{OFF}\left(1 - \frac{\mu_V R_{ON}}{D^2} q(t)\right) \tag{3.43}$$

It becomes apparent that this is a nanoscale phenomenon because the total thickness D must be very thin in order for the $q(t)$ term to become significant. The ability to control random networks of memristors will be an important advance in nanoscale and molecular networking.

3.5.2 Quorum sensing

Bacteria are independent organisms. However, colonies of bacteria appear to be able to coordinate their action. For example, not until there is a given density of bacteria will they begin to cooperatively produce tissue-degrading enzymes. Thus, some level of coordination must be taking place. Low molecular weight bacterial pheromones are the channels used to communicate with one another. The process of cell signaling with bacterial pheromones for coordinated behavior is known as quorum sensing.

The aspect of cooperative incentives as a control mechanism for signaling has been considered [110]. The idea is to find the individual cellular-level investment in signaling and cooperation that leads to optimal fitness. In other words, what keeps a cell from signaling all other cells to do what is beneficial for it alone, rather than the group, or bacterial colony?

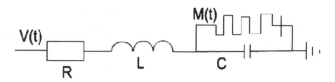

Figure 3.3 Schematic for the memristor model of bacterial learning.

There is much to be learned about communication and computation among supposedly simple, one-celled organisms. Here we explore quorum sensing, which is a form of cooperative behavior among one-celled organisms based upon collaborative communication. We also look at some intriguing results related to the nanoscale phenomenon of memristors and learning in one-celled organisms. Single-celled organisms, such as amoeba, have demonstrated both memory and learning. In particular, *Physarum polycephalum* displays remarkable intelligence. This one-celled organism is able to solve mazes and geometrical puzzles, control robots, and appears to learn and recall past events [111]. But where is the brain of the amoeba?

The amoeba has protoplasmic veins through which ectoplasm appears to flow in a pulsed pattern. Recall from Chapter 2 that actin-myosin motors provide the contractile force for muscles. The amoeba also has an actin-myosin system for contraction and movement. When contraction occurs, it pushes endoplasm in the direction of motion. The amoeba also contains a gel-sol solution; the sol moves through the gel like water through a sponge. The gel-sol solution is sensitive to pressure and can change to sol under pressure, that is, under the pressure caused by contraction. A key point is that the gel-sol solution may return to a prior state over time, but it is dependent on the underlying sol channels that were formed in the gel. There is a dependence on past state. This is also a key feature of the memristor, in which there is state, or memory, such that changes in voltage depend upon past electrical current levels.

A resistor R, capacitor C, inductor L, memory M circuit, shown in Figure 3.3, can mimic the learning capability of the amoeba in a simplified manner. A resistor, inductor, and capacitor are connected in series and a memristor is connected in parallel. This is a common harmonic oscillator circuit seen in communication equipment. Sometimes a variable capacitor or inductor will allow the circuit to be tuned. However, in this particular circuit, as proposed in [111], a memristor is connected in parallel with the capacitor. The memristor is modeled in Equation 3.44, where M is the resistance of the memristor, $f(V_C)$ is a function describing the change in memristor state, V_C is the voltage applied to the memristor, which is also the voltage drop across the capacitor, and $\theta()$ is a step function. M_1 and M_2 are the limits in resistance, between which the memristor will "learn" to operate.

$$\frac{dM}{dt} = f(V_C)\left[\theta(V_C)\theta(M - M_1) + \theta(-V_C)\theta(M_2 - M)\right] \tag{3.44}$$

The response of the RLC plus memristor circuit is shown in Equation 3.45 and Equation 3.46. Here I is the total current and $V(t)$ is the voltage applied across the entire RLC memristor circuit. Equation 3.45 simply models the voltage drops of each component of the circuit, whose sum must equal the voltage applied to the system. Equation 3.46 is Kirchoff's law applied to the capacitor, inductor, memristor connection.

$$V_C = L\dot{I} + IR = V(t) \tag{3.45}$$

$$C\dot{V}_C + \frac{V_C}{M} = I \tag{3.46}$$

Conditions that the amoeba enjoys are represented by positive voltages and conditions that the amoeba dislikes are negative voltages. When a long enjoyable (positive) voltage is applied, the memristor transitions to its M_1, or low resistance state [111]. The LC portion of the circuit is highly damped and oscillations decay quickly. Next, when irregular voltage pulses are applied, the voltage applied to the memristor is small and learning is slow. When periodic pulses are applied, particularly if they are close to the circuit resonant frequency, the memristor will experience a positive voltage and transition to resistance M_2. The LC circuit becomes less damped and oscillations survive longer. This represents learning, being able to predict the next positive, or enjoyable, event.

In summary, the responses of the actual amoeba as it learns to anticipate good and bad events in its environment show very strong similarities to the response of the RLC memristor circuit. This circuit is composed entirely of passive electronic components and appears to accurately model mechanisms occurring in the amoeba's protoplasm. This shows a nice transition in our understanding of cellular and nanoscale between biological and mechanical-electrical components.

3.5.3 Pheromone communication models and analysis

We have discussed the mechanics of how a pheromone is received and activates a signal. This section is concerned with the channel and how the signal is encoded. In natural conditions, there is a significant amount of noise. It's a continuous theme that noise, which human-engineered systems try so hard to avoid, is often beneficial, and sometimes necessary, in biological communication. A cloud of pheromone released into the air swirls into wisps and patches. An insect flying in a random pattern receives the pheromone as a periodic signal. In fact, it has been demonstrated that a certain type of moth cannot orient itself in a perfectly uniform cloud of pheromone, while it has no problem in naturally noisy patches [112]. As previously discussed, the mechanics of sensory signal transduction is a remarkable biochemical and electrical process. The presence of even a single ligand molecule at the cell surface can be detected and the corresponding cell signal amplified into an action potential through a nerve. As we saw earlier in this chapter with calcium signaling, it can be shown that the biological target is more effectively activated when the pheromone signal is oscillatory.

3.5.4 Neuronal communication

A biological mechanism that most people immediately think of as being a fast and efficient means of communication is a nerve impulse. This is more accurately called an *action potential* and is a brief change in the transmembrane voltage generated by the activity of voltage-gated ion channels embedded in the membrane. Action potentials are pulselike waves of voltage that travel along the axons of neurons. Perhaps partly influenced by our digital age, the action potential has been interpreted digitally, as spikes that are either on, or off; information should then be encoded with such spike trains, or sequences of spikes. The meaning and coding used within spike trains carried by the axon of the neuron is still largely an open research topic; different nerves appear to encode information in a different manner. We know from human-engineered communication systems that information that is highly compressed has a near-uniform symbol distribution; it looks similar to noise. There is a surprisingly high degree of variability in neural transmission, even when the input stimulation and experimental conditions remain constant [113]. As discussed shortly, noise can be beneficial in the form of stochastic resonance.

It should be mentioned at this point that the smallest axons are outside the nanoscale range; axons are on the order of 0.5 μm in diameter. It is the neurotransmitters within the synapse of a neuron that would be at the scale of interest to nanoscale networking. However, the concept of using the apparent "noise" within a neuron for useful purposes is mentioned here. First, the notion of stochastic resonance will be mentioned in the next section. Second, the idea of creating a subchannel through a nerve, in which the bandwidth currently used by apparent noise is instead used to communicate human-engineered signals, is suggested as a speculative concept.

As axons decrease in diameter, they become more susceptible to apparent randomly generated spikes. The effect of opening an ion channel increases with increasing membrane resistance, which increases significantly as the diameter of the axon decreases. Action potentials and spikes can be generated even when no synaptic input is present.

3.6 INFORMATION THEORETIC ASPECTS

Chapter 6 discusses information theory and nanoscale networks in general. Since this is the final chapter of the book devoted solely to biologically inspired nanoscale and molecular communication, it is a good point to discuss an interesting aspect of information detection and transmission known as stochastic resonance. In a nutshell, the right level of noise in a system can act as a means of improving signal detection, which can seem counterintuitive to human-scale engineers who go to great lengths to remove noise from communication systems. The direct connection to information theory is related to the analysis describing precisely under what conditions the mutual information transfer rate of a signal is maximized when stochastic resonance occurs.

3.6.1 Stochastic resonance

Biological systems, and in particular their cell signaling and networking, are very adaptive and robust in the presence of noise. Both organic and inorganic nanoscale networks will be faced with the similar problem of increased noise and stochastic operation, given their small size, the lack of precision in manipulating such small components, and their operation near the realm of thermal noise. Thus, understanding the concept of stochastic resonance [114], in which noise actually helps a system improve its performance, will be useful.

Stochastic resonance is the phenomenon that occurs when the correlation between a weak source signal and the output of a nonlinear system is improved by the presence of noise. It has been shown that biological systems use stochastic resonance to detect weak signals in a noisy environment. Stochastic resonance originated during research into the cause of the recurrence of ice ages. When examining the volume of ice sheets over a period of millions of years, it is difficult to see a pattern. When the power spectrum was examined, there were strong peaks at some cycles, which had very weak correlations with the earth's eccentricity. Stochastic resonance was used to suggest that strong "noise" factors, such as geothermal heat, when coupled with the earth's eccentricity, provide a strong enough correlation to be an explanation for the cause of the ice ages. A specific biological example of stochastic resonance is encompassed by the *Procambarus clarkii*, a crayfish. The tail of this crayfish contains mechano-receptor cells that are used to detect the presence of potential predators by means of small perturbations in water flow. The detection capability is amplified by noise in the surrounding water.

Let us consider a specific example of stochastic resonance applied to the Langevin equation. As discussed in Chapter 2, the Langevin equation, shown in Equation 3.47, is related to Brownian motion and played a key role in our discussion of the model that describes the movement of molecular motors. Here, $x(t)$ is the particle velocity at time t, γ is the diffusion coefficient, and $\eta(t)$ is a random noise term.

$$\frac{dx(t)}{dt} = -\gamma x(t) + \eta(t) \qquad (3.47)$$

A plot of $x(t)$ shows an oscillating signal around zero. The greater the variance σ of the noise term, the larger the random oscillations are around zero. On the other hand, the larger the value of γ, the more quickly $x(t)$ returns to a stable zero value.

Now, assume $x(t)$ is a normal distribution and that it is influenced in an arbitrary manner by a variable a. The average value of x when a is applied will be represented by $\langle x \rangle_a$. If a changes to $a + \Delta a$, then the result in Equation 3.48 can be shown. The change in the average value of x is proportional to its variance at a.

$$\frac{\langle x \rangle_{a+\Delta a} - \langle x \rangle_a}{\Delta a} \propto \langle (\delta x)^2 \rangle \tag{3.48}$$

The nature of stochastic resonance can be modeled by means of a modified Langevin equation as shown in Equation 3.49. The position of a particle at time t is $x(t)$. The base potential is defined as a function of the position $U(x)$. The normally distributed noise is represented by $\eta(t)$. Finally, the weak signal from the source is represented by $\epsilon \sin(\omega t)$.

$$\frac{dx(t)}{dt} = -\frac{dU(x(t)))}{dx} + \epsilon \sin(\omega t) + \eta(t) \tag{3.49}$$

In a specific example, $U(x) = -\alpha x^2 + \beta x^4$ [114]. The parameters α and β are set to 0.5 and 0.25, respectively. A bistable model can be assumed in which the output is binary; it is determined by partitioning the output into two ranges of values, a high and low value range separated by a threshold value. When the noise value is set appropriately, a weak high or low signal will yield an output in the appropriate high or low value range.

3.6.2   Towards human-engineered nanoscale biological communication networks

We end this chapter with a proposed nanoscale in vivo communication architecture [115], which brings together many of the components explained in this chapter and the last. The vision is to create a bio-transmission control protocol and a bio-user datagram protocol by careful interaction with cell signaling. A simple two-layer protocol stack is defined with a low-level encoding layer and a transport and error recovery layer. The encoding layer enables the device to encode the data bits into biomolecules; the transport and error recovery layer handles reliable transmission of the molecular information. The encoding mechanism is based upon phosphorylation of kinase, which provides a convenient on/off switch for representing binary values. Basic binary functions can be implemented using this technique [98, 116–118]. Note that there is a feedback loop between the activity occurring in the cells due to the release of kinase and the logic circuit controlling the network operation. More on the architecture of this nanoscale communication system is described in Chapter 7.

3.7   SUMMARY

Nature has evolved many forms of nanoscale and molecular communication beyond the molecular motors of the first chapter. All of these mechanisms are awe-inspiring; there is much to learn from how they operate. They range from relatively simple gap junctions and vesicles, to general biochemical cell signaling, to the fascinating calcium concentration channel, to neurons and olfactory communication. Each of these mechanisms has obvious pros and cons; however, they all inspire new ways to engineer communication at the molecular scale.

Now that we have drawn inspiration from biological approaches to nanoscale and molecular networking, what are some of the inorganic approaches? How has the ubiquitous carbon nanotube been utilized to construct information networks? What makes carbon nanotube networks unique and what are the challenges in utilizing them? These questions are addressed in the next section.

3.8 EXERCISES

Exercise 17   Diffusion

The diffusion constant D is approximately 10^{-6} cm^2/s for signaling molecules. Assume a cell radius ρ of 5 μm and a production rate F_0 *molecules/area · time* of 1,000.

1. What is the signal concentration at distances of 75, 150, and 235 μm from the cell?

Exercise 18   Gap Junction Signaling

Answer the following questions.

1. Do gap junction channels require cells to be in direct contact?

2. What are the signaling molecules that pass through gap junction channels?

3. What is phosphorylation and how is it used as a signaling mechanism?

4. Is the Ca^{2+} concentration higher or lower outside the cell and why?

Exercise 19   General Cell Signaling

Answer the following questions.

1. What is quorum sensing?

2. How do the concentrations of a neurotransmitter and an endocrine signal differ? How does their affinity to their receptors differ? Why is this difference important to the rate of transmission in a neuron?

3. What is the difference between autocrine and paracrine signals?

Exercise 20   Calcium Signaling

Use the parameters in Table 3.2 to construct a biochemical kinetic model as discussed in the section on chemical stores and relays and generate the Ca^{2+} waveform. (Hint: See the appendix for additional information about biochemical simulators and networks of differential equations.)

1. What does the waveform look like?

Table 3.2
Exercise 4 Parameters

Parameter	Value
v_0	1
v_1	7.3
B	0.01
k	3.5
k_f	1
V_{M2}	50
V_{M3}	500
K_2	1
K_R	2
K_A	0.9
m, n	2
p	4
d	0.5
n	50

Exercise 21   Calcium Signal Waveform

Consider the waveform for calcium signaling and the parameters describing its properties. Note that the frequency sensitivity of the target to calcium oscillations is maximized when the solution to Equation 3.50 is found.

$$\max_{\gamma} \left| \bar{X}(\omega \to \infty) - \bar{X}(\omega \to 0) \right|$$

$$\sigma = \text{constant} \tag{3.50}$$

1. How does the effective activation rate depend on the binding affinity K_S?

2. How does the effective activation rate depend on the Hill coefficient n?

3. What does a low value of the relative oscillation frequency ω indicate?

4. The text took the limit of the target activity as $\omega \to \infty$. What is the limit of the target activity as $\omega \to 0$?

5. Using Equation 3.50, what is the optimal duty ratio (as a function of σ)?

6. Using the answer to the previous question, what are the optimal values for the duty ratio and what does this tell us about the calcium signal?

Exercise 22   Memristor

1. What is the impact of the dopant mobility μ_V and the total doped and undoped distance D on the memristor?

Exercise 23   Stochastic Resonance

Consider the stochastic resonance example given in this chapter in Equation 3.49 with $U(x) = -\alpha x^2 + \beta x^4$.

1. Plot the solution for $x(t)$.

2. Choose an appropriate binary threshold for the output.

3. Plot the output for various levels of noise and weak signal strength.

Exercise 24   Network Coding

The max-flow min-cut theorem specifies that the maximum flow through a network is equal to the minimum cut of the network. In other words, the minimum number of edge flow rates that, when removed, partition the network into two disconnected regions is equal to the maximum flow possible through the entire network.

1. Recall that the information flow rate through a communication network using network coding is defined by the dimension of the vector space at the receiving node. As networks scale down to the nanoscale and become more interconnected, what is the gain in bandwidth when using network coding for uniformly connected random networks?

2. Consider the previous question, but for scale-free networks.

Chapter 4

Carbon Nanotube-Based Nanonetworks

In nature we never see anything isolated, but everything in connection with something else which is before it, beside it, under it and over it.

Johann Wolfgang von Goethe

In this section we examine nanoscale networks of carbon nanotubes. Much of the research in this area is focused upon using carbon nanotubes as interconnects within computer chips. While this chapter occasionally touches upon this aspect, our interest is in ad hoc approaches to nanoscale networks; approaches in which precise layout of nanotubes is not required. The carbon nanotube is a relatively new nanoscale structural discovery and we begin by reviewing some of its fundamental properties. It has properties that lend itself not only to nanonetworks but to a wide variety of applications. For example, sensor coverage will benefit from finding better ways to communicate among ever smaller sensing elements. As development in nanotechnology progresses, the need for low-cost, robust, reliable communication among nanomachines will become more urgent. Communication and signaling within newly engineered inorganic and biological nanosystems will allow for extremely dense and efficient distributed operation. We examine these potential challenges and benefits from the perspective of individual nanotubes within networks of nanotube structures, including both microtubules and random carbon nanotube (CNT) networks. In Chapter 1 we examined molecular motors that ride on microtubule network rails. While we focus on the inorganic carbon nanotube in this chapter, it should be noted that there are similarities between these molecular tube structures; much of the analysis of networks of carbon nanotubes will also apply to nanoscale networks of microtubules as well.

One may imagine small nanotube networks with functionalized tubes sensing multiple targets inserted into a cell in vivo. Information from each nanotube sensor can be fused within the network. This is clearly distinct from traditional, potentially less efficient, approaches of using CNT networks to construct transistors for wireless devices that perform the same task. This chapter also briefly considers nanotube networks, with random and semirandom network topologies, within field effect transistors (FET). FETs with nanoscale components are very important in their own right, due to their superior switching performance, but discussing this also exercises our understanding of the electrical and information transmission properties of such random and semirandom nanoscale network structures.

One of the main differences between human-scale and nanoscale networking is the difference in the underlying physics, one component of which is the quantum nature of matter which becomes more pronounced at the nanoscale. Therefore, the next section discusses nanotubes, quantum computing, and quantum networking. While the next chapter focuses on quantum networking, we introduce some of the topics here as they pertain to nanotubes and nanotube network structures.

One of the most direct applications of nanoscale networking has been the design of the single carbon nanotube radio. Thus, we next discuss this device. Although it has been promoted as a molecular radio, it is actually limited to being a radio receiver. However, this provides an excellent example of the wide versatility of the carbon nanotube, all aspects of which are leveraged, in this particular application, for wireless reception of information at the molecular level.

Following this, we will discuss nanotube networks and graph theory. Analysis of nanoscale network structure requires one to be facile in understanding and manipulating the mathematics behind networks. Eigenvalue analysis of network structures is introduced, also known as graph spectra, which will be required in the next chapter on quantum nanoscale networking.

Finally, manipulating a nanoscale network, both in terms of its construction and maintenance will require the ability to manipulate nanoscale components. The previous chapters have considered biological nanoscale network components. Nature has enabled biological systems to self-assembly components, so that for the most part, we simply leverage the preassembled nanoscale components. A completely human-engineered system, from inorganic components will require a means of reaching downwards in scale and assemble the required parts. Thus, the next part of this chapter discusses self-assembly of nanoscale networks. Not surprisingly, most research and implementations of self-assembly revert back to biological approaches.

4.1 INTRODUCTION

Carbon nanotubes are allotropes of carbon. Allotropes are elements that can exist in multiple forms. In each allotrope, the element's atoms are bonded together in a different manner. Thus, allotropes are different structural modifications of the same element. Carbon nanotubes are formed into tubular molecule structures, giving them desirable properties. They are very strong and may conduct electricity and thermal energy. Carbon nanotubes come in several varieties, with single-walled (SWNT) and multiwalled (MWNT) being the most common. They are known for their high aspect ratio; in other words, they are very long and thin. Some have been observed with length-to-diameter ratios of up to 28,000,000:1.

The molecular structure of the carbon nanotube is significant, affecting everything from strength to conductivity. In general, the way single-walled nanotubes are classified is by a chiral vector (n, m), where n and m are integers of the vector equation $R = na_1 + ma_2$. a_1 and a_2 are unit vectors defined as shown in Figure 4.1. The vector, which can be thought of as residing upon an infinite graphene sheet, describes how to "roll up" the sheet to make the nanotube, where T denotes the tube axis. This vector can affect the properties of carbon nanotubes important to their usefulness in nanonetworks. Specifically, depending on the vector, the electrical conductivity can change. An equal number of unit vectors will result in a metallic property for the tube, but tubes where $n - m$ is a multiple of three results in a semiconducting nanotube. Nanotubes with differing structures can then be networked together to produce parts for circuits and for stand-alone networks. Also, by changing the structure of a nanotube at a specific position, say a bend, the conductive properties and flow from one end to the other can be changed, similar to a resistor. The fact that a slight change in the winding of the hexagons can transform the tube from metal into a large-gap semiconductor has significant implications, particularly when we wish to use nanotubes to form semiconductor devices. Unfortunately, up to one third of the tubes appear to end up being metallic. Carbon nanotubes also have excellent thermal conducting properties with an estimated thermal capacity of up to 6,000 watts/meter/kelvin, far better than the standard copper wire. This opens up yet more opportunities for the use of carbon nanotubes. Another property of multiwalled carbon nanotubes is the telescoping effect of an inner core of a nanotube that is able to slide almost frictionlessly through the outer core. This may result in yet another information transmittance method, using the inner core to transport or to signal information.

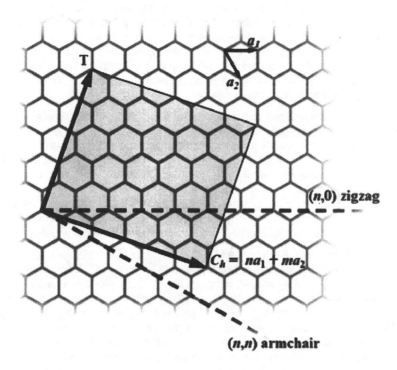

Figure 4.1 The chiral vector that defines the molecular structure of a carbon nanotube.

Yet another property of nanotubes, and possible method for transmitting information, is through vibration of carbon nanotubes. This will be discussed more in the section on the single carbon nanotube radio. Carbon nanotubes have been shown to be capable of oscillating in the terahertz range. In addition, researchers continue to find more uses for nanotubes. A functional radio receiver has been constructed from a single carbon nanotube, operating in the 40 to 400 MHz range [41]. The radio can receive and demodulate both AM and FM signals. This opens the amazing possibility of not only using radio communication inside nanonetworks but also using wireless radio communication to connect multiple nanonetworks and traditional networks. Carbon nanotubes show great potential for many uses. Their amazing properties, from conductivity to strength, make them a prime resource for realizing nanonetworks and implementing computer chips. Research continues to point towards additional possibilities to exploit these nanoscale structures for nanonetworks.

4.1.1 Comparison with microtubules

As we discussed in Chapter 2, microtubules are the railway for the transport of information throughout the cell. In a sense, microtubules are the cells' own version of a nanotube network. In fact, microtubules and carbon nanotubes have many similarities [119]. A clear similarity is their common structure; both are hollow, thin-walled tubes with a high aspect ratio and very efficient for bearing loads. Microtubules are cytoskeletal biopolymers that play a critical role in all phases of the cell's life cycle. They provide mechanical stability for the cell, including holding its shape during cell migration, and, in addition, provide tracks for intracellular transport. Microtubules are one hundred times stiffer than other cellular components and have a high degree of resilience. Carbon nanotubes are suggested to be the closest nonbiological counterpart of microtubules [119]. Carbon nanotubes are extremely stiff, with a Young's modulus five times higher than steel. Similar to microtubules, they are also highly resilient. While the chemical composition of microtubules comprised of proteins and noncovalent bonds differs from carbon nanotubes, which are comprised of carbon and covalent bonds, their mechanical behavior is quite similar. Both microtubules and carbon nanotubes spontaneously assemble into bundles.

In addition, microtubules and carbon nanotubes share electrical properties; namely both have conductances that have been carefully measured. The flow of current through microtubules and carbon nanotubes is a different process, namely microtubules use an ion channel while carbon nanotubes are either semiconducting or metallic. Current flow through microtubules was measured in [120] to be approximately 9 nS (nano-Siemens) at a rate of approximately 1.0 m/s and exhibits an amplification effect. Thus, we could even draw the analogy further to say that both microtubules and carbon nanotubes can act as transistors. Electric fields may be used to control the positioning of microtubules [28]. Also, both are impacted by magnetic fields; free-floating microtubules can be steered via a magnetic field. Microtubules naturally self-assemble while controlled self-assembly of carbon nanotubes is possible by coating them with amino acids.

One significant difference is that microtubules are more dynamic than currently engineered carbon nanotube networks. Microtubules switch between phases of assembly and disassembly on the time scale of seconds, thus constantly growing and shrinking in size. Also, microtubules commonly appear in a bent shape due to strong cytoskeletal forces. Persistence length quantifies the degree of bending in microtubules. The persistence length was discussed in detail in the previous chapter, but briefly, the persistence length is the rate at which tangents taken along each segment of a microtubule become decorrelated from one another. If $R(s)$ is a point on a segment s, then let $u(s)$ be the unit tangent vector, $u(s) = \frac{\partial R}{\partial s}$. The orientation of the unit tangent vectors for segment 0 and segment s is quantified by the inner product, $\langle u(s) \cdot u(0) \rangle = e^{-s/\xi_p}$, where ξ_p is the persistence length. For longer persistence lengths, or for shorter tubes, the microtubules will be straighter. For longer tubes and shorter persistence lengths, the impact of decorrelation along the chain's tangents becomes more significant. We can approximate the curved microtubules as many smaller random chains that happen to be connected end-to-end, but with decorrelated alignment. Thus, shorter persistence lengths will tend to decrease the percolation threshold, which is important in the explanation of network conductance discussed in a later section of this chapter.

4.1.2 Nanotubes and biology

Carbon nanotubes have been associated with biological applications for some time. They have been used to mimic neurons, carry and deliver drugs into cells, and they have been coupled with both DNA and molecular motors. The coupling between inorganic and organic nanoscale components is likely to be an important aspect in the advancement of nanonetworks and nanotechnology in general.

In one example, molecular motors have been use to position carbon nanotubes [121]. First recall the discussion of microtubule and molecular motors from Chapter 1. Carbon nanotubes have been functionalized with streptavidin, which forms a strong linkage with microtubules that have been functionalized with biotin. Streptavidin is a protein that comes from the bacterium *Streptomyces avidinii*. It finds wide use in molecular biology through its extraordinarily strong affinity for biotin. Specifically, the dissociation constant of the biotin-streptavidin complex is on the order of $\approx 10^{-15} mol/L$, ranking it among one of the strongest noncovalent interactions known in nature. This technique is used to attach microtubules to the nanotubes, or perhaps more precisely, attach the nanotubes to the microtubules, since the microtubules are the larger structure. Molecular motors are attached in fixed positions along gold-plated terminals. The motors then shuttle the molecular motors with their attached nanotubes across the terminals. As discussed in Chapter 1, ATP fuel is required in order to enable the transport to occur; adding or removing the fuel stops, starts, and allows control of the process.

Nanotubes can be functionalized in different ways to enable drug delivery [122]. They can be oxidized using acids, resulting in the reduction of their length to aid their ability to disperse within aqueous solutions. Also, chemical processing can be used to change the chemical structure of the carbon nanotube external walls and tips, making them soluble in water. Finally, functionalized carbon nanotubes (f-carbon nanotube) can be coupled to a variety of other biological molecules, including peptides, proteins, nucleic acids, and other therapeutic agents. Fluorescent markers attached to f-carbon nanotubes can be used to track their ability to easily penetrate the cell wall and into the cytoplasm. Evidence shows that macrophage cells could ingest significant amounts of nanotubes without apparent toxic effects [122]. In terms of drug delivery, one example is the transport and uptake of DNA by a cell. Injecting DNA alone, in the form of a plasmid, has difficulty in reaching the cell nucleus. When DNA is attached to the f-carbon nanotube, gene expression levels increased between five and ten times.

4.2 NANOTUBES AS FIELD EFFECT TRANSISTORS

To construct a nanoscale network, it is important to understand the medium with which the network will be built. As previously discussed, the chiral vector and angle describe the atomic structure of the tube and its electrical properties. The angle indicates to what degree the lattice is twisted. Multiple names exist for the most common of these structures; for example, armchair (n, n), zigzag $(n, 0)$, and chiral $(2n, n)$ [123, 124]. The nice thing about this structure is that depending how the atomic structure is arranged, the nanotube can have differing conducting properties [124–126]. The lattice has the property that it can backscatter electrons moving at the Fermi energy, which gives the nanotube the property of a semiconductor. If the electrons move along the y axis or at 60, 120, 180, or 240 degrees to the y axis of the lattice, the backscattering is suppressed [127, 128]. Thus, depending how the lattice is arranged, a carbon nanotube can be metallic or semiconducting.

More amazingly, semiconducting nanotubes can be made into transistors by applying either negative charge, which introduces holes that enable conductance, or a positive charge, which depletes the holes and reduces the conductance [127, 128]. These properties explain why carbon nanotubes are an active research area in nanonetworks and circuits [129]. Molecular dynamic simulations to test carbon nanotube structures for buckling and deformation have been carried out, finding differing buckling and deformation for tubes of differing diameters at differing energies [128]. An interesting result is that the simulations seem to indicate that the tubular cross-section can deform as much as 20% before buckling occurs. Both the mechanical properties and the energetic and vibrational properties of SWNTs have been intensely studied [130]; for example, the armchair (n, n), zigzag $(n, 0)$, and the chiral $(2n, n)$ nanotube structures.

Thermal conductivity is yet another area of research in carbon nanotubes. If nanotubes are to be used in very small nanosize devices such as nanonetworks and circuits, managing the heat created by high density electronics will be important. The exact thermal conductivity of carbon nanotubes is yet to be fully understood. There are results that show that the conductivity may be unusually high at approximately 6600 W/mK [131]. This result would be higher than previously reported results of approximately 2980 W/mK [132].

To overcome the difficulty of performing thermal conduction measurements due to problems with synthesizing and aligning high-quality nanotubes, molecular dynamic simulation models have been developed. For example, one particular model claims the advantage of not requiring high numbers of parameters to be deduced from experiments, such as the those based on the Boltzmann equation [133].

The model's result corresponds well with previous results, converging to a value of approximately 2,980 W/mk.

4.2.1 Electron transport

Electron transport is another key issue for carbon nanotubes that affect nanonetwork research. As stated previously, carbon nanotubes can be metallic or semiconducting, depending upon their chiral vector [127, 134]. [127] suggests that it is useful to employ the Landau-Buttiker formula, stated as: for a system with N one-dimensional channels in parallel: $G = (Ne^{\frac{2}{\hbar}})T$ where T is the transmission coefficient for the electrons through the tube. More about the derivation of this formula will be seen shortly.

For an SWNT with perfect contact, $T = 1.0$, which is approximately 155 μS. Resistance must be accounted for in imperfect contacts [127]. Modeling and simulating electron transport and conductivity will be important aspects for future nanonetworks. A simulation tool by [135, 136], simulates the electron transport in a single carbon nanotube. The simulator allows for changes in voltage temperature and chirality and simulates the effects on electron transport. A numerical algorithm for analyzing electron tunneling conduction through nanoparticle arrays is implemented in [137]. Using new experimental data, the nonlinear current-voltage characteristics were obtained and appear to largely depend on the structure of the network. The authors speculate that the measures of current flow can be used as a tool to characterize film structure conduction properties. Methods for interpreting both AC and DC conductivity results of SWNTs are examined in [138].

4.3 NANOTUBES AND QUANTUM COMPUTING

Nanotubes are a candidate for applications related to quantum computing, and thus, have a potential role in quantum networking. This can take a variety of forms, from traditional interconnects as quantum wire to a quantum dot substrate to form quantum dot networks. Chapter 5 introduces quantum computation and networking in detail. In this section we briefly mention the relationship between nanotubes and quantum computation and networking.

Nanotubes are one implementation of a quantum wire, a single long molecule with an extremely narrow diameter. The resistance along a traditional human-scale wire can be computed by Equation 4.1, where ρ is the resistivity, l is the length, and A is the area of a cross section of the wire.

$$R = \rho \frac{l}{A} \tag{4.1}$$

The equation above for resistance assumes a diffusive transmission of electrons through a material. The electrons scatter as they travel; bouncing off other particles in their path as they flow through the wire. Consider a hypothetical wire that is shrunk to the nanoscale, both in length and width. As the wire becomes shorter, approaching the nanoscale, the mean free path length of the electrons meets or exceeds the length of the distance traveled through the wire. Recall earlier discussions of Brownian motion and the Langevin equation from Chapter 2 that describes both directed and diffusive movement. As the width of the wire approaches the nanoscale, the only electron scattering that occurs is the electrons bouncing off the sides of the nanotube. The resulting conductance does not depend on the length of the wire, but instead becomes quantized in amounts that have been experimentally measured as shown in Equation 4.2 [139] where e is the electron charge and $\hbar$ is Planck's constant.

$$\frac{2e^2}{\hbar} = 12.9 k\Omega^{-1} \tag{4.2}$$

The quantization of conductance can be understood by the wave confinement that occurs as our hypothetical wire narrows to the point where it has only only one dimension, its length [140]. This is known as a quantum wire, and quantum mechanical effects dominate within a quantum wire. Let us see, in a little more detail, where this quantization originates. Assume as we just discussed, that the electron transport is ballistic, so that there is little or no scattering. If the channel is purely ballistic, the resistance of the channel is zero. All the dissipation of energy occurs in the contacts. The inverse of Equation 4.2 is sometimes called the contact resistance.

We can now start from the most fundamental nature of matter, its quantum mechanical description, as represented by the Schrödinger wave equation. The next chapter covers quantum computation and networking as it pertains to nanoscale communication networking. While this book assumes no prior knowledge of quantum mechanics, we need to introduce a key fundamental concept related to quantum computation at this point in order to gain insight into the nature of conductance and information flow through a carbon nanotube. Clearly this can be a challenging topic, so don't worry if it seems unclear during your first reading. At a minimum, the goal is simply to provide a qualitative understanding of why the carbon nanotube has some of its unique electronic characteristics. Understanding these characteristics allows one to better use these fundamental properties for nanoscale networking.

The wave equation is the fundamental representation of state in quantum mechanics. As introduced in Chapter 1, all matter has a waveform; this waveform may itself be used for nanoscale communication. Much more will discussed regarding quantum state in the next chapter. The Schrödinger wave equation is typically denoted by $\Psi(x,t)$, also known as the wavefunction, where x is the position of a particle and t is time. The assumption is that the particle has a mass m and is acted upon by a potential described by $V(x,t)$. The wavefunction describes where a particle may be in space and time; it represents a probability. More specifically, when squared, its value is the probability density of the particle's position in space at time t. In the case of the one-dimensional carbon nanotube, we are modeling the electron wave as shown in Equation 4.3. Here x is the direction along the length of the quantum wire L and y is the direction along the width of the wire W. Because we are concerned with the impact of the width, $V(y)$ is a potential that changes depending only upon the width of the wire.

$$-\frac{\hbar^2}{2m}\left(\frac{\partial^2\Psi}{\partial x^2} + \frac{\partial^2\Psi}{\partial y^2}\right) + V(y)\Psi = E\Psi \tag{4.3}$$

Equation 4.3 can be solved by separation of variables. Along the length L of the quantum wire, the result is a plane wave as one would expect. However, along the width of the quantum wire W, the solution is quantized depending upon the potential $V(y)$. The resulting energy is of the form shown in Equation 4.4. ϵ_n is the energy of the nth solution across the width of the wire. k_x is the x component of the wave vector. A wave vector has a magnitude equal to the wave number $\frac{2\pi}{\lambda}$ and a direction that indicates the direction of propagation of the wave.

$$E_n = \epsilon_n + \frac{\hbar^2 k_{xn}^2}{2m} \tag{4.4}$$

Here we take a quick diversion to mention the Fermi energy. Electrons are a class of particles known as fermions and obey the Pauli exclusion principle. This states that no two fermions can occupy the same quantum state. The quantum states are related to the energy levels of the particles. The Fermi energy is the energy of the highest occupied state. The ground state is the lowest energy level. Thus, in the solutions shown in Equation 4.4, it is only the solutions at the Fermi energy level that actually contribute to the conductance of the quantum wire. Each solution at the Fermi level is called a channel. Note that it is only electrons at the Fermi level that, with an infinitesimal addition of thermal energy, can jump to a free state and contribute to the conductance.

$$dI_n = ev_n\rho_n dE \tag{4.5}$$

Equation 4.5 contains a few symbols specific to a wavepacket. A wavepacket, which is the quantum mechanical instantiation of a particle, is a short oscillatory wave pattern. The short wave pattern has its own frequency; however, the wavepacket itself may be moving through space. Wavepackets can be constructed mathematically using the Fourier transform and integrating over a set of frequency components, or wave numbers k. Because wavepackets travel through space and the individual waves that comprise the wavepacket also have a frequency and travel through space, there is a distinction in how the wavepacket as a whole and individual waves are measured. The phase velocity is the velocity of the crest of a particular wave that comprises a wavepacket. The group velocity is the velocity of the overall pattern of the entire wavepacket as it travels through space. Here v_n is the group velocity in the x direction and ρ_n is the density of states in our one-dimensional wire. For wavepackets, the group velocity of a wave is the velocity with which the overall shape of the wave's amplitudes, known as the modulation or envelope of the wave, propagates through space. The density is $(\pi\frac{dE}{dk_x})^{-1}$ and the group velocity is $\frac{dE}{\hbar k_x}$. These values can be inserted into Equation 4.5 to yield Equation 4.6. Thus, the rate of change of current with regard to energy is constant for all channels.

$$\frac{dI_n}{dE} = \frac{2e}{\hbar} \tag{4.6}$$

Instead of writing differentials, we can consider larger differences and also replace the energy differential with the potential difference of $e\Delta V$ as shown in Equation 4.7.

$$\frac{\Delta I_n}{dE} = \frac{2e^2}{\hbar}\Delta V \tag{4.7}$$

Notice that $\frac{2e}{\hbar}$ in Equation 4.7 is also the value stated earlier in Equation 4.2. The total conductance is the sum of all existing channels, thus if there are N channels, then the total conductance is shown in Equation 4.8.

$$G = \frac{2e^2}{\hbar}N \tag{4.8}$$

This result is known as the Landauer equation, and it predicts the quantized conductance that has been experimentally measured.

4.4 A SINGLE CARBON NANOTUBE RADIO

The single carbon nanotube radio is an excellent example of leveraging many fundamental phenomena within a single carbon nanotube in order to implement components required for a radio receiver [41]. One of the physical differences at the nanoscale is that mechanical resonances are in the range of 50 MHz through 5 GHz. The fact that the nanotube exists at the nanoscale means that it has mechanical resonance that overlaps with the frequency of mobile phones, wireless local area networks, and global positioning system signals. Note that the resonance frequency is related to the tendency of a system to oscillate at a larger amplitude at some frequencies than others. At the resonant frequency, increases in the amplitude of oscillation do not change the frequency of oscillation. Thus, the energy is not being released through oscillation, but is being stored within the object. At these frequencies, even small periodic driving forces can produce large amplitude vibrations, because the system stores vibrational energy. The fact that this overlap between commercial human-scale radio communication frequency and nanoscale mechanical resonance occurs is one of the enabling aspects of the nanotube radio.

There are unique phenomena within the nanotube that correspond to the antenna, tuner, amplifier, and demodulator of a radio receiver. The complete system is comprised of a nanotube mounted in close proximity to another electrode. A direct current power supply is connected to the nanotube and electrode, which power the nano-radio receiver. This places a charge on the tip of the nanotube, which makes the nanotube tip very sensitive to changes in the surrounding electric field. Because, as previously mentioned, the resonant frequency of the nanotube is in the range of commonly used radio frequencies, the tube readily oscillates when a carrier frequency is detected.

The nanotube's resonance frequency can be tuned in at least two different ways. A rough and nonreversible tuning process can occur by making the voltage strong enough to burn off the tip of the tube, thus shortening the tube and increasing its frequency; just as a guitar string when shortened increases in pitch, the resonance frequency of the nanotube increases. Thus, the length of the tube is reduced. Finer resolution tuning of the radio is accomplished by increasing the bias voltage through the tube, which changes the tension of the nanotube based upon the electrostatic field. Thus, the nanotube acts both as an antenna, detecting the signal, and as a tuner, by varying its length. An amplifier exists within the nanotube radio because the current flowing through the nanotube is powered by the external battery. Thus, it is the power of the battery that is modulated and is much larger than the detected signal.

Because commonly used radio frequencies at the human-scale and mechanical resonance frequencies at the nanoscale overlap, much of the radio communication engineering is mechanical rather than electrical and requires a good understanding of how to control the resonance frequency. The resonance frequency is shown in Equation 4.9 where L is the length of the nanotube, Y is Young's modulus, I is the area moment of inertia (explained shortly), ρ is the density, and A is the cross-section area.

$$f_0 = \frac{0.56}{L^2}\sqrt{\frac{YI}{\rho A}} \qquad (4.9)$$

The area moment of inertia, also called the second moment of inertia, measures an object's ability to resist deflection. An I-beam used for constructing a building has higher area moment of inertia than a simple rod of the same mass and length. It can be thought of as the position along a lever required to stop a force from turning a lever. The farther one is from the pivot point, the more leverage one has and the easier it is to stop the object from turning. However, the notion of the area moment of inertia is to take into account the shape of the object. Each point within the object itself resists turning based upon its distance from the pivot point and, thus, the shape of the object. A simple example of an object that is symmetric along the x axis is shown in Equation 4.10, where A is the cross-section area, y is defined by the shape of the material, and I_x is the resulting second moment of inertia in units of length to the fourth power. Note that for a cylinder, such as a nanotube, the moment of inertia can be computed by $\frac{\pi}{4}(r_{outer}^4 - r_{inner}^4)$ where r_{outer} is the outer radii and r_{inner} is the inner radii; thus a thicker tube has a larger second moment of inertia.

$$I_x = \int y^2 dA \qquad (4.10)$$

From [141], the minimum detectable electric field amplitude is shown in Equation 4.11 where T is the temperature, k_B is Boltzmann's constant, $m_{eff} \approx 0.24m$ is the effective mass of the nanotube, Q is the quality factor, and B is the bandwidth of the oscillation. The oscillation of the nanotube depends upon the portion of the mass of the tube that is actually oscillating. Not all of the tube moves during the oscillation.

The effective mass of the tube is the mass effectively taking part in the oscillation, which correctly predicts the behavior of the system. ω_0 is the angular resonant frequency where $\omega_0 = 2\pi f$.

The quality factor Q is a measure of the degree to which an oscillating system is underdamped. An underdamped, or high Q, system will expend most of its energy at, or near, its resonant frequency. An overdamped, or low Q, system will spread its energy over a larger range of frequencies as it slowly comes to a halt. Thus, in general, high Q systems are most desirable as they pack the energy into a narrow, predictable frequency range.

$$E_{rad} = \frac{1}{Q}\sqrt{4k_B T m_{eff}\omega_0 \frac{B}{Q}} \qquad (4.11)$$

Demodulation of a signal from an amplitude-modulated carrier wave simply requires a rectifier in order to perform envelope detection. The carbon nanotube is capable of demodulating an amplitude-modulated RF signal due to its nonlinear current-voltage (I_{DS} vs V_{DS}) characteristics. These nonlinearities can rectify a portion of the applied RF current [142]. The current to voltage relationship is shown in Equation 4.12 where the voltage of the applied RF signal is V_{RF}, and the second derivative represents the nonlinear current-voltage (I_{DS} vs V_{DS}) characteristics of the carbon nanotube itself.

$$I = I_0 \frac{1}{4}\frac{d^2 I}{dV^2}V_{RF}^2 \qquad (4.12)$$

4.5 NANOTUBES AND GRAPH THEORY

Connectivity of multiple carbon nanotubes is a step towards realizing a full nanonetwork. Developing methods for connecting, evaluating, and modeling connections, how electrons move through the differing types of possible connections, and how this can be modeled and used to route information are all current research areas.

Another step in building nanonetworks of carbon nanotubes is to examine conductivity through the entire network. The area of random or partially aligned nanotubes is of interest given the difficulty in fabricating and placing individual nanotubes. The behavior of carbon nanotube networks differ from thin film networks [143]. Poorly connected networks have a relatively even conductivity through the network but decrease dramatically at certain points. At some points, resistance can increase on the order of $10^8\,\Omega$. Well-connected networks will also have nonmonotonically decreasing conductance but have areas of very high conductance. Random networks of carbon nanotubes and routing via controlled changes in resistance has been modeled [3] and will be discussed in more detail later in this chapter. The nanotube network capacity, as a function of area available for connectivity, results in much higher capacity over smaller distances than could be achieved through a wireless network. Transistors are one application for nanotube networks and nanotechnology in general [144–146]. The ability to use nanonetworks to shrink the size of current electronics is one of the driving forces behind nanotechnology research.

4.5.1 Eigensystem network analysis

Information can be communicated simply through a connected path of carbon nanotubes that are in direct contact with one another. Each contact point is a node, and the length of each carbon nanotube forms a link in the communication network. Electric current flowing into the network carries information; depending upon the network structure, current will flow to a destination point within, or on the opposite side of, the network. Let us examine what a channel looks like in such a wired network.

Any communication network can be described at its most general level by a set of nodes that either generate or forward information and a set of links, or channels, that enable the information to flow from one node to another. Thus, the network can be represented by an adjacency matrix A, a square matrix with the same number of rows and columns as there are nodes in the network. The entries in the adjacency matrix A_{ij} represent the presence (1) or absence (0) of a connection, or link from a node specified by the row i to a node specified by the column j.

In general, the Laplacian matrix L is defined as the difference between the degree matrix D and the adjacency matrix A. The degree matrix is simply a diagonal matrix in which the degree of each node is listed along the diagonal for that node. Thus, the Laplacian matrix will have a positive diagonal and negative or zero elements in all other locations. Following the derivation of [147], elements of the Laplacian matrix can represent the conductances. Thus, the conductances are the inverse of the resistances for each link $c_{i,j}$ as shown in Equation 4.13. A conductance of zero is the same as an infinite resistance, or no connection.

$$c_{ij} = r_{ij}^{-1} = c_{ji} \qquad (4.13)$$

Kirchoff's law must apply at each node in the network. This means that there is no charge storage within a node; the sum of the ingoing and outgoing currents must be zero as shown in Equation 4.14. Here I_i is the current flowing into node i.

$$\sum_{i=1}^{N} I_i = 0 \qquad (4.14)$$

The electric potential at each node is represented by V_i. A simple application of both Ohm's law and Kirchoff's law yields Equation 4.15. This shows $I = V/R$, or $I = Vc$, where c is the conductance, for each node.

$$\sum_{j=1}^{N} c_{ij}(V_i - V_j) = I_i, i \neq j, i = 1, 2, 3 ..., N \qquad (4.15)$$

Simply put, the conductance of the network, represented by the Laplacian conductance matrix L multiplied by a vector of the voltages for each node $\vec{V}$, yields a vector of currents entering each node $\vec{I}$ as shown in Equation 4.16.

$$L\vec{V} = \vec{I} \qquad (4.16)$$

L is shown explicitly in Equation 4.17.

$$L = \begin{pmatrix} c_1 & -c_{12} & ... & -c_{1N} \\ -c_{12} & c_2 & ... & -c_{2N} \\ ... & ... & ... & ... \\ -c_{1N} & -c_{1N} & ... & -c_N \end{pmatrix} \qquad (4.17)$$

The diagonal is shown explicitly in Equation 4.18.

$$c_i = \sum_{j=1}^{N} c_{ij}, i \neq j \qquad (4.18)$$

As a side note related to the network structure, the Laplacian matrix also shows the number of spanning trees of the resulting network. A spanning tree is a subset of the network graph that forms a tree and contains all the nodes within the network. In other words, it has a root node that reaches all the nodes of the original graph. Specifically, if λ_n is the nth eigenvalue of the Laplacian matrix for graph G, then the number of spanning trees is shown in Equation 4.19. Spanning trees are not only important for theoretical reasons, but they also have practical application in the setup and maintenance of routing within communication networks.

$$trees(G) = \frac{1}{n}\lambda_1\lambda_2\lambda_3\ldots\lambda_n \tag{4.19}$$

It's not hard to see now that the resistance between any two nodes α and β can be found by applying battery leads to the nodes and measuring the resistance, which will be as shown in Equation 4.20.

$$R_{\alpha\beta} = \frac{V_\alpha - V_\beta}{I} \tag{4.20}$$

Recall that the current entering node i is I_i. If $\bar{I}$ is the vector of node currents, then the current into node i is shown in Equation 4.21 where δ_{ij} is the Kronecker delta function. This function is one when $i = j$ and zero otherwise.

$$I_i = \bar{I}(\delta_{i\alpha} - \delta_{i\beta}) \tag{4.21}$$

Now, following the work in [147], we can solve for the network conductance in terms of the eigensystem of the Laplacian matrix L. Let Ψ_i and λ_i be the eigenvectors and eigenvalues of L, which, by definition, are shown in Equation 4.22.

$$L\Psi_i = \lambda_i\Psi_i, i = 1, 2, \ldots, N \tag{4.22}$$

In general, if there are N nodes, then there will be N eigenvectors and eigenvalues, and the eigenvector will have N elements. Let each element of the eigenvector be labeled α, so that we can index the ith eigenvector as $\Psi_{i\alpha}$.

$$(\Psi_i^*\Psi_j) = \sum_\alpha \Psi_{i\alpha}\Psi_{j\alpha}^* = \delta_{ij} \tag{4.23}$$

A Hermitian matrix is a square matrix with complex entries. A Hermitian matrix is equal to its own complex transpose. In other words, the element in the ith row and jth column is equal to the complex conjugate of the element in the jth row and ith column, for all indices i and j as shown in Equation 4.24.

$$a_{i,j} = \bar{a}_{j,i} \tag{4.24}$$

The conjugate transpose, or matrix adjoint, is often represented as $A^\dagger$. Thus, a Hermitian matrix is one in which $A = A^\dagger$. Hermitian matrices have many interesting properties. One important property is that the eigenvalues are orthogonal. The eigen-decomposition of a Hermitian matrix A is shown in Equation 4.25 where $UU^\dagger = I$.

$$A = U\Sigma U^\dagger \tag{4.25}$$

Since the left and right inverse are the same, $U^\dagger U = I$. Thus, $A = \sum_i \sigma_i u_i u_i^\dagger$. Here, σ_i are the eigenvalues and u_i are the eigenvectors.

Returning to the Laplacian matrix with conductances L, since it is Hermitian, Equation 4.26 holds. Here δ_{ij} is the Kronecker delta function. It has a value of one when i and j are equal and zero otherwise.

$$\Psi_i^\dagger \cdot \Psi_j = \sum_\alpha \Psi_{i\alpha}^\dagger\Psi_{j\alpha} = \delta_{ij} \tag{4.26}$$

It's important to note that one of the eigenvalues is zero, because the sum of any row, or column, of L is zero, $l_1 = 0$ and the corresponding eigenvector is shown in Equation 4.27.

$$\Psi_{1\alpha} = \frac{1}{\sqrt{N}}, \alpha = 1, 2, \ldots, N \tag{4.27}$$

Following the development in [147], we can solve for the resistance between any two points of an arbitrarily complex network using a simple formula involving the eigensystem, that is the eigenvalues and eigenvectors.

First, we need an inverse of the Laplacian matrix L, which we will call G. However, to avoid problems in the analysis due to the zero eigenvalue, a small value ϵ is added to the matrix values, $L(\epsilon) = L + \epsilon I$. This adds ϵ to the diagonal values changing them from c_i to $c_i + \epsilon$. Since this only effects the diagonals the eigenvalues are simply modified by $\lambda_i + \epsilon$. The inverse of $L(\epsilon)$ is shown in Equation 4.28.

$$G(\epsilon) = L^{-1}(\epsilon) \tag{4.28}$$

Now that we have an inverse of L, we can go back to the fundamental relationship in Equation 4.16 and replace it with $L(\epsilon)\bar{V}(\epsilon) = \bar{I}$. This can be the inverse of $L(\epsilon)$, namely $G(\epsilon)$ to yield $V(\epsilon)G(\epsilon) = \bar{I}$ which is shown in Equation 4.29 where $i = 1, 2, 3, \ldots, N$.

$$V_i(\epsilon) = \sum_{j=1}^{N} G_{ij}(\epsilon) I_j \tag{4.29}$$

Now determine $G(\epsilon)$. First, note that L and $L(\epsilon)$ can be diagonalized as shown in Equation 4.30 as was previously mentioned for Hermitian matrices. The matrix U is comprised of $U_{ij} = \Psi_{ji}$, Λ contains the eigenvalues λ_i along the diagonal, or $\lambda_i \delta_{ij}$ in each position of the matrix, and $\Lambda(\epsilon)$ contains the eigenvalues $\lambda_i + \epsilon$ along its diagonal, or $(\lambda_i + \epsilon)\delta_{ij}$ in each position of its matrix.

$$U^\dagger L U = \Lambda$$
$$U^\dagger L(\epsilon) U = \Lambda(\epsilon) \tag{4.30}$$

The inverse of $\Lambda(\epsilon)$ from Equation 4.30 is shown in the top of Equation 4.31. A simple rearrangement yields the bottom line of the same equation.

$$U^\dagger G(\epsilon) U = \Lambda^{-1}(\epsilon)$$
$$G(\epsilon) = U \Lambda^{-1}(\epsilon) U^\dagger \tag{4.31}$$

Changing the matrix form into a summation form, Equation 4.32 and Equation 4.33 show the operations on the explicit elements of the matrices. Note carefully the reason for the term $\frac{1}{N\epsilon}$ in Equation 4.32. This term incorporates the first eigenvalue, which we already mentioned is zero back in Equation 4.27 and has eigenvectors of $\frac{1}{\sqrt{N}}$. Thus, the summation in Equation 4.33 begins from $i = 2$.

$$
\begin{aligned}
G_{\alpha\beta}(\epsilon) &= \sum_{i=1}^{N} U_{\alpha i} \frac{1}{\lambda_i + \epsilon} U_{\beta i}^* \\
&= \frac{1}{N\epsilon} + g_{\alpha\beta}(\epsilon)
\end{aligned} \tag{4.32}
$$

$$g_{\alpha\beta} = \sum_{i=2}^{N} \frac{\Psi_{i\alpha} \Psi_{i\beta}^*}{\lambda_i + \epsilon} \tag{4.33}$$

The previous equations describing G, namely Equations 4.32 and 4.33, can be substituted into $V(\epsilon)G(\epsilon) = \bar{I}$, which yields the top line of Equation 4.34. Now that the inverse has been accomplished, the

ϵ can be removed by taking the limit as ϵ goes to zero as shown in the bottom line of Equation 4.34.

$$V_i(\epsilon) = \sum_{j=1}^{N} g_{ij}(\epsilon) I_j$$

$$V_i = \sum_{j=1}^{N} g_{ij}(0) I_j \tag{4.34}$$

Going back to Equation 4.20, and using Equations 4.34 and 4.21, yields Equation 4.35.

$$R_{\alpha\beta} = g_{\alpha\alpha}(0) + g_{\beta\beta}(0) - g_{\alpha\beta}(0) - g_{\beta\alpha}(0) \tag{4.35}$$

Next, Equation 4.33, which defines $g_{\alpha\beta}$, can be inserted into Equation 4.35. After some final manipulation, the useful result in Equation 4.36 is obtained.

$$R_{\alpha\beta} = \sum_{i=2}^{N} \frac{1}{\lambda_i} |\Psi_{i\alpha} - \Psi_{i\beta}|^2 \tag{4.36}$$

The resistance between any two points in an arbitrarily connected network can be computed. In this case, we are considering a nanoscale network, in which there is a random network of carbon nanotubes, which have all the resistance characteristics previously mentioned, namely one-dimensional quantum wire characteristics and quantized conductances. Now that we can compute the resistance of a random network, we need to address how to assemble such a network.

4.6   NANOTUBES AND SELF-ASSEMBLY

One can imagine two very general approaches to nanoscale networking with a nanotube or nanostructure substrate. The first approach is to attempt to assemble the network substrate into a precise pattern, using self-assembly techniques. The alternative approach is to allow the tube positions to be semirandom or random and adapt the communication to handle the random infrastructure. First, we discuss the aspect of controlled placement of nanostructures using self-assembly taken from the biological realm. In the next section, we will discuss the impact of communication with randomized placement of nanostructures.

The choice to not use a random network requires that the network be assembled in some way. Currently, lithographic printing methods have difficulty achieving the desired accuracy. One proposed solution, is to use DNA self-assembly in a bottom up approach [148]. This technique is applied prima facia for computer chip design, but as we will see, it has many of the elements of computer networking to make it work properly. Before explaining this nanoscale networking approach, it is necessary to be familiar with active networking.

4.6.1   Active networking

Active networking is a novel approach to network architecture in which network nodes—switches, routers, hubs, bridges, gateways, and so forth—perform customized computation upon packets flowing through them. The packets that the network is carrying can change the underlying operation of the network itself. Thus, this type of network is called an "active network" because new computations are injected into the nodes dynamically, thereby altering the behavior of the network on-the-fly. Packets in an active network can carry fragments of program code in addition to data. Customized computation is embedded within the packet's code, which is executed on the intermediate network nodes through which the packet flows [1,56]. Active networks allow processing to be distributed within the network; each packet can process information that it is carrying on each node along its path. In fact, active networks are an alternative to the current Internet architecture, whose protocols were built upon the unquantifiable notion of simplicity. The current Internet, with its carefully crafted and static protocols, requires a massive amount of documentation comprised of $O(6,000)$ Request for Comments and is growing rapidly.

Active networking at the nanoscale is an ideal mechanism for handling both the low processing power of individual nanoscale components and the variability in their connectivity due to the imprecise nature of self-assembly. Each node, or minimal nanoscale processing element, contains just enough capability to communicate as well as perform a simple processing task. The individual nanoscale components are connected using DNA hybridization, or the natural joining of complementary DNA sequences, to guide connections into position. Processing occurs by creating an active packet; a communication packet that contains the processing commands as well as the data. The packet flows from one nanoscale processing unit to another; each node performs its limited processing upon the packet and then forwards it to the next appropriate node. This requires that the resource-constrained nanoscale nodes must be capable of routing. Let us see how this works in more detail next.

4.6.2 Nanoscale active networking and routing

A *via* in an electronic circuit is essentially a hole filled with conductive material between layers. Vias are used to connect the traditional microelectronics world with the nanoscale implementation [149]. The via is large compared to the individual nanoscale processing components, so one via may overlap many nanoscale components. However, one of the components is designated to be the interface for a via. A reverse path forwarding technique is used to initialize routing at the nanoscale. A specially identified broadcast packet is injected into the nanoscale network through a via. The packet is then forwarded by each node along all its connected links, except for the link upon which the packet entered the node. Thus, the packet will propagate in a continuous wave, or front, away from the point of entry into the nanoscale network with the goal of establishing the "interest" of different vias to receive packets. The goal is to establish directions back to the responsible nanoscale elements for each via; the paths of the reverse flows look like gradients of a vector field. Thus, the packet is called a gradient packet. When a node receives such a packet, the node keeps track of the link upon which the the packet arrived so that it knows where to route packets to reach back to the appropriate via. The process will terminate, because a node stops forwarding gradient packets from the same via once it has forwarded one. Upon termination, a tree is created with the via's nanoscale processing element as the root. Multiple vias, one from each of the four edges of the nanoscale system, are used to establish north, south, east, west gradients. It is easy to see that defective nodes and links will not remain isolated from the final connected system. Thus, this simple route establishment process also removes defective elements.

A very simple routing technique is used known as up/down routing. Direction is assigned to the operational links based upon the gradient tree formation that was just described. Each link that is closer to the root of the tree is considered to be "up." If there are ties, they are broken by comparing node identifiers. A route is simply a path such that a packet never crosses a link in the "up" direction after it has used one in the "down" direction. This simple scheme avoids deadlocks and any potential loops as well as ensures that all connected nodes are reachable. Because of their simplicity, both the gradient tree establishment and this up/down routing technique are ideal for the limited operational capability of nanoscale processing elements.

The trade-off between node complexity and control of the self-assembly process has been examined [52]. Using a custom network topology generator, a variety of networks that have a varying degree of control over how nodes are placed and how links are created during self-assembly are examined. System performance is measured by the effects of topology on system performance. The running time of matrix multiplication is used as a benchmark. If the processing elements are placed correctly, there is little variation in performance due to changes in topology. Controlling placement and orientation has a greater effect on network connectivity than link sharing. Another custom event simulator was used to evaluate defect tolerance mechanisms [51]. The reverse path forward algorithm is applied to map out the defective node in the network at startup. This allows the defective nodes to be identified and connective paths to be found. Using the custom simulator to test the algorithm for varying defect rates of 0% to 50% and grids of size 30×30 to 100×100, for a system with no defective nodes, the time needed to complete the broadcast is a linear function of the square root of the number of nodes. The time taken to complete the broadcast actually decreases as the number of defective nodes increases, due to the fact there are fewer reachable nodes. Gradients are not forwarded through defective nodes so links built using gradients will not have links from defective nodes.

4.7   SEMIRANDOM CARBON NANOTUBE NETWORKS

Nanoscale sensors exist. However, human-engineered communication mechanisms to transport information from such sensors are not yet common. In addition to this, development of both biological and engineered nanomachines is progressing; such machines will need to communicate. Conceptually, information systems in living cells and computer networks have much similarity. We focus on the analogy between living, self-organizing microtubules and engineered carbon nanotube networks. We note the strong similarities and differences between microtubules and carbon nanotubes. Recent work has shown that both microtubules and carbon nanotubes can form random networks capable of transporting information at the nanoscale. The similarity is strong enough that in the following text, we use the word nanotube to refer to both microtubules and carbon nanotubes. Subsequently, we will discuss a technique based upon graph spectral analysis for analyzing a nanotube network. A network graph is extracted from the layout of the tubes and the ability to route information at the level of individual nanotubes is considered. The impact of random tube characteristics, such as location and angle, upon the corresponding network graph and its impact are examined. The nanotube network and routing of information play an integral part of the physical layer in the emerging field of nanoscale networks.

Due to their small size, nanotubes can reach deep into their environment without affecting their natural behavior. For example, a single carbon nanotube is small enough to penetrate a cell without triggering the cell's defensive responses. Individual nanotubes can be used to construct a network of sensing elements [20, 21]. The depth and coverage provided by such a network of sensing elements are greater than today's sensor networks. From a medical standpoint, the use of wireless implants using current techniques is unacceptable for many reasons, including their bulky size, inability to use magnetic resonance imaging after implantation, potential radiation damage, surgical invasiveness, need to recharge/replace power, postoperative pain and long recovery times, and the reduced quality of life for the patient [22–24]. Better, more humane implant communication is needed. Development of both biological and engineered nanomachines is progressing as well; such machines will need to communicate [25, 26]. Unfortunately, networking vast collections of nanoscale sensors and robots using current techniques, including wireless techniques, is not possible without the communication hardware exceeding the nanoscale. A solution is to use randomly aligned nanotubes, as discussed in [27] as the communications media, thus bringing the scale of the communications network down to the scale of the sensing elements. The biological approach has been well established in [28–36].

Current technology is focused on utilizing an entire carbon nanotube network as semiconducting material to construct a single transistor or field-effect transistor (FET). Many such transistors are required to build legacy network equipment. The result is that nanoscale networks are embedded within each device that might be otherwise more effectively and directly utilized for communication. Consider rethinking the communication architecture such that the carbon nanotube network itself is the communication media and individual nanotubes are the links.

Much research has gone into understanding how to align tubes. Unfortunately, cost and separation of impurities (metallic tubes) are still unsolved problems. Lower-cost, randomly oriented tubes may be directly utilized as a communication medium. Figure 4.2 illustrates a sample communication network where users at a molecular level simultaneously share random carbon nanotube network bandwidth. The randomly oriented lines depict the nanotubes and the thicker lines directly connected to the molecular users show input/output channels (or probes) into the network media. Because of the random nature of the nanotubes and the difference in distance between each user, there is a distinct difference in resistance between any sender and receiver node. Each sender has a distinguishable impact on a given receiver via network interaction as shown conceptually by the range of resistances from each user as shown along the bottom of the figure.

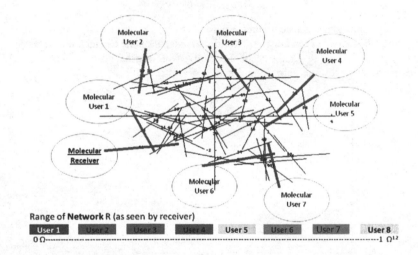

Figure 4.2   Routing through an embedded random carbon nanotube network.

To analyze such graphs, spectral analysis is used. Spectral analysis reveals the topological properties of a graph such as its patterns of connectivity and typically involves computing the eigenvalues and eigenvectors of the Laplacian matrix of a graph. The Laplacian matrix of a graph is an undirected, unweighted graph without loops or multiple edges from one node to another. The eigenvalues are related to the topology of the graph and represent specific features. For instance, the second smallest eigenvalue represents the measure of compactness of a graph. A large value implies a compact graph whereas a small value represents an elongated graph. Such an analysis is often used for relative comparisons of graphs; however, only graphs with the same number of vertices can be meaningfully compared. Spectral analysis has been used in a wide variety of applications, including semantic analysis of documents to cluster documents into areas of interest, comparing structural, functional, and evolutional similarity in RNA molecules, and connectivity on the Internet. Though very elegant there are some limitations to this technique. For instance, Mihail and Papadimitriou [150] argue that for randomly generated graphs that satisfy a power law distribution, spectral analysis of the adjacency matrix will simply produce the neighborhoods of the high degree nodes as its eigenvectors and miss any embedded structure. However, for graphs that do not have skewed degree distributions, spectral analysis is an efficient tool that reveals inherent embedded structures. Even for graphs with skewed degree distributions transformations can be applied to enable spectral analysis in recovering the latent structure with a high probability [151]. A graph spectral technique precisely solves the problem of determining resistance through the random tube layouts and is applied to the analysis of the nanotube graphs to reveal their inherent structural properties.

The impact of scale on a traditional communication network is considered as the network is scaled down to the size of a carbon nanotube network. An obvious consideration from a network perspective is the change in capacity, specifically in bandwidth. Simple harmonic oscillation, which provides bandwidth, increases with reduction in scale; thus potential bandwidth increases dramatically [152]. The increase is $1/L$ where L is a linear scale dimension. The capacity, C_{ij}, of a link from a transmitter at j to a receiver at i is given by Shannon's famous formula (4.37). Considering all possible multilevel and multiphase encoding techniques, the theorem states that the theoretical maximum rate of clean (or arbitrarily low bit error rate) data with a given average signal power that can be sent through an analog communication channel subject to additive, white, Gaussian-distribution noise interference is:

$$C_{ij} = BW \ln(1 + (S/N)_{ij}) \qquad (4.37)$$

The term BW is the bandwidth of the communication, and $(S/N)_{ij}$ is the signal-to-noise ratio (SNR) of the link. SNR measures the ratio between noise and an arbitrary signal on the channel, not necessarily the most powerful signal possible. In Figure 4.3, the channel capacity, assuming that noise is minimal ($SNR = 1/2$), rises as the scale is reduced towards zero using Equation 4.37.

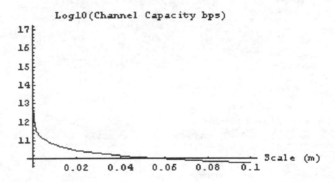

Figure 4.3 The approximate capacity increase with reduced scale is shown assuming that bandwidth varies as discussed in Equation 1.3 from Chapter 2. As the scale becomes smaller, the potential channel capacity grows significantly.

In addition to the increased bandwidth potential, the nanotube density allows for an increase in the number of bits per square meter. Consider a wireless network of today. A typical bit-meters per second capacity is limited in a traditional wireless network [58]. The maximum wireless capacity approximation in a wireless broadcast medium can be used to determine the collective capacity. Assume a perfect distribution mechanism in which all links are used as efficiently as possible to disseminate route update information [57]. Assume a network of n nodes is spread over an area A and each possible connection has capacity W. Also, assume Δ is a guard distance to ensure channel transmissions do not overlap. The maximum wireless capacity in bit-meters per second is shown in Equation 4.38, which was derived in Chapter 1.

$$C_{max} = \sqrt{\frac{8}{\pi} \frac{W}{\Delta}} \sqrt{n} \qquad (4.38)$$

Generalizing to a uniformly random distribution of n sensors over a circular area A, the density is A/n, and the expected nearest-neighbor distance is $\sqrt{A}/n$. The total distance that data must travel is shown in Equation 4.39.

$$E[d] = \sum_{k=1}^{n} \frac{\sqrt{A}}{n} \qquad (4.39)$$

Now consider a nanotube network. A point source radiates information omnidirectionally via a tube structure limited by the degree of compactness of the tube network. If tubes could be well aligned, then the notion of a guard distance would be unnecessary. A macroscopic source is assumed to generate data omnidirectionally. A microtubule is on the order of 8 to 25 nm in diameter and a carbon nanotube is on the order of 1.4 nm in diameter. The communication capacity in the case of carbon nanotubes radiating compactly from a circular source is shown in Figure 4.4. Essentially, the limit is reached as an extremely large number of tubes are joined to the source without overlapping. Unfortunately, current technology cannot align tubes with this degree of accuracy.

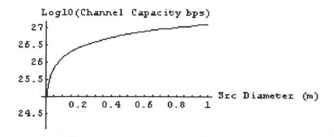

Figure 4.4   The capacity of a carbon nanotube network is shown as a function of the area available for nanotube connectivity. This is significantly higher capacity over much smaller distances than could be achieved with a wireless network.

A Mathematica [153] framework for evaluating random nanotube networks has been developed and used to verify design characteristics of carbon nanotubes. The framework relates tube placement characteristics comprised of tube center location t_{xy}, tube angle θ, and tube density d_t. The intersections of tubes form vertices V and tubes form edges E of a graph $G(V, E)$. In the specific instance of FET mobility, the graph structure impacts the mobility μ of the FET. Thus, a goal has been to find the relationship among tubes, the network topology, and mobility of the network. Let $f(t_{xy}, \theta, d_t)$ be a function of the physical tube characteristics that yields the network graph. As shown in Equation 4.40, we want to know how the nanotube network graph is affected by the individual tube parameters and how the nanotube network impacts the mobility, as a function of source-drain resistance R_{sd}.

$$f(t_{xy}, \theta, d_t) \to G(V, E) \to \mu(\Delta R_{sd}) \qquad (4.40)$$

We let I_{on} and I_{off} be the FET gate "on" and "off" currents that are determined by the resistance of the underlying nanotube network; w and L_{sd} are the gate width and length, respectively. Mobility is then approximated in Equation 4.41. Assuming L_{sd} and V_{sd} are predefined, only I_{on} and I_{off} are a function of the nanotube network structure. We can focus on determining how the nanotube network impacts the on and off currents, which allows us to find the mobility.

$$\mu = \frac{L_{sd}(I_{on} - I_{off})}{w} \frac{t_{ox}}{20\epsilon V_{sd}} \qquad (4.41)$$

We now consider how the individual tube characteristics impact the nanotube network. A key component of the tube layout is the overall directionality of the tubes; that is, the angle of each tube relative to all other tubes. Isotropy means uniformity in all directions; anisotropy is dependency upon direction. Anisotropy quantifies the directionality of the nanotubes and is defined in Equation 4.42, where l is the tube length and a is the tube angle. If nanotubes are aligned, then material properties depend upon direction and the properties are anisotropic.

$$\frac{\sum l \cos(a)}{\sum l \sin(a)} \tag{4.42}$$

Tubes that are nearly aligned have a high anisotropy and tubes that are randomly oriented have a low anisotropy. Figure 4.5 shows the anisotropy of a set of nanotube networks with constrained tube angles. The tube density is 1.2 per micron and lengths are constant at 3 microns. The angles range from −90 through +90 degrees.

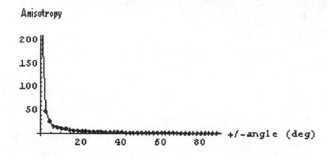

Figure 4.5 Anisotropy decreases with greater variance in nanotube angles. Isotropy is the invariance to rotation of the nanotube structure, which occurs when the tubes are less aligned relative to one another.

The angle of each tube can be considered to encode information. Entropy, from an information theoretic viewpoint, measures the amount of information. Angle entropy is defined as $- \sum Pr(a) \log_2(Pr(a))$ where a is the tube angle and Pr is the probability of a tube of angle a given the network under analysis. The angle entropy of the network analyzed in Figure 4.5 is shown in Figure 4.6. The more random the angle, the more angular entropy exists and thus there should be a relationship between anisotropy, angular entropy, the type of networks that are formed, and, ultimately, their performance and resilience to metallic tubes.

Clearly, there is a relationship between anisotropy and angle entropy as shown in Figure 4.7 for the same networks analyzed in the previous figure. High angle entropy implies that the directionality and thus the anisotropy are low. Information can be stored in nanotube angles; reading the information from the change in resistance of nanotube angles is discussed in [154].

As anisotropy increases and angle entropy decreases, the density of nanotube intersections decreases as shown in Figure 4.8. Greater angular variation enables the tubes to intersect nearer to one another. The intertube contact resistance has a greater impact as intersection density increases.

4.7.1   Characteristics of a semirandom nanotube network

For any given orientation of nanotubes, the corresponding network $G(V, E)$ is extracted and resistances are assigned based upon the probability that a tube is either a semiconducting carbon nanotube of 10^6 ohms when the gate is "on" (10 volts) and 10^{12} ohms when the gate is "off." Impure, that is, metallic, nanotubes

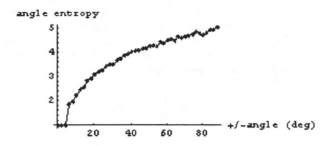

Figure 4.6 Angle entropy increases with greater variance in nanotube angles. In other words, greater angle entropy increases with greater uncertainty in nanotube angles and greater isotropy, or randomness.

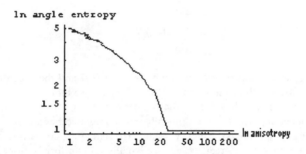

Figure 4.7 The natural logarithm of nanotube angle entropy decreases as the natural logarithm of anisotropy increases. As nanotubes become more anisotropic, or aligned, there is less uncertainty in the angle variance.

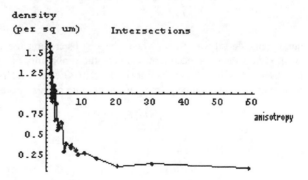

Figure 4.8 The density of nanotube intersections (nodes) varies inversely with anisotropy. An isotropic, or randomized, orientation will have a greater probability of contact.

remain at 10^6 ohms regardless of gate voltage and the probability of a metallic tube is 0.33. The network formed by the overlapping nanotubes is extracted by determining the location of junctions. The FET gate area is overlaid on this network and virtual vertices are added as source and drain; the virtual vertices are assigned edges with no resistance to each nanotube that is adjacent to the source or drain edge of the layout, respectively. The equivalent resistance of the network of resistors across the virtual source and drain is determined by Equation 4.43 where λ_i is the ith eigenvalue of the graph Laplacian matrix and ϕ_{ix} is the xth component of the ith eigenvector of the graph Laplacian matrix.

$$R_{sd} = \sum_{1}^{N} \frac{1}{\lambda_i} |\phi_{is} - \phi_{id}| \qquad (4.43)$$

Consider the relation in Equation 4.40, namely, $f(t_{xy}, \theta, d_t) \rightarrow G(V, E)$. Note that tube center locations t_{xy} and tube angles θ, are random variables. Tube density d_t is the number of tubes per unit area and is not considered a random variable in this analysis. Intuitively, one would expect the anisotropy, Equation 4.44, to have an impact on vertex density d_v.

$$\frac{\sum_{i=1}^{N} \cos(\theta_i)}{\sum_{i=1}^{N} \sin(\theta_i)} \qquad (4.44)$$

In Equation 4.44, the x component of each tube is $L_i \cos(\theta_i)$ and the y component of each tube is $L_i \sin(\theta_i)$ where L_i is the length of the ith tube. If tube lengths are infinite, the number of vertices in $G(V, E)$ is defined as in Equation 4.45 where t is the number of intersections among tubes t. The intuition is that each new tube will overlap with $t - 1$ existing tubes assuming no tubes are exactly parallel, yielding additional $t - 1$ nodes. Figure 4.9 shows the number of the extracted graph nodes versus tubes. Tube angles are uniformly distributed from $-\frac{\pi}{4}$ to $\frac{\pi}{4}$ radians and tube lengths vary uniformly from 3 to 10 microns in a 5 by 5 micron area. The simulation has a lower number of nodes because it assumes infinite tube lengths. The simulated nodes were finite; as the tube lengths increase, it is expected that the actual number of nodes would approach the analytical result.

$$|V| = n_i = \left(\sum_{i=1}^{i-1} t \right) \approx \frac{t^2}{2}, n_1 = 0, n_2 = 1 \qquad (4.45)$$

Determining the number and density of vertices when tube lengths are finite becomes more complex. Equation 4.45 needs to be modified such that each term includes the probability of overlap between tube pairs as shown in Equations 4.46–4.48 where the probability of overlap is defined in terms of the probability of overlap in both the x and y components of tube pairs. o_t is the number of overlaps among t tubes.

$$P_{ij}(O_2) = P_{ij}(O_x)P_{ij}(Oy) \qquad (4.46)$$

$$P_{ij}(o_y|y, L, \theta) = P_{ij}(|y_i - y_j| < \qquad (4.47)$$
$$(L_i \sin(\theta_i) + L_j \sin(\theta_j)))$$

$$P_{ij}(o_x|y, L, \theta) = P_{ij}(|x_i - x_j| < \qquad (4.48)$$
$$(L_i \cos(\theta_i) + L_j \cos(\theta_j)))$$

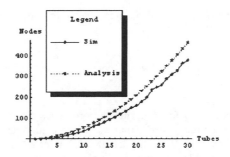

Figure 4.9 Simulated versus analytical results for tube angles uniformly distributed from 0 to $\frac{\pi}{2}$ radians. Analytical results assume overlaps from infinite tube lengths, thus the analytical results are an upper bound on the actual number of nodes.

Combining Equations 4.46–4.48 yields Equation 4.49. The analysis from Equation 4.49 is plotted versus simulated results in Figure 4.10.

$$|V| = P_{ij}(O_2) \binom{t}{2}$$ (4.49)

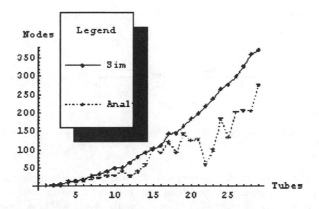

Figure 4.10 Probabilistic analyses versus simulation results. Because finite nanotube lengths are taken into account, the number of nanotube intersections (nodes) is less than with the assumption of infinite length nanotubes.

A maximum number of vertices are generated when the difference between the x, y values is small, that is, when there is a high concentration of tubes, when L is large, and when θ is $\pm\pi/4$ radians or $\pm3\pi/4$ radians. The concentration of tubes required for a connected network across the gate increases at these angles. The relation between L and θ to create a connected network for a given concentration of tubes in area wL_{sd} also needs to be determined. If tube lengths are held constant and each tube center is located farther apart, then tube angles must be reduced in order to achieve a connected graph, which will reduce the number of vertices. Thus, there is an optimal range of θ for a given area that meets the requirement for a connected graph, but that also maximizes (or minimizes) the number of vertices in the nanotube network $G(V, E)$. The probability of a connected network comes from Equation 4.46. The requirement for a network reaching from source to drain is the probability that tubes i and j are connected and that they cover the required distance. The expected distance covered that meets or exceeds the source to drain distance is shown in Equation 4.50.

$$\sum_{i=1}^{N} \sum_{j=1, j\neq i}^{N} P_{ij}(o_2) \left(|x_i - x_j| + \left(\frac{L}{2}\cos(\theta_i) + \frac{L}{2}\cos(\theta_j) \right) \right) \geq L_{sd} \qquad (4.50)$$

Graph spectral analysis is used for analyzing the properties of the nanotube graph. Figure 4.11 shows the conductance of a nanotube network that has random tube layouts with a gate area 2×2 microns, tube density 1.5 tubes/sq micron, tube length of 2 microns, and samples at tube angles of $\pm\{0, 15, 30, 45, 60, 75, 90\}$ degrees. The second lowest eigenvalue indicates graph connectivity (Fiedler value). These results show that there is a very strong correlation between the second lowest eigenvalue and both the percolation threshold and the conductance (in units of Siemens which is the inverse of resistance). Note that at zero degree angles, single tubes spanned the source drain allowing for a relatively high conductance.

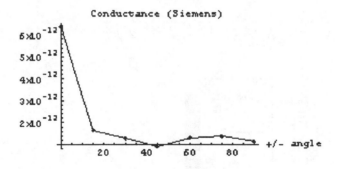

Figure 4.11 Network conductance versus range of nanotube angles in degrees. Zero degrees is perfect alignment directly from the source to drain contacts. As the nanotubes lose this anisotropy, the conductance decreases.

In Figure 4.12, the second smallest eigenvalue of the graph Laplacian is plotted for each of the same tube layouts as in the previous figure. Note the similar trend in this plot and the previous plot of conductance. The probability of percolation is the probability that a given tube layout will have connectivity from source to drain. It is determined by using the same network parameters as in the previous figures, namely tube density, area, and length and analyzing many layouts with each set of tube angles. The results are shown in Figure 4.13. The most important point of these results is that the second smallest eigenvalue of the graph Laplacian does indeed correlate strongly with both total conductance of the network and the probability of percolation as shown in Figure 4.14.

4.7.2 Data transmission in a semirandom nanotube network

Data transmission occurs via modulated current flow through the nanotube network guided towards specific nanodestination addresses. The addresses identify spatially distinct areas of the nanotube network. Since gate control is used to induce routes through the nanotube network, nanoaddresses are directly mapped to combinations of gates to be turned on that induce a path from a source to a destination. Figure 4.15 shows a conceptual view of the nanotube network infrastructure. Note that in addition to gate routing control,

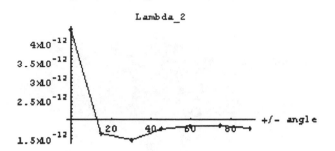

Figure 4.12 The second smallest eigenvalue of the graph Laplacian versus the range of tube angles. As the isotropy increases, the second-smallest eigenvalue decreases. Note the similar trend of this graph with the conductance from Figure 4.11.

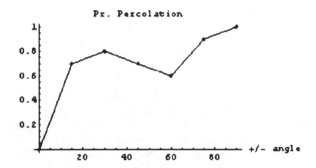

Figure 4.13 The probability of achieving percolation as a function of ranges of tube angles. A larger range of tube angles (higher nanotube entropy) increases the probability of percolation.

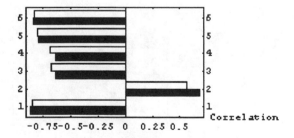

Figure 4.14 The Pearson's correlation of the lowest six graph Laplacian eigenvalues with both the probability of percolation and the network resistance over many simulated layouts. In each pair of bars, the upper bar is the network resistance correlated with the graph Laplacian eigenvalues and the lower bar is the probability of percolation correlated with the graph Laplacian eigenvalues. A correlation value of one indicates a strong linear correlation and negative one indicates a strong linear anti-correlation. The second-lowest eigenvalue stands out as being correlated with both network resistance and probability of percolation.

sensors are often constructed directly from nanotubes in such a manner as to change the resistance based upon the amount and specificity of the material being sensed [153]. Thus, the act of sensing may change the routing through the network.

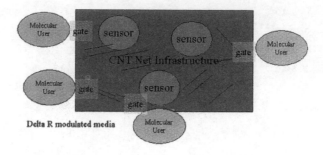

Figure 4.15   The carbon nanotube network infrastructure is comprised of resistance-controlled media routing information among molecular-level addresses. Sensors are nanotubes within the network whose resistance changes upon sensing a target, thus altering the pattern of current flow.

4.7.3   Routing in a semirandom nanotube network

Given a nanotube network, the mechanism used to route data through such a network must be considered. Consider a random nanotube network with a matrix of gates as shown in Figure 4.16. The gates are identified by number and when turned on, change the resistance of the semiconducting nanotubes within its area. Most nanotube sensing devices operate by changing tube resistance. A gate that is turned on, for any reason, may be used to route data through the network. Thus, the sensing elements, which sense by variation in resistance, may act simultaneously as routing elements. When a gate is turned on, the nanotubes within the gate area become conducting. Properly choosing gates to turn on also changes the current flow to the edges of the nanotube network, effectively creating a controlled network, which may act as a communication network.

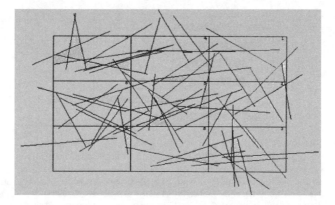

Figure 4.16   A matrix of gates superimposed on a random carbon nanotube network. Turning the gates on or off changes the conductance of the corresponding portion of the nanotube network.

The potential for such routing capability is simulated using a specific nanotube network shown in Figure 4.17. Nanotubes labeled 31, 5, 9, 50, 8, 39, and 35 are considered the outputs of this switch. Tube 52 is considered the input. The hypothesis is that in this isotropic media, tubes are randomly dispersed at all possible angles providing an approximately equal propagation of current in all directions. Activating gates appropriately serves to channel the flow into desired directions.

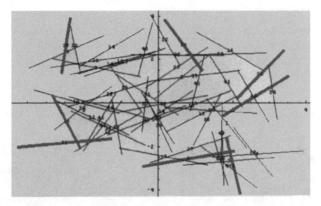

Figure 4.17 Nanotubes used for input/output to the network extend farthest from the random nanotube network for easy access. The input/output nanotubes in the network above are nanotubes labeled 31, 5, 9, 50, 8, 39, 35, and 52. These tubes can act as contacts, distributed evenly around the network (areas shown in previous figure).

Using a relatively small 3×3 gate matrix, we choose 3 possible combinations of gates turned on and compute the impact on the predefined output tubes. In this simple demonstration, we are checking the impact of every possible gate state on the flow of information through the network. The ratio of the resistances from tube 52 to all output tubes when no gates are turned on R_{off}, to the resistance between same tube pair when combinations of the gates are turned on R_{on} for selected I/O tubes is plotted on bar charts as shown in Figure 4.18 and Figure 4.19. The gates turned on that generate the bar chart values are shown beneath each bar chart. The last number in the list below each bar graph is the resistance threshold distinguishing the output resistance ratio from the next highest ratio. The effectiveness of the routing capability is measured by the difference between the resistance ratio at each output and the expected resistance ratio at all outputs (Equation 4.51); only the most effective gate combination is shown for each output, where *to* refers to the target outputs and E_o is the expected output resistance ratio.

$$\max_{on gates} [(R_{off}/R_{on})_{to} - E_o(R_{off}/R_{on})] \qquad (4.51)$$

Information flow through a nanotube network may be controlled in spite of the random nature of tube alignment. The same technique used for sensing in nanotube networks, namely, change in resistance of semiconducting material, may be used to effectively route information. The traditional networking protocol stack is compressed because, rather than the network layer being logically positioned above the physical and link layers, the nanotube network and routing of information is an integral part of the physical layer. The potential benefits of better utilizing individual nanotubes within random carbon nanotube networks to carry information is distinct from traditional, potentially less efficient and wasteful, approaches of using nanotube networks to construct transistors which are then used to implement communication networks. There are some theoretical questions with significant practical impact, namely, (1) whether one might achieve an information rate through the nanotube network that approaches the maximum flow through the equivalent network graph, in other words, network coding at the level of individual nanotubes [155],

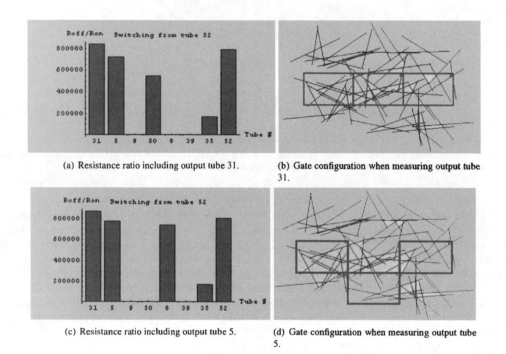

(a) Resistance ratio including output tube 31.   (b) Gate configuration when measuring output tube 31.

(c) Resistance ratio including output tube 5.   (d) Gate configuration when measuring output tube 5.

Figure 4.18   (a–d) Gates and routing from tube 52 to tubes 31 and 50. The ratio of resistance with no gates turned on to the resistance with the indicated gates turned on is shown in the bar graphs for selected I/O tubes.

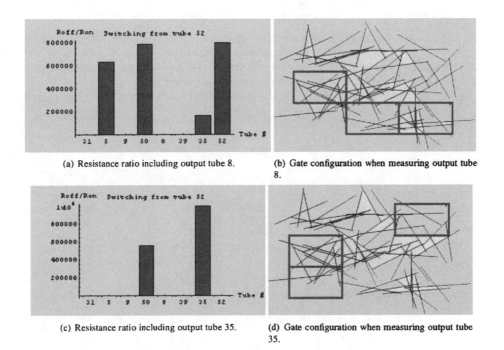

(a) Resistance ratio including output tube 8.

(b) Gate configuration when measuring output tube 8.

(c) Resistance ratio including output tube 35.

(d) Gate configuration when measuring output tube 35.

Figure 4.19 (a–d) Gates and routing from tube 52 to gates 8, 39, and 55. The ratio of resistance with no gates turned on to the resistance with the indicated gates turned on is shown in the bar graphs for selected I/O tubes.

and (2) whether given a network resistance, one could generate the underlying tube layout. That is, given a network resistance (as well as minimal information regarding tube characteristics), one can generate a set of feasible tube layouts with the given resistance. This may be approached via the inverse eigenvalue problem.

4.8 SUMMARY

This chapter focused upon the use of carbon nanotubes to implement communication channels. Carbon nanotubes have many unique properties that may be leveraged for improved communication. This includes their one-dimensional quantum aspects and conductance and their thermal and vibrational characteristics. One of the main takeaways from this chapter should be how to analyze random networks of such structures.

In this chapter, we touched upon the quantum effect of a one-dimensional nanowire. As the components for communication networks are reduced to reach the goal of nanoscale and molecular networking, quantum effects will become more significant and may be harnessed for useful purposes. What is quantum networking and how much quantum computation is required for quantum networking? What are the benefits and challenges of quantum network channels? These questions will be explored in the next chapter.

4.9 EXERCISES

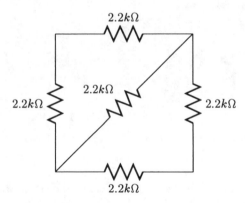

Figure 4.20   Random resistor network for Exercise 25.

Exercise 25   Random Network Resistance

Consider the derivation of random network resistance in terms of the Laplacian matrix.

1. In Equation 4.30, why is ϵ introduced?

2. In Equation 4.33, why does the subscript of the summation begin with two, instead of one?

3. In the derivation of Equation 4.36, why is it important that the Laplacian matrix is Hermitian?

4. What is the Laplacian matrix for the resistor network shown in Figure 4.20?

5. What are the eigenvalues and eigenvectors for the resistor network shown in Figure 4.20?

6. What is the resistance between every pair of points in network shown in Figure 4.20?

Exercise 26   Graph Spectra

Assume the following random network structure.

1. What is the matrix that represents the conductance of the network (e.g., the graph Laplacian)?

2. What are the eigenvalues of the matrix?

3. What are the eigenvectors of the matrix?

4. What is the resistance from point A to B?

Exercise 27   Persistence Length

Consider the discussion of persistence length from Chapter 1.

1. What is the impact of persistence length on the conductance of a network with a constant density of nanotubes?

Exercise 28   Single Nanotube Radio

Consider the minimum detectable amplitude wave that the carbon nanotube radio can detect.

1. Specify and show analytically all the ways and conditions under which sensitivity can be improved. (Hint: see Equation 4.11.)

Exercise 29   Interconnects and Graphs

Assume a random network of nanotubes has high isotropy.

1. What can you say about its second-largest eigenvalue and why?

Exercise 30   Percolation

1. How does persistence length impact percolation?

Exercise 31   Mobility and Carbon Nanotubes

1. What is the benefit of high mobility semiconductors?

2. How does the difference in "on" and "off" current impact the mobility? (Hint: see Equation 4.41.)

Exercise 32   Node Density

Assume a carbon nanotube network with tubes of infinite length.

1. What are the number of vertices (intersecting tubes)? (Hint: see Equation 4.45.)

Chapter 5

Nanoscale Quantum Networking

The mathematical framework of quantum theory has passed countless successful tests and is now universally accepted as a consistent and accurate description of all atomic phenomena.

Erwin Schrödinger

This chapter introduces quantum networking. It assumes no prior knowledge in quantum computation or networking. There is a short introduction to the concept of quantum networking and how it relates to nanoscale networking. Then there is a more detailed introduction to quantum computation, which is required in order to better understand quantum networking. This includes concepts related to quantum network security. This is followed by a discussion on entanglement and a simple overview of teleportation, which is one of the main forms of quantum networking. Armed with the background in quantum computation, a more detailed discussion of quantum networking follows. Finally, the chapter analyzes the latency and bandwidth of two quantum networking approaches, swapping channels and teleportation.

5.1 INTRODUCTION

Classical mechanics deals with the physical laws of bodies in motion based up the forces applied to them. Quantum mechanics is a set of physical principles that operate at the atomic and subatomic scale. When these principles are applied to implement computation, we have quantum computation. One of the key principles that quantum mechanics addresses is the particle-wave duality of matter, the fact that at the scale at which quantum mechanics applies, matter has characteristics of both solid particles and waves. Since quantum mechanics becomes an accurate representation of matter at the atomic and subatomic scale, it will naturally be a significant part of nanoscale networking. This section begins with a simple explanation of quantum networking and lays the foundation for delving gradually deeper into the mathematical description as we proceed through this chapter.

5.1.1 The nature of quantum networks

The most prevalent forms of quantum networking utilize both the wave nature of matter and the notion of entanglement to transmit quantum state from one location to another. One of the simplest experiments that demonstrates the particle wave nature is the beam splitter experiment. A beam splitter is a device that allows a portion of light to pass directly through and reflects the remainder of the light, as shown in Figure 5.1. An example is a half-silvered mirror, in which a very thin coating of aluminum is applied to a plate of glass. This allows some of the light to pass through and the remainder to be reflected. Imagine that the reflective coating is such that precisely one half of the light is reflected and one half travels directly through the mirror to detectors. Photons of light cannot be split, they must be either reflected or pass through the splitter. With classical physics, one would expect to detect 50% of the photons at one detector and 50% at the other detector. This is precisely what happens in experiments.

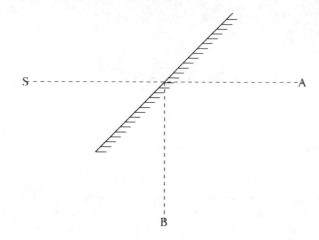

Figure 5.1   Beam splitter experiment one: 50% photon count from source S to detectors A and B.

However, the photons cannot be considered particles that happen to either bounce off when reflected or pass through. Instead, all photons take both directions simultaneously. It is only when they are detected, or "measured," that the particles choose which path they have taken. This is known as single particle interference and can be seen in another experiment.

Consider a new configuration as shown in Figure 5.2 in which light passes through a splitter, then bounces off two fully reflective mirrors, and then both beams pass through another beam splitter from two different directions. If the photons were particles, one would expect to see 50% of the light coming from each direction at the final desination. Instead, 100% of the light comes from only one or the other direction. This is because the light is taking both paths simultaneously and canceling out at the point of intersection near the destination. This is single particle interference, in which particles act like waves whose troughs cancel one another.

If one of the paths in this experiment is blocked as shown in Figure 5.3, 50% of the detections will occur at both destinations, similar to the first experiment. There is a superposition of photon paths through the system. Superposition plays a key role in enabling quantum computation to theoretically have significantly more processing power than existing computers. More details on quantum computation and networking in particular will be provided in the next section. For now, the goal is to become exposed to some of the unique properties of quantum systems. Note that in these simple experiments, which demonstrate quantum phenomena, the information is being transmitted optically. Also note that these experiments are dealing with subnanoscale particles, photons. Thus, we already have seen some of the elements required to implement a nanoscale network.

Before discussing the analytical nature of quantum networking, let's take an introductory look at another phenomenon that appears to have a relationship with nanoscale networking, namely, plasmonics. When light strikes a metal, just like people on a trampoline, the metal's surface electrons oscillate; they

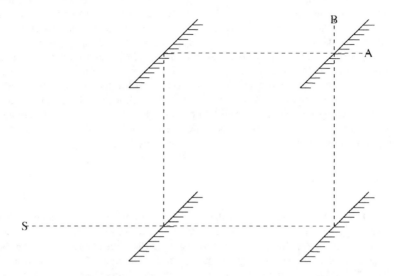

Figure 5.2   Beam splitter experiment two: 100% photon count from source S to either detector A and B.

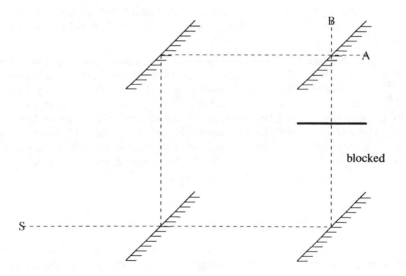

Figure 5.3   Beam splitter experiment three: 50% photon count from source S to detectors A and B when interferer is blocked.

absorb and scatter the incident photons from the light source. The oscillating surface electrons establish their own small electric field. This small field propagates along the surface of the thin metal film. The propagation occurs along the surface, typically for 10 to 100 μm, while the electromagnetic wave reaches 200 nm to 300 nm above the surface of the metal. In essence, a nanoscale wireless transmission propagates at the nanoscale along the metal film. To be more specific, a plasmon refers to a quantum of plasma oscillation. It is a quasiparticle resulting from the quantization of plasma oscillations, free metal electrons in our example, just as photons and phonons are quantizations of light and sound waves. The surface plasmon waves typically occur at optical frequencies. Surface plasmons allow smaller subwavelength structures to be utilized than with typical optical systems. Thus, surface plasmonics is being explored for nanoscale chip design in which computation is performed by operating directly with surface plasmons. In addition, recall the previous examples of the quantum matter-wave nature of light using a beam splitter. It has been suggested that plasmonics may be a useful interface to quantum systems [156].

5.1.2   Forms of quantum networking

Here we discuss a few of the tangible directions in which quantum networking is headed, or that are likely to have an influence on quantum networking. The first two chapters of this book focused heavily on biological mechanisms for nanoscale networking. Erwin Schrödinger, in 1946, clearly suggested that biology would need to be studied from the perspective of quantum mechanics. The field of quantum biology, which does precisely as Erwin Schrödinger suggested, seems to be gathering momentum as of the date of the writing of this book. The goal is to study biological processes and structures in terms of the unique properties of quantum mechanics. Clearly, all matter is described by quantum mechanics at the atomic and subatomic scale. Thus, a somewhat controversial issue in quantum biology is to find nontrivial, that is, nonobvious, roles and effects of quantum phenomena in biology. Quantum biology is a large and growing field that is outside the scope of this book. However, we note that several of the biological aspects mentioned related to the biological implementation of nanoscale networks, such as molecular motors [157], enzyme reactions, and transcription control [158] have been studied and experimentally verified from a quantum mechanics perspective. Thus, quantum computation and networking may be a fundamental and universal framework from which to study all forms of nanoscale networking.

From a practical perspective, as of December 2004, a quantum communication network demonstrating quantum security protocols has been operational. At that time the network was comprised of two weak coherent BB84 transmitters (the concepts underlying BB84 will be discussed later in this chapter), two compatible receivers, and a switch that could couple any transmitter to any receiver. The network connected BBN's laboratory, Harvard University, and Boston University. The Harvard-BBN connection is approximately 10 km and the BU-BBN strand is approximately 19 km. The full Harvard-BU path, through a switch at BBN, is approximately 29 km. The link was delivering about 1,000 privacy-amplified secret bits/second at an average 3% quantum bit error rate. Thus, human-engineered quantum networking has been a reality for some time.

5.2   PRIMER ON QUANTUM COMPUTATION

Quantum networking is implemented using techniques developed for quantum computation. Thus, an understanding of quantum computation is required in order to understand quantum networking. In this section we review just enough of the basics required to understand quantum networking [159].

Because quantum computation relies heavily on using matrices and matrix operations to represent the superposition of quantum states, we begin with a very brief overview of matrices and linear algebra.

In linear algebra, a row vector or row matrix is a $1 \times n$ matrix; that is, a matrix consisting of a single row $\mathbf{x} = \begin{bmatrix} x_1 & x_2 & \ldots & x_m \end{bmatrix} = \langle x |$. The transpose of a matrix changes the rows to columns and the columns to rows. Thus, the transpose of a row vector is a column vector as shown in Equation 5.1. Note that the complex conjugate is applied to each element to obtain $|x\rangle$ in addition to the transpose.

$$\begin{bmatrix} x_1 \\ x_2 \\ \vdots \\ x_m \end{bmatrix} = \begin{bmatrix} x_1 & x_2 & \ldots & x_m \end{bmatrix}^{\mathrm{T}} = |x\rangle \tag{5.1}$$

The dot product of two vectors a and b is equivalent to multiplying the row vector representation of a by the column vector representation of b as shown in Equation 5.2.

$$\mathbf{a} \cdot \mathbf{b} = \begin{bmatrix} a_1 & a_2 & a_3 \end{bmatrix} \begin{bmatrix} b_1 \\ b_2 \\ b_3 \end{bmatrix} = \langle a \, | \, b \rangle \tag{5.2}$$

Matrix multiplication involves the action of multiplying each row vector of one matrix by each column vector of another matrix as shown in Equation 5.3.

$$\begin{bmatrix} 1 & 0 & 2 \\ -1 & 3 & 1 \end{bmatrix} \times \begin{bmatrix} 3 & 1 \\ 2 & 1 \\ 1 & 0 \end{bmatrix} = \begin{bmatrix} (1 \times 3 + 0 \times 2 + 2 \times 1) & (1 \times 1 + 0 \times 1 + 2 \times 0) \\ (-1 \times 3 + 3 \times 2 + 1 \times 1) & (-1 \times 1 + 3 \times 1 + 1 \times 0) \end{bmatrix} \tag{5.3}$$

Thus, matrix multiplication is an inner product, or dot product, of all the rows of one matrix with all columns of the other matrix. This means that the order of matrix operations is extremely important.

As previously mentioned, the transpose of an m-by-n matrix A is the n-by-m matrix A^T (also sometimes written as A^{tr}) formed by turning rows into columns and columns into rows (i.e., $A^T[i, j] = A[j, i]$ for all indices i and j) and is also shown in Equation 5.4.

$$(A + B)^T = A^T + B^T \text{ and } (AB)^T = B^T A^T \tag{5.4}$$

Quantum state is represented by vectors in Dirac notation; the state is represented symbolically as $|\psi\rangle$. We will discuss how the $|\psi\rangle$ relates to a superposition of basis states later; for now, it is perfectly fine to think of $|\psi\rangle$ and $\langle\psi|$ as column and row vectors. There is a special condition that is very important when observing quantum state, in which $A |\psi\rangle = \lambda |\psi\rangle$. Here λ is the eigenvalue of A and ψ is the eigenvector of A. The matrix A becomes equivalent to a scalar λ in this special case. As an example, consider the eigenvalue shown in Equation 5.5, where the juxtaposition of matrices indicates matrix multiplication.

$$\begin{bmatrix} a_{11} & a_{12} \\ a_{21} & a_{22} \end{bmatrix} \begin{bmatrix} x \\ y \end{bmatrix} = \lambda \begin{bmatrix} x \\ y \end{bmatrix} \tag{5.5}$$

For any given matrix A, one can solve for the eigenvalues using simple algebra and obtain $det(A - \lambda I) = 0$, solve for λ. Recall that the matrix A as shown in Equation 5.6 has determinant $det(A) = ad - bc$.

$$A = \begin{bmatrix} a & b \\ c & d \end{bmatrix} \tag{5.6}$$

For a 3×3 matrix as shown in Equation 5.7, the cofactor expansion can be used to find the determinant as shown in Equation 5.8.

$$A = \begin{bmatrix} a & b & c \\ d & e & f \\ g & h & i \end{bmatrix} \tag{5.7}$$

$$
\begin{aligned}
\det(A) &= a \begin{vmatrix} e & f \\ h & i \end{vmatrix} - b \begin{vmatrix} d & f \\ g & i \end{vmatrix} + c \begin{vmatrix} d & e \\ g & h \end{vmatrix} = \\
&\quad aei - afh - bdi + bfg + cdh - ceg = \\
&\quad (aei + bfg + cdh) - (gec + hfa + idb)
\end{aligned}
$$

As an example, we find the eigenvalue of $A = \begin{vmatrix} 5 & 3 \\ 2 & 10 \end{vmatrix}$ as shown in Equation 5.8.

$$
\begin{aligned}
det(A - \lambda I) &= det(\begin{vmatrix} 5 & 3 \\ 2 & 10 \end{vmatrix} - \lambda \begin{vmatrix} 1 & 0 \\ 0 & 1 \end{vmatrix}) = \\
&\quad det(\begin{vmatrix} 5 & 3 \\ 2 & 10 \end{vmatrix} - \begin{vmatrix} \lambda & 0 \\ 0 & \lambda \end{vmatrix}) = \\
&\quad det(\begin{vmatrix} 5 - \lambda & 3 \\ 2 & 10 - \lambda \end{vmatrix}) = (5 - \lambda)(10 - \lambda) - 6 = \lambda^2 - 15\lambda + 44 \tag{5.8}
\end{aligned}
$$

Solving for λ, $(5 - \lambda)(10 - \lambda) - 6 = \lambda^2 - 15\lambda + 44 = 0 \rightarrow (\lambda - 11)(\lambda - 4) = 0$. Thus, $\lambda = 11$ and $\lambda = 4$, so this matrix has two eigenvalues.

Another notation that we will see are the Dirac and Kronecker delta functions. The Dirac delta function is infinity when its parameter is zero as shown in Equation 5.9.

$$\delta(x) = \begin{cases} \infty, & x = 0 \\ 0, & x \neq 0 \end{cases} \tag{5.9}$$

The Kronecker delta function is one when its two parameters are equal, as shown in Equation 5.10.

$$\delta_{ij} = \begin{cases} 1, & \text{if } i = j \\ 0, & \text{if } i \neq j \end{cases} \tag{5.10}$$

5.2.1   What is quantum mechanics?

Quantum mechanics is a framework in which the details of small-scale physical interactions can be worked out. The keyword "framework" is used because it is not, by itself, a complete physical theory, but rather the rules within which details are added to describe specific small-scale phenomena. Quantum mechanics consists of four rather simple-looking mathematical postulates. It is important to note that quantum mechanics, through this set of small postulates, has been very successful in describing the physical nature of small systems and will play a large role in the understanding and development of nanoscale networks.

5.2.1.1 The structure of quantum mechanics

Before continuing to the postulates of quantum mechanics, it's important to understand the language of matrices. This is how quantum mechanics, computation, and networking are most easily understood and analyzed. A very specific form of matrix notation is used in quantum mechanics that may seem odd at first, but is very convenient, concise, and natural to use with a little practice. It is called Dirac notation, in which angle brackets are used to indicate common quantum mechanical operations on vectors. If ψ, ϕ, and A are vectors, then $\langle\psi|$, $\langle\phi|$, $\langle A|$ indicate information about the quantum state within those vectors, which will be described shortly. For now, consider that $|\psi\rangle$ represents a column vector.

The postulates of quantum mechanics describe four fundamental notions regarding the framework in which quantum state exists and evolves. The first postulate describes quantum states of a closed system, specifically "state vectors" and "state space." The second postulate describes quantum dynamics, in particular, the nature of "unitary evolution." The third postulate describes the nature of measurements of a quantum system, specifically "projective measurements." And finally, the fourth postulate describes the quantum state of a composite system using the notion of "tensor products," which will be described shortly.

5.2.2 The nature of qubits

A nanoscale quantum network will most likely be transporting quantum bits, or qubits. Just like a classical digital system, the qubit is a two-state representation of information, either high or low, or one or zero. These may be implemented by photons, electron spin, or any small particle and its associated property.

However, there is at least one significant difference between a qubit and classical digital bit. The qubit can be in both of its states simultaneously. It is not until the qubit is actually observed, or measured, that it collapses firmly into one state or the other. There is a well-defined method for determining the probability of which state the qubit will take when measured, which we will discuss soon.

An easy way to think about a qubit is to consider two one-element vectors, $|0\rangle$ and $|1\rangle$. Let these be the computational basis states, either the qubit is all zero or all one, or it may be any linear combination of zero and one. This means that if we think of $|0\rangle$ as the y-axis and $|1\rangle$ as the x-axis of a circle, then an arbitrary qubit is, in general, comprised of α parts $|0\rangle$ and β parts $|1\rangle$, where α and β are called amplitudes, in keeping with the underlying waveform theme. If we normalize in such a manner to obtain probabilities, that is, the probability of obtaining one or zero ranges between zero and one, then Equation 5.11 must be true. It's also worth pointing out that the concept of projecting onto basis states is the mechanism behind quantum measurement. More will be said about this later.

$$|\alpha^2| + |\beta^2| = 1 \qquad\qquad (5.11)$$

Table 5.1 provides a concise summary of quantum computation vector and matrix notation. Quantum states can have complex values, thus, the complex conjugate is needed when squaring such values. As previously mentioned, $|\psi\rangle$ is the quantum state ψ, pronounced "ket" ψ. The complex conjugate transpose of the ket is the "bra" ψ, or $\langle\psi|$. This is the dual form of the ket for a quantum state. When the "bra" and the "ket" are placed together, we get a "bra-ket" or "bracket," $\langle\phi|\psi\rangle$. This is, or will come to be, esthetically pleasing as well as meaningful. This is the multiplication of a vector with its complex conjugate transpose, which yields a real scalar value. It is also known as the inner product. Note that both $|\phi\rangle \otimes |\psi\rangle$ and $|\phi\rangle\,|\psi\rangle$ are different notations for the same thing, namely, row vectors multiplied by row vectors, which assumes that both vectors must be the same length. The result is a matrix and this is known as a tensor product, or outer product. The tensor product will be explained later. Whereas the inner product collapsed the vectors into a single real scalar value, the outer product expands the vector representation of two states into a full

Table 5.1
A Brief Summary of Quantum Mechanics Vector and Matrix Notation

z^*	Complex conjugate of the complex number z, $(1+i)^* = 1 - i$
$\lvert \psi \rangle$	Vector or *ket*
$\langle \psi \rvert$	Vector dual to $\lvert \psi \rangle$ or *bra*
$\langle \phi \mid \psi \rangle$	Inner product between the vectors $\lvert \phi \rangle$ and $\lvert \psi \rangle$ (vector product)
$\lvert \phi \rangle \otimes \lvert \psi \rangle$	Tensor product of the vectors $\lvert \phi \rangle$ and $\lvert \psi \rangle$
$\lvert \phi \rangle \lvert \psi \rangle$	Alternative notation for tensor product
A^*	Complex conjugate of the A matrix
A^T	Transpose of the A matrix
$A^\dagger$	Hermitian conjugate or adjoint of the A matrix, $A^\dagger = (A^T)^*$
	$\begin{bmatrix} a & b \\ c & d \end{bmatrix}^\dagger = \begin{bmatrix} a^* & c^* \\ b^* & d^* \end{bmatrix}$
$\langle \phi \mid A \mid \psi \rangle$	Inner product between $\lvert \phi \rangle$ and $A \lvert \psi \rangle$
	Also, inner product between $A^\dagger \lvert \phi \rangle$ and $\lvert \psi \rangle$

matrix. A^* is the complex conjugate of a matrix. Note this simply means taking the complex conjugate of all the matrix values. A^T is simply the transpose of a matrix; switch rows to columns and columns to rows. $A^\dagger$ is a combination of the previous two operations. It indicates taking the complex conjugate of a matrix, and then taking the transpose. This is also known as the Hermitian conjugate of the adjoint of a matrix. Another combination operation that is often seen is in the final row of Table 5.1, namely, the inner product of a quantum state vector with a matrix operator. First, note that we just used the term *matrix operator*. Operators in quantum computation are in the form of matrices that act upon quantum state vectors and yield new quantum state vectors as a result. More will be said about this later. Also, note that, as indicated in the table, the order in which the matrix and vector operations occur is important.

5.2.3 Postulate one

The first postulate of quantum mechanics states that a complex vector space is associated with any quantum system. This is known as the state space. More specifically, the state of a closed quantum system is a unit vector in state space. Suppose we have a state space C^2, C for complex and comprised of two states. Then the state space looks like Equation 5.12.

$$\alpha \lvert 0 \rangle + \beta \lvert 1 \rangle \equiv \begin{bmatrix} \alpha \\ \beta \end{bmatrix} \tag{5.12}$$

Note that, as we mentioned previously, quantum mechanics is a framework rather than a specific theory for a specific system. It does not describe the state space of specific systems, such as electrons. As an example, quantum electrodynamics provides the details for this type of system within the quantum mechanics framework.

As previously mentioned, vectors are written as $\lvert \psi \rangle \equiv \vec{\psi}$ which is known as the ket notation. Discussion will be limited to systems that have finite dimensional state spaces. As shown in Equation 5.13, we are not necessarily limited to qubits, there can be an arbitrarily finite number of quantum states. This is known as a qudit.

$$|\psi\rangle = \alpha_0 |0\rangle + \alpha_1 |1\rangle + \alpha_2 |2\rangle \ldots \alpha_{d-1} |d-1\rangle = \begin{bmatrix} \alpha_1 \\ \alpha_2 \\ \vdots \\ \alpha_{d-1} \end{bmatrix} \quad (5.13)$$

5.2.3.1 Dynamics: quantum logic gates

As we alluded to previously, quantum operators, or quantum logic gates, are represented in the form of matrices. One of the simplest nontrivial gates to implement is the quantum NOT gate. For example, suppose we would like to cause a change in state that flips a one to zero and vice versa. Consider the operation as shown in Equation 5.14. Operator X is expected to perform this operation on quantum state vector $|0\rangle$ or $|1\rangle$. We need a matrix for X that performs the desired operation.

$$X |0\rangle = |1\rangle$$
$$X |1\rangle = |0\rangle \quad (5.14)$$

The more general case of a NOT gate is shown in Equation 5.15.

$$\alpha |0\rangle + \beta |1\rangle \rightarrow \alpha |1\rangle + \beta |0\rangle \quad (5.15)$$

The matrix that accomplishes this is shown in Equation 5.16. Working through the matrix multiplication of the qubit quantum state vector with the X matrix yields a vector in which the zero and one positions are swapped.

$$X = \begin{array}{cc} & \begin{array}{cc} |0\rangle & |1\rangle \end{array} \\ \begin{array}{c} |0\rangle \\ |1\rangle \end{array} & \begin{array}{cc} 0 & 1 \\ 1 & 0 \end{array} \end{array} \quad (5.16)$$

A question at this point may be: Can one use any matrix as an operator? With some careful thought, we can work out the answer to this question based upon the material that we have already covered. First, the operator matrix must be square and of the same size as the quantum state vectors being acted upon. Second, we mentioned earlier that results must be normalized such that valid probabilities are obtained after application of the operator as stated back in Equation 5.11. This places a critical constraint on the values that a legitimate operator can have. An operator that satisfies this second constraint is called a *unitary* matrix.

5.2.3.2 Unitary matrices

Consider the matrix shown in Equation 5.17.

$$A = \begin{bmatrix} a & b \\ c & d \end{bmatrix} \quad (5.17)$$

Now consider taking the Hermitian conjugation or adjoint as shown in Equation 5.18.

$$A^\dagger = (A^*)^T = \begin{bmatrix} a^* & c^* \\ b^* & d^* \end{bmatrix} \quad (5.18)$$

Now, we can state that the matrix is unitary if $AA^\dagger = A^\dagger A = I$, where I is the identity matrix, namely, all ones along the diagonal. Unitary matrices are often denoted as U. An example is shown in Equation 5.19.

$$XX^\dagger = \begin{bmatrix} 0 & 1 \\ 1 & 0 \end{bmatrix} \begin{bmatrix} 0 & 1 \\ 1 & 0 \end{bmatrix} = \begin{bmatrix} 1 & 0 \\ 0 & 1 \end{bmatrix} = I \qquad (5.19)$$

There is one last point regarding types of matrices as quantum operators. The following enumerated list of types of matrices is given from most common to least common. In other words, the largest set of matrices are normal. Hermitian matrices are a subset of normal matrices. Unitary matrices are also a subset of normal matrices and also overlap with Hermitian matrices. Finally, positive matrices are a subset of both Hermitian and Unitary matrices.

1. Normal $AA^\dagger = A^\dagger A$

2. Hermitian $A = A^\dagger$

3. Unitary $AA^\dagger = I$

4. Positive $\langle v\,|\,A\,|\,v \rangle \geq 0$

Remember that we are still discussing postulate one of the four postulates of quantum mechanics in order to understand quantum networking. Quantum networking will require quantum operators in order to implement transport of information. We can see now that a quantum operator is a matrix that enables a linear transformation of state. The requirement that the matrix be unitary enforces the normalization required to achieve meaningful probabilities of being in each state. Now we can move on to the next postulate.

5.2.4 Postulate two

Postulate two follows directly from the previous discussion. The evolution of a closed quantum system is described by a unitary transformation as shown in Equation 5.20.

$$|\psi'\rangle = U\,|\psi\rangle \qquad (5.20)$$

As mentioned, unitary matrices are the only linear matrices that preserve normalization. In other words, $|\psi'\rangle = U\,|\psi\rangle$ implies $\|\,|\psi'\rangle\,\| = \|U\,|\psi\rangle\,\| = \|\,|\psi\rangle\,\| = 1$. So now, we can examine a few more commonly used quantum gates.

5.2.4.1 Pauli gates

The Pauli gates are denoted by matrices X, Y, and Z. Each Pauli matrix represents an observable describing the spin of a spin one-half particle in three spatial directions.

The X gate (AKA σ_x or σ_1) is shown in Equation 5.21.

$$X\,|0\rangle = |1\rangle\,,\, X\,|1\rangle = |0\rangle\,,\, X = \begin{bmatrix} 0 & 1 \\ 1 & 0 \end{bmatrix} \qquad (5.21)$$

The Y gate (AKA σ_y or σ_2) is shown in Equation 5.22.

$$Y\,|0\rangle = i\,|1\rangle\,,\, Y\,|1\rangle = -i\,|0\rangle\,,\, Y = \begin{bmatrix} 0 & -i \\ i & 0 \end{bmatrix} \qquad (5.22)$$

The Z gate (AKA σ_z or σ_3) is shown in Equation 5.23.

$$Z|0\rangle = |0\rangle, Z|1\rangle = -|1\rangle, Z = \begin{bmatrix} 1 & 0 \\ 0 & -1 \end{bmatrix} \qquad (5.23)$$

Another almost trivial operator is I, which is often represented by the notation σ_0. This is simply the identity matrix. As its name suggests, applying the operator is applying the identity operation, which returns the original quantum vector. It may seem useless to have such a no-operation operator, however, it plays an important role, particularly in isolating and extracting state from composite state systems.

This concludes the explanation of the second postulate. You should now have an introductory understanding of the first two postulates and gained some familiarity with the Dirac notation for quantum mechanics and quantum computation. This is a good point to explore some of the exercise questions and solidify your understanding before continuing. Exercises 33 and 34 cover the topics discussed so far.

Now, we will review the last two postulates of quantum mechanics and introduce two fundamental but difficult issues, measurement and entanglement. Measurement is required in order to determine the result of a transmission over a quantum network and entanglement is one of the key mechanisms behind quantum teleportation, which *is* quantum networking in action.

5.2.5 Measuring a qubit

Recall that the first postulate of quantum mechanics describes the state space of a closed system. A system that is being measured is, by definition, being measured by some other system, and thus is open, at least temporarily. The third postulate of quantum mechanics explains how this measurement process may be described.

Let's begin with our general description of a quantum state as: $|\psi\rangle = \alpha|0\rangle + \beta|1\rangle$. A critical point regarding quantum mechanics is that it does not allow us to determine α and β directly. This is actually a useful feature for quantum network security. However, it means that we have to be clever about determining these values when we need them. It is possible to read some information about α and β.

As we know from earlier in this chapter, measuring in the computational basis is simply done by projecting onto the computational basis axes of a unit circle for a qubit. Thus, Equation 5.24 shows the probability of obtaining either quantum state when measured.

$$P(0) = |\alpha|^2$$
$$P(1) = |\beta|^2 \qquad (5.24)$$

Another key point that has a crucial impact on quantum networking is that measurement unavoidably disturbs the system, leaving it in a state $|0\rangle$ or $|1\rangle$ after the measurement operation. Notice that this is called a "measurement operation," measurement can be considered another quantum operator.

Consider an example in which we let $|e_1\rangle \dots |e_d\rangle$ be an orthonormal basis for C^d. A measurement of $|\psi\rangle$ in the basis $|e_1\rangle \dots |e_d\rangle$ gives result j with probability $P(j) = ||e_j\rangle \cdot |\psi\rangle|^2$. The symbol $\cdot$ is the usual vector inner product operation. Measurement unavoidably disturbs the system, leaving it in a state $|e_j\rangle$ determined by the outcome.

Here is another specific qubit example: $|\psi\rangle = \alpha|0\rangle + \beta|1\rangle$. What we can do now is to show that is possible to measure by projecting on a different set of axes. Therefore, consider the orthonormal basis described by Equation 5.26. If we think of the basis as the axes of a unit circle, then these new bases are similar to the computational basis, except that they are rotated slightly on the unit circle forming something like an "x" shape instead of a "+" shape.

$$|+\rangle = \frac{|0\rangle + |1\rangle}{\sqrt{2}}$$

(5.25)

$$|-\rangle = \frac{|0\rangle - |1\rangle}{\sqrt{2}}$$

Now we can use the technique just mentioned to determine the probability of obtaining one of the bases as shown in Equation 5.27.

$$Pr(+) = \left| \frac{1}{\sqrt{2}} \begin{bmatrix} 1 \\ 1 \end{bmatrix} \cdot \begin{bmatrix} \alpha \\ \beta \end{bmatrix} \right|^2 = \left| \frac{\alpha + \beta}{\sqrt{2}} \right|^2 = \frac{|\alpha + \beta|^2}{2}$$

(5.26)

$$Pr(-) = \frac{|\alpha - \beta|^2}{2}$$

When performing the analysis for measurements, it's useful to have facility with the inner product and the dual of a vector. The inner product is used to define the dual of a vector $|\psi\rangle$. If $|\psi\rangle$ lives in C^d then the dual of $|\psi\rangle$ is a function $\langle\psi| : C^d \rightarrow C$ defined by $\langle\psi| (|\phi\rangle) = |\psi\rangle \cdot |\phi\rangle$. The simplified notation is $\langle\psi | \phi\rangle$. As an example: $\langle 0| (\alpha |0\rangle + \beta |1\rangle) = \begin{bmatrix} 1 \\ 0 \end{bmatrix} \cdot \begin{bmatrix} \alpha \\ \beta \end{bmatrix} = \alpha$. Properties of the dual are: $\langle a|b\rangle = \langle b | a\rangle^*$, since $(|a\rangle, |b\rangle) = (|b\rangle, |a\rangle)^*$ $A |b\rangle \leftrightarrow \langle b| A^\dagger$, since $(A |b\rangle, |c\rangle) = (|b\rangle, A^\dagger |c\rangle) = \langle b| A^\dagger |c\rangle$.

Now we can look at the dual as a row vector. Suppose $|a\rangle = \sum_j a_j |j\rangle$ and $|b\rangle = \sum_j b_j |j\rangle$. Then $\langle a|b\rangle = (|a\rangle, |b\rangle) = \sum_j a_j^* b_j = [a_1^* a_2^* ...] \begin{bmatrix} b_1 \\ b_2 \\ \vdots \end{bmatrix}$. This suggests the very useful identification of $\langle a|$ with the row vector $\begin{bmatrix} a_1^* & a_2^* & ... \end{bmatrix}$.

5.2.6 Postulate three

Postulate three deals with the crucial operation of measurement of quantum state. If we measure $|\psi\rangle$ in an orthonormal basis $|e_1\rangle, ..., |e_d\rangle$, then we obtain the result j with probability $P(j) = |\langle e_j | \psi\rangle|^2$. The measurement disturbs the system, leaving it in a state $|e_j\rangle$ determined by the outcome.

5.2.6.1   Irrelevance of "global phase"

There is an interesting simplification that can take place during quantum operations including measurement. Suppose we measure $|\psi\rangle$ in the orthonormal basis $|e_1\rangle, ..., |e_d\rangle$. Then the probability of outcome j is $Pr(j) = |\langle e_j | \psi\rangle|^2$. Now suppose we measure $e^{i\theta} |\psi\rangle$ in the orthonormal basis $|e_1\rangle, ..., |e_d\rangle$. Then the probability of outcome j is $Pr(j) = |\langle e_j | e^{i\theta} | \psi\rangle|^2 = |\langle e_j | \psi\rangle|^2$. The $e^{i\theta}$ term disappeared! Let's see why this happened. Note that the $Pr(j)$ can also be stated as $\langle e_j | e^{i\theta} | \psi\rangle \cdot \langle e_j | e^{i\theta} | \psi\rangle$ where the term to the right of the dot product symbol is the complex conjugate of the term on the left side. The simplification occurs because the complex conjugate of $e^{i\theta}$ is given by $e^{-i\theta}$ and the product of these two values is one. Thus, the global phase factor $e^{i\theta}$ is unobservable, and we cannot differentiate the states $|\psi\rangle$ and $e^{i\theta} |\psi\rangle$. The global phase factor is not the same as a "relative" phase factor, or amplitude, which *is* important and cannot be removed the analysis.

Now we can go back and refine postulate one. Associated to any quantum system is a complex *inner product space* known as state space. The state of a closed quantum system is a unit vector in state space. These inner product spaces are often called Hilbert spaces.

5.2.6.2 Multiple-qubit systems

We can also consider multiple-qubit systems. Amplitudes are required for each element of the multiple-qubit vector, for example: $\alpha_{00}|00\rangle + \alpha_{01}|01\rangle + \alpha_{10}|10\rangle + \alpha_{11}|11\rangle$. All the same rules apply as for a single qubit vector, except that we are now dealing with multiple dimensions. In particular, the computational basis states must form an orthonormal basis, so that we can project onto it as per postulate three. The corresponding probability for determining the bit string (x, y) as the measurement outcome is just the modulus squared of the amplitude α_{xy}. Thus, measurement in the computation basis looks like $P(x, y) = |\alpha_{xy}|^2$ which is simply the probability of obtaining the bit string (x, y). In other words, we can deal with bit strings. The general state of n qubits is $\sum_{x \in \{0,1\}^n} \alpha_x |x\rangle$. This means that the state of n qubits can be written as a superposition over computational basis states corresponding to every possible n-bit string. Compare this to a classical state, which requires $O(2^n)$ bits to describe.

Simply put: it appears to take much more classical information to describe a quantum state than it does qubits. Thus, it seems that much more information can be stored and manipulated within a quantum system than a classical system. Unfortunately, it would seem that some of this information is lost during measurement; however, one of the keys to leveraging the benefits of quantum computation is to utilize the additional information that is present to the maximum extent possible before measurement.

5.2.7 Postulate four

The next and final postulate of quantum mechanics deals with composite quantum systems. This also will bring us into the notion of entanglement, which is a key mechanism in quantum networking.

Postulate four can be stated simply as the fact that the state space of a composite physical system is the tensor product of the state spaces of the component systems. Refer back to Table 5.1 regarding the notation and description of the tensor product. The tensor product is typically represented by the symbol $\otimes$.

Some properties of the tensor products are shown in Equation 5.28. The key to deriving these properties is simply to remember that these are vectors, and the outer product of vectors is a matrix. Thus, it is important to follow the rules regarding the order of matrix and vector multiplication.

$$z(|\nu\rangle \otimes |w\rangle) = (z|\nu\rangle) \otimes |w\rangle = |\nu\rangle \otimes (z|w\rangle) \tag{5.27}$$
$$(|\nu_1\rangle + |\nu_2\rangle) \otimes |w\rangle = |\nu_1\rangle \otimes |w\rangle + |\nu_2\rangle \otimes |w\rangle$$
$$|\nu\rangle \otimes (|w_1\rangle + |w_2\rangle) = |\nu\rangle \otimes |w_1\rangle + |\nu\rangle \otimes |w_2\rangle$$

5.2.7.1 Outer product notation

Tensors and the outer product are related as we will see in this section. Consider states $|\psi\rangle$ and $|\phi\rangle$ as vectors. We can define a linear operator, namely a matrix, formed by $|\psi\rangle \langle\phi|$. Using the Dirac notation, we can apply the operator to a state $|\gamma\rangle$ as follows: $|\psi\rangle \langle\phi| (|\gamma\rangle) \equiv |\psi\rangle \langle\phi|\gamma\rangle$.

Suppose that γ is $\alpha|0\rangle + \beta|1\rangle$. Then as a specific example, the previous operator can be applied as follows: $|1\rangle \langle0| (\alpha|0\rangle + \beta|1\rangle) \equiv |1\rangle \alpha = \alpha|1\rangle$. Note that $\langle0|$ is orthogonal to $|1\rangle$, so that $\langle0|\beta|1\rangle$ is zero. One way to become familiar with Dirac notation and the outer product is to look at some examples as shown in Equations 5.28–5.31.

$$|0\rangle\langle0| = \begin{bmatrix} 1 \\ 0 \end{bmatrix} \begin{bmatrix} 1 & 0 \end{bmatrix} = \begin{bmatrix} 1 & 0 \\ 0 & 0 \end{bmatrix} \tag{5.28}$$

$$|1\rangle\langle1| = \begin{bmatrix} 0 \\ 1 \end{bmatrix} \begin{bmatrix} 0 & 1 \end{bmatrix} = \begin{bmatrix} 0 & 0 \\ 0 & 1 \end{bmatrix} \tag{5.29}$$

$$Z = \begin{bmatrix} 1 & 0 \\ 0 & -1 \end{bmatrix} = (|0\rangle\langle0|) - (|1\rangle\langle1|) \tag{5.30}$$

$$|0\rangle\langle1| = \begin{bmatrix} 1 \\ 0 \end{bmatrix} \begin{bmatrix} 0 & 1 \end{bmatrix} = \begin{bmatrix} 0 & 1 \\ 0 & 0 \end{bmatrix} \tag{5.31}$$

5.2.8   The tensor product

Let's consider the tensor product in more detail. The tensor product ($\otimes$) may be applied in different contexts to vectors, matrices, tensors, vector spaces, algebras, and topological vector spaces. As previously mentioned, it is also referred to as the *outer product*. A representative case is the Kronecker product of any two rectangular arrays, considered as matrices as shown in Equation 5.32. Note the pattern of the columns of the resulting matrix.

$$\begin{bmatrix} b_1 \\ b_2 \\ b_3 \\ b_4 \end{bmatrix} \otimes \begin{bmatrix} a_1 & a_2 & a_3 \end{bmatrix} = \begin{bmatrix} a_1b_1 & a_2b_1 & a_3b_1 \\ a_1b_2 & a_2b_2 & a_3b_2 \\ a_1b_3 & a_2b_3 & a_3b_3 \\ a_1b_4 & a_2b_4 & a_3b_4 \end{bmatrix} \tag{5.32}$$

A vector tensor product, which we use for quantum states, is of the form shown in Equation 5.33. Again, notice the pattern in the resulting matrix.

$$\begin{bmatrix} 1 \\ 2 \end{bmatrix} \otimes \begin{bmatrix} 2 \\ 3 \end{bmatrix} = \begin{bmatrix} 1 \times 2 \\ 1 \times 3 \\ 2 \times 2 \\ 2 \times 3 \end{bmatrix} \tag{5.33}$$

Recall the operation of the tensor product. Another example is shown in Equation 5.34.

$$|\psi\rangle|\phi\rangle \rightarrow \begin{bmatrix} \psi_0 \begin{bmatrix} \phi_0 \\ \phi_1 \end{bmatrix} \\ \psi_1 \begin{bmatrix} \phi_0 \\ \phi_1 \end{bmatrix} \end{bmatrix} = \begin{bmatrix} \psi_0\phi_0 \\ \psi_0\phi_1 \\ \psi_1\phi_0 \\ \psi_1\phi_1 \end{bmatrix} \tag{5.34}$$

Equation 5.36 shows the result of applying the tensor to two matrices.

$$U1 \otimes U2 = \begin{bmatrix} U1_1 & U1_2 \\ U1_3 & U1_4 \end{bmatrix} \otimes \begin{bmatrix} U2_1 & U2_2 \\ U2_3 & U2_4 \end{bmatrix} = \tag{5.35}$$

$$\begin{bmatrix} U1_1 \begin{bmatrix} U2_1 & U2_2 \\ U2_3 & U2_4 \end{bmatrix} & U1_2 \begin{bmatrix} U2_1 & U2_2 \\ U2_3 & U2_4 \end{bmatrix} \\ U1_3 \begin{bmatrix} U2_1 & U2_2 \\ U2_3 & U2_4 \end{bmatrix} & U1_4 \begin{bmatrix} U2_1 & U2_2 \\ U2_3 & U2_4 \end{bmatrix} \end{bmatrix}$$

5.2.8.1 Conventions implicit in postulate four

For readers not already familiar with literature related to quantum computation, the names Alice, Bob, and Carol, as well as other names proceeding down the alphabet, are used frequently. These names are archetypes, or prototypes, used to ease the explanation when many, potentially confusing, interactions are involved. Instead of saying "Person A transmits a message to person B," we simply say "Alice sends a message to Bob." There is no deep meaning to the names other than that they are used by convention, are meant to be easy to remember, and can sometimes offer some humor into a potentially tedious and dull description. Finally, it should be understood that the operations by these named archetypes are not necessarily performed by people, but by computational devices or agents. So let us begin with some operations between Alice and Bob.

If Alice prepares her system in state $|a\rangle$, and Bob prepares his in state $|b\rangle$, then the joint state is $|a\rangle \otimes |b\rangle$. Conversely, if the joint state is $|a\rangle \otimes |b\rangle$ then we say that Alice's system is in the state $|a\rangle$, and Bob's system is in the state $|b\rangle$. This is shown in Equation 5.36. Note the addition of the global phase $e^{-i\theta}$ has no effect on the outcome.

$$|a\rangle \otimes |b\rangle = (e^{i\theta} |a\rangle) \otimes (e^{-i\theta} |b\rangle) \tag{5.36}$$

"Alice applies the gate U to her system" means that $(U \otimes I)$ is applied to the joint system. I leaves Bob unchanged. Consider the tensor manipulation shown in Equation 5.37.

$$(A \otimes B) |v\rangle \otimes |w\rangle = A |v\rangle \otimes B |w\rangle \tag{5.37}$$

As another example, suppose a NOT gate is applied to the second qubit of the state shown in Equation 5.38.

$$\sqrt{0.4} |00\rangle + \sqrt{0.3} |01\rangle + \sqrt{0.2} |10\rangle + \sqrt{0.1} |11\rangle \tag{5.38}$$

The resulting state is shown in Equation 5.39. The identity matrix I is a place holder that keeps the first qubit constant while X applies the NOT operation to the second qubit only. Compare carefully the last line of Equation 5.39 with Equation 5.38.

$$\begin{aligned}(I \otimes X)(\sqrt{0.4} |00\rangle + \sqrt{0.3} |01\rangle + \sqrt{0.2} |10\rangle + \sqrt{0.1} |11\rangle) &= \\ \sqrt{0.4} |01\rangle + \sqrt{0.3} |00\rangle + \sqrt{0.2} |11\rangle + \sqrt{0.1} |10\rangle \end{aligned} \tag{5.39}$$

5.3 QUANTUM ENTANGLEMENT

Now we can begin to address one of the key topics related to quantum networking, namely, entanglement. Entanglement between quantum states can be leveraged in order to aid in conveying information. Before we can discuss this, we need to gain at least an intuitive understanding of entanglement, which we can do with an example.

Suppose that Alice has a qubit in state $|a\rangle = \alpha |0\rangle + \beta |1\rangle$. Also suppose that Bob also has a qubit and its state is $|b\rangle = \gamma |0\rangle + \delta |1\rangle$. Here comes the key point: suppose the composite state of both qubits is as shown in Equation 5.40.

$$|\psi\rangle = \frac{|00\rangle + |11\rangle}{\sqrt{2}} \tag{5.40}$$

This is an entangled state as will be shown shortly. First recall that a composite system state is a tensor product of the component states. The reason it is entangled is that $|\psi\rangle \neq |a\rangle |b\rangle$. In other words, the

Table 5.2

The Four Postulates

Postulate 1	A closed quantum system is described by a unit vector in a complex inner product space known as state space.
Postulate 2	The evolution of a closed quantum system is described by a unitary transformation $\lvert \psi' \rangle = U \lvert \psi \rangle$.
Postulate 3	If we measure $\lvert psi \rangle$ in an orthonormal basis $\lvert e_1 \rangle, ..., \lvert e_d \rangle$, then we obtain the results j with probability $P(j) = \lvert \langle e_j \vert \psi \rangle \rvert^2$. The measurement disturbs the system, leaving it in a state $\lvert e_j \rangle$ determined by the outcome.
Postulate 4	The state space of a composite physical system is the tensor product of the state spaces of the component systems.

composite state cannot be written as a tensor product of the component states. But let us try to do so in order to demonstrate why it is not possible. First, form the product of the component states as shown in Equation 5.41.

$$\lvert \psi \rangle = (\alpha \lvert 0 \rangle + \beta \lvert 1 \rangle)(\gamma \lvert 0 \rangle + \delta \lvert 1 \rangle) \tag{5.41}$$

Next, expand the product form as shown in Equation 5.42.

$$\alpha\gamma \lvert 00 \rangle + \beta\gamma \lvert 10 \rangle + \alpha\delta \lvert 01 \rangle + \beta\delta \lvert 11 \rangle \tag{5.42}$$

Now comes another key point, namely a contradiction showing that the simple tensor product of the individual states cannot exist for the particular composite state in this example. Look back at the composite state in Equation 5.40. Now compare that equation with Equation 5.42; consider how Equation 5.42 could be equivalent to 5.40. Simple inspection shows that the $\lvert 10 \rangle$ term needs to drop out. The only way that could happen is if either $\beta = 0$ or $\gamma = 0$. However, in either case, one of the terms necessary for the composite state in Equation 5.40 would be lost as well. This contradiction demonstrates the nature of entanglement. Thus, it is impossible to find a solution. In other words, ψ cannot be written as a tensor product in this situation.

We have now quickly and hopefully easily learned the four postulates of quantum mechanics, which are summarized in Table 5.2. Our goal in the next section is to apply the principles introduced in this section to some illustrative examples: superdense coding and a simplified quantum cryptography example. We will also refine postulate two, which covers quantum state evolution and postulate three, which covers measurement.

5.3.1   Superdense coding

Describing the state of even a single qubit means specifying two complex amplitudes, to arbitrary precision, which requires an infinite amount of classical information. Transmitting a qubit should result in the transport of this huge amount of information. Can we utilize this for improved communication? If so, how? Let's start with a simple overview. Suppose Alice has two classical bits: a and b. Alice also has a qubit. Can Alice transmit two classical bits with one qubit that Bob can discover by measurement? Unfortunately, the short answer is no, but suppose the protocol starts with Bob and two qubits. Bob prepares them in a joint state and sends one to Alice. Alice applies a local operation that changes the joint state. Alice sends the qubit to Bob who measures it and thus determines the value of a and b.

Let's consider this in more detail. Bob starts with a Bell state qubit pair. A Bell state is one of the simplest examples of entanglement. Recall that entanglement is not affected by the distance with which the qubits are separated and is not limited by the speed of light. A Bell state is a *maximally* entangled state between two qubits as shown Equation 5.43, where Alice's qubit has subscript A and Bob's has subscript B. If Alice measured her qubit, the outcome would be perfectly random. If Bob measured his qubit after Alice, then his outcome would be the same as Alice's. To put it from Bob's perspective, his measurement looks random to him, but if he later spoke with Alice, they would find that their measurements were mysteriously similar. It turns out, as we will discuss later, there are no predefined agreements, or hidden variables, controlling the correlation between the outcomes.

$$|\Phi^+\rangle = \frac{1}{\sqrt{2}}(|0\rangle_A \otimes |0\rangle_B + |1\rangle_A \otimes |1\rangle_B) \tag{5.43}$$

Alice wants to transmit two classical bits a and b to Bob. This is the main goal. Bob has entangled bits as shown in Equation 5.44.

$$\frac{|00\rangle + |11\rangle}{\sqrt{2}} \tag{5.44}$$

Bob sends one of the entangled qubits to Alice. Recall again, that the goal is for Alice to transmit her two classical bits. Also recall that entangled qubits require both states in order to fully describe their joint state. Thus, if Alice changes the qubit that is entangled with Bob, the quantum state of Bob's entangled qubit will "mysteriously" change as well, regardless of time or distance. Alice modifies the qubit based on the classical bits that she wishes to transmit, a and b. If $a = 0$, $b = 0$, then Alice applies I, since there is no change. Recall that I is the identity matrix operator and has no effect.

$$\frac{|00\rangle + |11\rangle}{\sqrt{2}} \rightarrow \frac{|00\rangle + |11\rangle}{\sqrt{2}} \tag{5.45}$$

If $a = 0$, $b = 1$, then apply Z. Recall what Z does as shown in Equation 5.46. Keep in mind that Alice is only changing her entangled qubit. However, the overall effect is changing the joint state of the entangled qubit pair as shown in Equation 5.47.

$$Z|0\rangle = |0\rangle$$
$$Z|1\rangle = -|1\rangle \tag{5.46}$$

$$\frac{|00\rangle + |11\rangle}{\sqrt{2}} \rightarrow \frac{|00\rangle - |11\rangle}{\sqrt{2}} \tag{5.47}$$

If $a = 1$, $b = 0$, then apply X. Recall what X does as shown in Equation 5.48. The joint quantum state changes as shown in Equation 5.49.

$$X|0\rangle = |1\rangle$$
$$X|1\rangle = |0\rangle \tag{5.48}$$

$$\frac{|00\rangle + |11\rangle}{\sqrt{2}} \rightarrow \frac{|10\rangle + |01\rangle}{\sqrt{2}} \tag{5.49}$$

Alice changes the pair as follows. If $a = 1$, $b = 1$, then apply XZ as shown in Equation 5.50. The overall joint quantum state changes as shown in Equation 5.51.

$$XZ|0\rangle = |1\rangle$$
$$XZ|1\rangle = -|0\rangle \tag{5.50}$$

$$\frac{|00\rangle + |11\rangle}{\sqrt{2}} \rightarrow \frac{|10\rangle + |01\rangle}{\sqrt{2}} \tag{5.51}$$

Bob measures his qubit and from the result of the measurement learns the two classical bits that Alice has. These four states are, in fact, orthonormal to one another, and thus can be distinguished by a quantum measurement. Superdense coding is a sibling of teleportation which we will see later. Teleportation is essentially the reverse of the superdense coding process. In superdense coding, one entangled bit plus one qubit of communication is equivalent to two bits of classical communication.

5.3.2 Measurement of composite quantum states

In order to better understand entangled states, we need to understand composite systems and the impact of measurement within composite systems. As explained previously, if we observe, that is, measure, the state $|\psi\rangle$ in an orthonormal basis $|e_1\rangle, ..., |e_d\rangle$, the result, or outcome, j is observed with probability $P(j) = |\langle e_j|\psi\rangle|^2$. To put it another way, the amplitudes that compose state $|\psi\rangle$ are projected onto the orthonormal basis. Recall that the probability of a state in superposition is simply the square of the amplitude. Thus, the probability of observing an outcome is the square of the amplitudes of the state projected onto one of the orthonormal bases. The measurement irreversibly disturbs the system, leaving it in state $|e_j\rangle$ determined by the outcome.

Now assume we measure a quantum system A in the orthonormal basis $|e_1\rangle, ..., |e_d\rangle$. When entanglement occurs, there are are two quantum systems involved, which together form a larger composite quantum state. Thus, suppose system A is part of a larger system, consisting of two components, A and B. If only A is measured, what is the effect of the measurement on the larger system? Ignore entanglement for now and focus on any composite system, not necessarily one that is entangled. We can replace the orthogonal states $|e_1\rangle, ..., |e_d\rangle$ with a complete set of orthogonal subspaces $V_1, ..., V_m$. Thus, the complete composite system state is V as shown in Equation 5.52.

$$V = V_1 \oplus V_2 \oplus ... \oplus V_m \tag{5.52}$$

Let V be a vector space and let $\{v_1, v_2, ... v_n\}$ be a set of elements in V. If every vector in V can be expressed as a linear combination of $\{v_1, v_2, ... v_n\}$ then $\{v_1, v_2, ... v_n\}$ spans V. We can consider the example state $|\psi\rangle = (\alpha |e_1\rangle + \beta |e_2\rangle) + \gamma |e_3\rangle$. Notice the components of the state that are in parentheses. The notion is that even though the full state consists of basis vectors $|e_1\rangle, |e_2\rangle, |e_3\rangle$, we can consider the state space to be separated into $|e_1\rangle + |e_2\rangle$ and $|e_3\rangle$. Consider it is as the space spanned by $|e_1\rangle + |e_2\rangle$ and the space spanned by $|e_3\rangle$ as shown in Equation 5.53.

$$sp(|e_1\rangle, |e_2\rangle, |e_3\rangle) = sp(|e_1\rangle, |e_2\rangle) \oplus sp(|e_3\rangle) \tag{5.53}$$

Thinking of measurement in this manner, it becomes possible to proceed as follows. We can break the state space up into two parts, one part spanned by $|e_1\rangle$ and $|e_2\rangle$, and the other part spanned by $|e_3\rangle$ alone. A general measurement can be thought of as asking the question of how to partition the subspaces more appropriately. Projection can be used to isolate a subspace of interest. We can describe the subspaces $V_1, ..., V_m$ in terms of their corresponding projectors, $P_1, ..., P_m$, thereby removing orthogonal components. For example, a projector P may be chosen such that when projected onto $sp(|e_1\rangle, |e_2\rangle)$ it acts as $P(\alpha |e_1\rangle + \beta |e_2\rangle + \gamma |e_3\rangle) = \alpha |e_1\rangle + \beta |e_2\rangle$. In other words, the projector onto a subspace acts as the identity on that subspace, and annihilates everything orthogonal to V.

Consider a set of projectors $P_1, ..., P_m$ onto a complete set of orthogonal subspaces of a state space as shown in Equation 5.54. Then, as shown in Equation 5.55, we note that the entire set of projectors P_j is chosen so that the sum is the identity matrix. This means that our set of projectors covers the entire space. Also, the projectors are orthogonal, so that multiplying any two returns zero, unless they happen to be the same projector. This is represented by the delta function, which returns one only when j and k are equal and zero otherwise. We could also state in Dirac notation that $\langle i \mid j \rangle = \delta_{ij}$.

$$\sum_j P_j = I \tag{5.54}$$

$$P_j P_k = \delta_{jk} P_j \tag{5.55}$$

Essentially, these sets of projections define bases upon which measurements can be made, just as measurement was first presented. Recalling that probability depends upon the amplitude squared, then measuring $|\psi\rangle$ with this set of projectors yields outcome j with probability $Pr(j) = \langle \psi \mid P_j \mid \psi \rangle$. As always, the measurement unavoidably disturbs the system, leaving it in the postmeasurement state $P_j |\psi\rangle$. However, this final state is not completely correct, as it simply applied the projector to the initial state, but did not normalize the amplitude. The correct measurement value must be divided by the square of the projected amplitudes, yielding Equation 5.56.

$$\frac{P_j |\psi\rangle}{\sqrt{\langle \psi \mid P_j \mid \psi \rangle}} \tag{5.56}$$

5.3.2.1 An example measuring the first of two qubits

Suppose that we have a two-qubit system. Now assume that we have a measurement basis $|e_1\rangle$, $|e_2\rangle$ and we want to measure only the first qubit of the two-qubit system. As we discussed, we can form projectors P_1, P_2 onto the state space of the target qubit, qubit one in this case. We want to leave qubit two alone, so we use the identity on the second qubit. Recall that a multiple qubit system is described by the tensor product of each of the component states. Thus, we want $P_1 \otimes I$ for projector one applied to the first qubit only and $P_2 \otimes I$ for the second projector applied to the first qubit only. As an example, assume the state of two qubits is as shown in Equation 5.57.

$$|\psi\rangle = \alpha_{00} |00\rangle + \alpha_{01} |01\rangle + \alpha_{10} |10\rangle + \alpha_{11} |11\rangle \tag{5.57}$$

Measuring the first qubit in the computational basis yields the result zero with the probability as shown in Equation 5.58.

$$
\begin{aligned}
Pr(0) \;=\; & \langle \psi \mid (P_0 \otimes I) \mid \psi \rangle = \\
& (\alpha_{00} |00\rangle + \alpha_{01} |01\rangle + \alpha_{10} |10\rangle + \alpha_{11} |11\rangle) \cdot (\alpha_{00} |00\rangle + \alpha_{01} |01\rangle) = \\
& |\alpha_{00}|^2 + |\alpha_{01}|^2
\end{aligned}
\tag{5.58}
$$

5.3.3 The Bell inequality

Recall the discussion of entanglement; it will be explained as a key resource in quantum communication later. It is a key resource, not only in communication, but also in quantum security. This section discusses the Bell inequality, which is a tangible measure of correlation due to entanglement. Entanglement violates the principle of locality. Simply put, the principle of locality is that objects can only be directly influenced by their immediate surroundings. The opposing principle is that of nonlocality, a somewhat mysterious principle in which objects can directly influence other objects at a distance, and is incompatible with relativity. The principle of nonlocality only exists in quantum mechanics.

The Bell inequality is a manifestation of nonlocality. To begin, ignore quantum mechanics for the first part of this explanation. The goal is to establish a measure of correlation between events as we commonly perceive it in our everyday classical world.

The classical example is to have our two favorite people, Alice and Bob, each independently measure two properties. Let us assume that the measurements yield either $+1$ or -1 for each property. Let us assume that Alice and Bob each receive an object from a third party. They each measure one of the two properties independently and randomly. Let us establish a specific relationship that measures the correlation in the results. Let Alice's two properties be called Q and R. Let Bob's two properties be called S and T. We can look at the product of each of the measurement outcomes between Alice and Bob as shown in Equation 5.59.

$$QS + RS + RT - QT = (Q + R)S + (R - Q)T \tag{5.59}$$

Both R and Q have values of ± 1. Thus, no matter what combinations of ± 1 are assumed by R and Q, either $(Q + R)S = 0$ or $(R - Q)T = 0$. Whichever term goes to zero leaves the remaining term to represent the value of Equation 5.59. Running through the combinations of ± 1 shows that the final result can only be ± 2 as shown in Equation 5.60.

$$QS + RS + RT - QT = \pm 2 \tag{5.60}$$

It's possible to frame the above equation in terms of probabilities. Thus, let $p(q, r, s, t)$ be the probability of the system to be in state Q, R, S, T before measurements are performed. Then Equation 5.62 shows the expected value of being in the state specified in Equation 5.59.

$$E(QS + RS + RT - QT) = \tag{5.61}$$

$$\sum_{qrst} p(q, r, s, t)(qs + rs + rt - qt) \leq \sum_{qrst} p(q, r, s, t) \times 2$$

$$= 2$$

The probabilities can be multiplied through the terms as shown Equation 5.63.

$$E(QS + RS + RT - QT) = \tag{5.62}$$

$$\sum_{qrst} p(q, r, s, t)qs + \sum_{qrst} p(q, r, s, t)rs +$$

$$\sum_{qrst} p(q, r, s, t)rt - \sum_{qrst} p(q, r, s, t)qt =$$

$$E(QS) + E(RS) + E(RT) - E(QT)$$

Finally, since the probabilities must be less than or equal to one and we already established that the sum of the value of the terms can be at most two, the result must be less than or equal to two as shown in Equation 5.63.

$$= E(QS) + E(RS) + E(RT) - E(QT) \leq 2 \tag{5.63}$$

5.3.3.1   Bell inequality with quantum mechanics

Next we will consider what happens when quantum mechanics are involved. But first, we need to review how to find the expected value of a quantum state. The expectation of operator A is $\langle A \rangle = \langle \psi | A | \psi \rangle$ with reference to state ψ. Now reconsider the situation in which Alice measures two properties, but this time they are properties of the perfectly entangled qubit state. Specifically, Alice measures $Q = Z_1$ and $R = X_1$. In other words, Alice measures only the first qubit of the entangled pair after applying a Z operator and an X operator to it. Bob measures only the second qubit of the entangled pair, $S = \frac{-Z_2 - X_2}{\sqrt{2}}$ and $T = \frac{Z_2 - X_2}{\sqrt{2}}$.

It turns out that the expected value of QS is $\langle QS \rangle = \frac{1}{\sqrt{2}}$ and RS is $\langle RS \rangle = \frac{1}{\sqrt{2}}$. For RT and QT it is $\langle RT \rangle = \frac{1}{\sqrt{2}}$ and $\langle QT \rangle = -\frac{1}{\sqrt{2}}$. Inserting these values into Equation 5.59 yields Equation 5.64.

$$\langle QS \rangle + \langle RS \rangle + \langle RT \rangle - \langle QT \rangle = 2\sqrt{2} \tag{5.64}$$

But we now have contradictory results, the classical result gave an expected value ≤ 2 and the quantum value > 2. Quantum experimental validation has always shown an expected value > 2. This clearly distinguishes classical events from quantum ones.

5.3.4 Quantum cryptography example

One of the advantages of working with nanoscale networks and being able to harness quantum networking is that quantum security may come along with it. For this example, assume there is a simple quantum communication mechanism, that is, no teleportation or superdense coding. We are assuming a quantum wire, for example, which was described in Section 4.3. Let us assume now that Alice generates pairs of entangled qubits: $|\psi_{ab}\rangle = \frac{1}{\sqrt{2}}(|0_a 1_b\rangle - |1_a 0_b\rangle)$. Alice sends the second qubit of each pair to Bob, so the qubit with subscript a belongs to Alice and the one with subscript b belongs to Bob.

Notice that the qubits are entangled in a Bell state, also called a singleton. Assume that Alice and Bob now perform measurements in the computational basis. When measured on the computational basis, there is a 50% probability of obtaining a one and a 50% probability of obtaining a zero. Thus, the stream of bits that are observed will appear to be random. However, because of the entanglement, Bob will observe the complement of the bits that Alice observes. Thus, Bob simply needs to complement his bits in order to recover Alice's observed bits. Since the bit stream is random, a secret crypotgraphic key can be generated and transmitted in this manner.

5.3.4.1 A cryptography example with an eavesdropper

Now consider what happens when an eavesdropper, Eve, tries to discretely listen to the message. Suppose that she intercepts the qubits from Alice, measures, and copies the state, then passes it along to Bob. In a classical system, no one would realize the compromised communication. However, the quantum channel is significantly more secure; the eavesdropping will be detected immediately as follows.

Suppose that Alice and Bob allow multiple measurement bases to be used. The bases are the computational basis, $0/1$, and the $+/-$ basis. As long as Alice and Bob measure in the same basis, the bits are random and perfectly anticorrelated. Alice and Bob randomly and independently change their basis for each entangled. Eve does the same. After n measurements, Alice and Bob exchange the bases only, not the measurements that they had used. They reveal the measurements when different bases were used and some when the same bases were used. The Bell correlation that was recently described can then be used to find out whether Eve was listening. The only way that Eve could eavesdrop would be to measure the qubits on some basis herself. This would collapse the qubits based upon the outcome of the measurement. If Eve were to happen to guess the correct basis for each qubit, she could discretely create new qubits with the same state that Alice had sent and inject them into the network to Bob. However, once the qubits touched the classical world, by Eve's measurement, the quantum Bell correlation would be lost between Alice and Bob and the tampering would be immediately identified.

5.4 TELEPORTATION

Now we discuss one of the main quantum networking approaches, teleportation. Just like superdense coding, which we discussed earlier, teleportation involves quantum computation in order to transmit information from one location to another. In teleportation, the same quantum state is reconstructed at a destination as was transmitted at the source. Note that the famous no-cloning theorem of quantum mechanics does not allow for a quantum state to be copied, or directly reproduced. However, teleportation prescribes a set of quantum operations, using entanglement, that allows a quantum state to be inferred and reconstructed at the destination. Let us describe the operation in general terms.

First, Alice has a qubit and she wants to transmit it to Bob. Recall that a quantum state for a qubit, in general, is a superposition of zero and one with the degree of being in zero or one specified by the amplitudes. Unfortunately, as we have seen with measurement, there is no possible manner in which Alice can directly examine the amplitudes. Any attempt to examine the amplitudes will result in a collapse of the superposition into one of the discrete states, assuming measurement is with respect to the computational bases. The amplitudes, in general, can be complex numbers with an arbitrary amount of precision. So the amount of information in a single qubit can be enormous; transmission of the state is a significant feat. Now assume that there is a third party named Victor who prepares the obligatory entangled qubit, providing one of the entangled pair to Alice and the other to Bob. Note that now Alice has two qubits: the original qubit that she wished to transmit and the entangled one from Victor.

Now comes a key point, Alice treats her qubit and the entangled one as a single composite system. She has a pair of qubits that she measures on the Bell basis, which will be described in more detail shortly. Alice transmits the measurement result, namely the outcome of her measurement, to Bob using any classical means. Bob applies an operation, to be explained shortly, based upon Alice's measurement. After Bob applies the operation, his state becomes exactly the same as Alice's original qubit.

Now let's look at the details of how teleportation works following the above exchange between Alice and Bob. Let's specify the state that Alice wants to transmit to Bob as shown in Equation 5.65.

$$\alpha |0\rangle + \beta |1\rangle \tag{5.65}$$

As we mentioned, Victor prepares his qubits in the entangled Bell state as shown in Equation 5.66.

$$\frac{1}{\sqrt{2}}(|00\rangle + |11\rangle) \tag{5.66}$$

Next, Victor sends one of his qubits to Alice and the other to Bob. We see the joint state of Alice's qubits in Equation 5.67.

$$(\alpha |0\rangle + \beta |1\rangle)\left(\frac{(|00\rangle + |11\rangle)}{\sqrt{2}}\right) \tag{5.67}$$

We can expand the joint state of Alice's qubits via the distributive property of the outer product as shown in Equation 5.68.

$$\frac{\alpha |000\rangle + \alpha |011\rangle + \beta |100\rangle + \beta |111\rangle}{\sqrt{2}} \tag{5.68}$$

Alice's two qubits, in state $|00\rangle$, can be rewritten as a linear combination of two elements of the Bell basis as shown in Equation 5.69.

$$|00\rangle = \frac{1}{\sqrt{2}}\left(\frac{|00\rangle + |11\rangle}{\sqrt{2}}\right) + \frac{1}{\sqrt{2}}\left(\frac{|00\rangle - |11\rangle}{\sqrt{2}}\right) \tag{5.69}$$

Alice's two qubits, in state $|01\rangle$, can be written in the Bell basis as shown in Equation 5.70.

$$|01\rangle = \frac{1}{\sqrt{2}}\left(\frac{|01\rangle + |10\rangle}{\sqrt{2}}\right) + \frac{1}{\sqrt{2}}\left(\frac{|01\rangle - |10\rangle}{\sqrt{2}}\right) \tag{5.70}$$

Alice's two qubits, in state $|10\rangle$, can be written in the Bell basis as shown in Equation 5.71.

$$|10\rangle = \frac{1}{\sqrt{2}}\left(\frac{|01\rangle + |10\rangle}{\sqrt{2}}\right) - \frac{1}{\sqrt{2}}\left(\frac{|01\rangle - |10\rangle}{\sqrt{2}}\right) \tag{5.71}$$

Alice's two qubits, in state $|11\rangle$, can be written in the Bell basis as shown in Equation 5.72.

$$|11\rangle = \frac{1}{\sqrt{2}}\left(\frac{|00\rangle + |11\rangle}{\sqrt{2}}\right) - \frac{1}{\sqrt{2}}\left(\frac{|00\rangle - |11\rangle}{\sqrt{2}}\right) \tag{5.72}$$

The next step is to substitute these expressions for the computational basis states to rewrite this state in terms of the Bell basis on the first two qubits, which is pure algebra as shown in Equation 5.73.

$$\begin{aligned}
= \quad &\frac{1}{\sqrt{2}}\left(\frac{|00\rangle + |11\rangle}{\sqrt{2}}\right)(\alpha|0\rangle + \beta|1\rangle) \\
&\frac{1}{\sqrt{2}}\left(\frac{|00\rangle - |11\rangle}{\sqrt{2}}\right)(\alpha|0\rangle - \beta|1\rangle) \\
&\frac{1}{\sqrt{2}}\left(\frac{|01\rangle + |10\rangle}{\sqrt{2}}\right)(\alpha|1\rangle + \beta|0\rangle) \\
&\frac{1}{\sqrt{2}}\left(\frac{|01\rangle - |10\rangle}{\sqrt{2}}\right)(\alpha|1\rangle - \beta|0\rangle)
\end{aligned} \tag{5.73}$$

Now suppose that Alice measures a 01, as shown in Equation 5.75.

$$\begin{aligned}
= \quad &\frac{1}{\sqrt{2}}\left(\frac{|00\rangle + |11\rangle}{\sqrt{2}}\right)(\alpha|0\rangle + \beta|1\rangle) \\
&\frac{1}{\sqrt{2}}\left(\frac{|00\rangle - |11\rangle}{\sqrt{2}}\right)(\alpha|0\rangle - \beta|1\rangle) \rightarrow (Z)(\alpha|0\rangle + \beta|1\rangle) \\
&\frac{1}{\sqrt{2}}\left(\frac{|01\rangle + |10\rangle}{\sqrt{2}}\right)(\alpha|1\rangle + \beta|0\rangle) \\
&\frac{1}{\sqrt{2}}\left(\frac{|01\rangle - |10\rangle}{\sqrt{2}}\right)(\alpha|1\rangle - \beta|0\rangle)
\end{aligned} \tag{5.74}$$

All probabilities in the Bell basis are $\frac{1}{4}$, independent of the state being teleported. Suppose Alice finds her qubits are in the second state of the Bell basis. Now suppose that Bob applies the Z gate. The result is the original qubit state of Alice's system. Thus, Bob has reconstructed Alice's original qubit at his location.

If Alice were to obtain the first Bell state, then no change is required and I may be applied as shown in Equation 5.76 to represent no change.

$$= \frac{1}{\sqrt{2}} \left(\frac{|00\rangle + |11\rangle}{\sqrt{2}} \right) (\alpha |0\rangle + \beta |1\rangle) \rightarrow (I)(\alpha |0\rangle + \beta |1\rangle) \qquad (5.75)$$

$$\frac{1}{\sqrt{2}} \left(\frac{|00\rangle - |11\rangle}{\sqrt{2}} \right) (\alpha |0\rangle - \beta |1\rangle) \rightarrow (Z)(\alpha |0\rangle + \beta |1\rangle)$$

$$\frac{1}{\sqrt{2}} \left(\frac{|01\rangle + |10\rangle}{\sqrt{2}} \right) (\alpha |1\rangle + \beta |0\rangle)$$

$$\frac{1}{\sqrt{2}} \left(\frac{|01\rangle - |10\rangle}{\sqrt{2}} \right) (\alpha |1\rangle - \beta |0\rangle)$$

For the third state, X may be applied as shown in Equation 5.77.

$$= \frac{1}{\sqrt{2}} \left(\frac{|00\rangle + |11\rangle}{\sqrt{2}} \right) (\alpha |0\rangle + \beta |1\rangle) \rightarrow (I)(\alpha |0\rangle + \beta |1\rangle)$$

$$\frac{1}{\sqrt{2}} \left(\frac{|00\rangle - |11\rangle}{\sqrt{2}} \right) (\alpha |0\rangle - \beta |1\rangle) \rightarrow (Z)(\alpha |0\rangle + \beta |1\rangle)$$

$$\frac{1}{\sqrt{2}} \left(\frac{|01\rangle + |10\rangle}{\sqrt{2}} \right) (\alpha |1\rangle + \beta |0\rangle) \rightarrow (Z)(\alpha |0\rangle + \beta |1\rangle) \qquad (5.76)$$

$$\frac{1}{\sqrt{2}} \left(\frac{|01\rangle - |10\rangle}{\sqrt{2}} \right) (\alpha |1\rangle - \beta |0\rangle)$$

For the fourth state, ZX may applied; in each case as shown in Equation 5.77.

$$= \frac{1}{\sqrt{2}} \left(\frac{|00\rangle + |11\rangle}{\sqrt{2}} \right) (\alpha |0\rangle + \beta |1\rangle) \rightarrow (I)(\alpha |0\rangle + \beta |1\rangle)$$

$$\frac{1}{\sqrt{2}} \left(\frac{|00\rangle - |11\rangle}{\sqrt{2}} \right) (\alpha |0\rangle - \beta |1\rangle) \rightarrow (Z)(\alpha |0\rangle + \beta |1\rangle)$$

$$\frac{1}{\sqrt{2}} \left(\frac{|01\rangle + |10\rangle}{\sqrt{2}} \right) (\alpha |1\rangle + \beta |0\rangle) \rightarrow (X)(\alpha |0\rangle + \beta |1\rangle)$$

$$\frac{1}{\sqrt{2}} \left(\frac{|01\rangle - |10\rangle}{\sqrt{2}} \right) (\alpha |1\rangle - \beta |0\rangle) \rightarrow (ZX)(\alpha |0\rangle + \beta |1\rangle) \qquad (5.77)$$

For each possible measurement outcome that Alice can observe, Bob can apply a predetermined operator in order to change his entangled qubit state to become that of the qubit that Alice originally intended to transmit. The qubit and thus the information that it encodes have been effectively transmitted. Notice that two classical bits had to be transmitted by Alice to Bob in order for the quantum information to be properly decoded. Also note that Alice did *not* have to know the quantum state of the qubit she intended to transmit, in fact, she could not know it. The transmission succeeded, using just two bits of classical information and a shared Bell state.

To recap, superdense coding used one entangled bit and one qubit to transmit two bits of classical communication. Teleportation used one entangled bit and two classical bits to transmit one quantum state. It seems that in both cases, one qubit of information is equivalent to two classical bits.

Table 5.3

Continuous Versus Discrete Forms of the Gradient and Laplacian Matrix

Continuous	Discrete
$grad(f) = \nabla f = \partial f/\partial x, \partial f/\partial y, \ldots$	$I_m = \text{edges} \times \text{nodes} = \begin{bmatrix} i_1 & i_2 \\ i_3 & i_4 \end{bmatrix}$
$laplacian(f) = \nabla^2 f = \partial^2 f/\partial^2 x, \partial^2 f/\partial^2 y, \ldots$	$L = \text{nodes} \times \text{nodes} = \begin{bmatrix} v_1 & v_2 \\ v_3 & v_4 \end{bmatrix}$

5.4.1 The spectral theorem

Recall the discussion from the last chapter regarding eigensystem network analysis. In particular, recall Equation 4.25. Here we discuss the theorem behind that equation, namely the spectral theorem. Suppose that A is an arbitrary Hermitian matrix, $A\dagger = A$. Then A is diagonalizable as shown in Equation 5.78, where U represents a unitary matrix, and $\lambda_1, \ldots, \lambda_d$ are the eigenvalues of A.

$$A = U \operatorname{diag}(\lambda_1, \ldots, \lambda_d) U^\dagger \tag{5.78}$$

Recall that $|j\rangle \langle j|$ is an outer product. It returns a matrix with ones in position (j, j) along the diagonal of the matrix. Thus, the matrix with lambdas along the diagonal, $\operatorname{diag}(\lambda_1, \ldots, \lambda_d) = \sum_j \lambda_j |j\rangle \langle j|$.

Now, consider Equation 5.78. Let $|e_j\rangle \equiv U |j\rangle$ be the λ_j eigenvector of A. In other words, $A |e_j\rangle = \lambda_j |e_j\rangle$. Then $A = \sum_j \lambda_j |e_j\rangle \langle e_j|$. The eigenvectors are always orthogonal, so $A = \sum_k \lambda_k P_k$, where P_k is a projector onto the λ_k eigenspace of A. An eigenspace of a given transformation is the set of all eigenvectors of that transformation that have the same eigenvalue, together with the zero vector. An eigenspace is an example of a subspace of a vector space.

In the eigensystem analysis of carbon nanotube networks from the previous chapter we used the Laplacian matrix to represent the nanotube network and solved for the network conductance by finding the eigenvalues and eigenvectors. In essence, the network, represented by the Hermitian Laplacian matrix, was analogous to a quantum system and by finding the eigensystem, we collapsed the system onto it along various orthogonal bases as one would when performing a measurement operation. There are many practical uses of the spectral theorem, from finding graph cliques and image segmentation to document search, for example Web searches and latent semantic analysis. One final word on the carbon nanotube network is the relationship of the Laplacian matrix to the simple incidence matrix for the underlying network graph. As shown in Table 5.3, the Laplacian matrix is the square of the incidence matrix just as the continuous Laplacian is the derivative of the gradient.

Another operator that will be used shortly is the trace. The trace of an $n \times n$ square matrix A is defined to be the sum of the elements along the main diagonal as shown in Equation 5.79.

$$tr(A) \equiv \sum_j A_{jj} \tag{5.79}$$

It is important to note that the order of the matrices does matter in taking traces: in general, $tr(ABC) \neq tr(ACB)$. However, the trace is invariant under cyclic permutations; that is, $tr(ABCD) = tr(BCDA) =$

$tr(CDAB) = tr(DABC)$. Some simple examples are shown in Equation 5.80.

$$
\begin{aligned}
X &= \begin{bmatrix} 0 & 1 \\ 1 & 0 \end{bmatrix} \\
tr(X) &= 0 \\
I &= \begin{bmatrix} 1 & 0 \\ 0 & 1 \end{bmatrix} \\
tr(I) &= 2
\end{aligned}
\tag{5.80}
$$

As previously mentioned, the trace is cyclical, $tr(AB) = tr(BA)$. Given that $tr(AB) = \sum_j (AB)_{jj} = \sum_{jk} A_{jk} B_{kj} = \sum_{jk} B_{kj} A_{jk} = tr(BA)$, then, $tr(ABC) = tr(BCA) = tr(CAB)$. Finally, a useful identity to note is that $tr(|a\rangle \langle b|) = \langle a \,|\, b\rangle$. An exercise at the end of the chapter asks for a proof of this identity. This would be a good time to think through some of the exercises to solidify the material before proceeding.

5.4.2 An alternative form of postulate two

Recall that postulate two stated that the evolution of a closed quantum system is described by a unitary transformation. In other words, a new quantum state ψ' is derived from the application of the unitary operator U as shown in Equation 5.81.

$$
|\psi'\rangle = U |\psi\rangle
\tag{5.81}
$$

However, the above operation is a discrete one while quantum operations are in reality continuous. The evolution of a closed quantum system is described by Schrödinger's equation as shown in Equation 5.82, where H is a constant Hermitian matrix known as the Hamiltonian of the system.

$$
i\frac{d|\psi\rangle}{dt} = H |\psi\rangle
\tag{5.82}
$$

The eigenvectors of H are known as the energy eigenstates of the system, and the corresponding eigenvalues are known as the energies. The Hamilton equations for a closed system are the sum of the kinetic and potential energy in the system as represented by a set of differential equations. A discussion of the Hamiltonian would take us beyond the scope of this book. As an example, $H = \omega X$ has energy eigenstates $(|0\rangle + |1\rangle)/\sqrt{2}$ and $(|0\rangle - |1\rangle)/\sqrt{2}$, with corresponding energies $\pm\omega$. The solution of Schrödinger's equation is $|\psi(t)\rangle = e^{(-iHt)} |\psi(0)\rangle$ where $\psi(0)$ is the initial state and $\psi(t)$ is the state at time t. Then $U = e^{(-iHt)}$ and $|\psi'\rangle = U |\psi\rangle$.

One other useful operator to be aware of is the Hadamard operator H, also called the square-root of NOT gate. It is useful because it turns a $|0\rangle$ or $|1\rangle$ into a state "halfway" between either one; it puts them into a perfect superposition. Thus, it is useful for placing inputs into superimposed states. The gate is shown in Equation 5.83.

$$
H = \frac{1}{\sqrt{2}} \begin{bmatrix} 1 & 1 \\ 1 & -1 \end{bmatrix}
\tag{5.83}
$$

Thus, when the Hadamard operator is applied to $|0\rangle$, the result is shown in Equation 5.84.

$$
|0\rangle H = \frac{(|0\rangle + |1\rangle)}{\sqrt{2}}
\tag{5.84}
$$

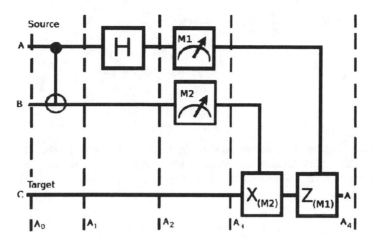

Figure 5.4   A quantum circuit for teleportation.

Similarly, when the Hadamard operator is applied to $|1\rangle$ the result is shown in Equation 5.85.

$$|1\rangle H \;=\; \frac{(|0\rangle - |1\rangle)}{\sqrt{2}} \tag{5.85}$$

5.4.3   Building a quantum communication network

Teleportation is a key mechanism for nanoscale network protocols. Let us examine a quantum teleportation circuit as shown in Figure 5.4. Inputs are on the left side of the circuit and flow proceeds to the right. $|\psi\rangle$ is Alice's qubit that is to be transported. The entangled qubit, shared by Alice and Bob, is β_{00}. The standard practice of using single lines for quantum information flow and double lines for classical information flow is used here.

We show "slices" (A_0 through A_4) through the quantum circuit in the figure to indicate the operation of the circuit one step at a time. The joint state of Alice's qubit and the entangled state is denoted by slice $A_0 \, |\psi_0\rangle$ shown in Equation 5.86.

$$
\begin{aligned}
|\psi_0\rangle \;&=\; |\psi\rangle \, |\beta_{00}\rangle \\
&=\; \frac{1}{\sqrt{2}} \left[\alpha \, |0\rangle \, (|00\rangle + |11\rangle) + \beta \, |1\rangle \, (|00\rangle + |11\rangle) \right]
\end{aligned} \tag{5.86}
$$

Next, the controlled-NOT gate implements a flip of the qubits in Alice's shared entangled qubit in slice A_1 as shown in Equation 5.87. If a $|1\rangle$ is applied, a controlled-NOT gate implements the qubit flip operation, otherwise the state passes though unchanged.

$$|\psi_1\rangle \;=\; \frac{1}{\sqrt{2}} \left[\alpha \, |0\rangle \, (|00\rangle + |11\rangle) + \beta \, |1\rangle \, (|10\rangle + |01\rangle) \right] \tag{5.87}$$

In the next step, slice A_2, $|\psi_2\rangle$, the Hadamard gate is applied to Alice's qubit as shown in Equation 5.88.

$$|\psi_2\rangle \;=\; \frac{1}{2} \left[\alpha(|0\rangle + |1\rangle)(|00\rangle + |11\rangle) + \beta(|0\rangle - |1\rangle)(|10\rangle + |01\rangle) \right] \tag{5.88}$$

The result can be expanded via simple algebra as shown in Equation 5.89.

$$
\begin{aligned}
|\psi_2\rangle \ = \ \tfrac{1}{2} \ [& |00\rangle \left(\alpha \, |0\rangle + \beta \, |1\rangle \right) + \\
& |01\rangle \left(\alpha \, |1\rangle + \beta \, |0\rangle \right) + \\
& |10\rangle \left(\alpha \, |0\rangle - \beta \, |1\rangle \right) + \\
& |11\rangle \left(\alpha \, |1\rangle - \beta \, |0\rangle \right)]
\end{aligned}
\tag{5.89}
$$

Next, Alice performs measurement on her qubit system as indicated by slice $A_3 \, |\psi_3\rangle$. The result is sent classically to Bob. Look carefully at Equation 5.89; essentially, every possible permutation of Alice's qubit amplitudes are inside that equation. Equation 5.90 shows how the measured result in classical bits should map into the appropriate quantum state that yields the state of Alice's original qubit.

$$
\begin{aligned}
00 &\mapsto |\psi_3(00)\rangle \equiv [\alpha \, |0\rangle + \beta \, |1\rangle] \\
01 &\mapsto |\psi_3(01)\rangle \equiv [\alpha \, |1\rangle + \beta \, |0\rangle] \\
10 &\mapsto |\psi_3(10)\rangle \equiv [\alpha \, |0\rangle - \beta \, |1\rangle] \\
11 &\mapsto |\psi_3(11)\rangle \equiv [\alpha \, |1\rangle - \beta \, |0\rangle]
\end{aligned}
\tag{5.90}
$$

We see that a quantum network using teleportation will not work without the necessary entangled qubits. Yet, a teleportation network could be used to transmit the entangled qubits. This sounds like the chicken and egg problem: which came first? As we see with teleportation and many other applications, entangled qubits are a consumable resource. Can we teleport entangled qubits? More generally, can we create a teleportation network for transmitting quantum information? Entanglement swapping appears to be a key to the solution of this problem.

5.4.4   Sharing entanglement in a quantum network

Recall that the Bell basis is a basis of a two-qubit system, where the basis vectors are defined in terms of the computational basis and are as shown in Equation 5.91.

$$
\begin{aligned}
\Phi^+\rangle &= \frac{1}{\sqrt{2}}(|00\rangle + |11\rangle) \\
\Psi^+\rangle &= \frac{1}{\sqrt{2}}(|01\rangle + |10\rangle) \\
\Psi^-\rangle &= \frac{1}{\sqrt{2}}(|10\rangle - |01\rangle) \\
\Phi^-\rangle &= \frac{1}{\sqrt{2}}(|00\rangle - |11\rangle)
\end{aligned}
\tag{5.91}
$$

The quantum states represented by these vectors are called Bell states and are maximally entangled. Density matrices that are diagonal in this basis are called Bell-diagonal. Let us generalize the Bell basis as a symbolic function $|\Psi(i,j)\rangle_\pm$ where $|\Psi(i,j)\rangle_\pm = |u_i, u_j\rangle \pm |u_i^c, u_j^c\rangle$ [160]. The state $u_i \in \{0,1\}$ and its complement is u_i^c is $1 - u_i$. Then, for a pair of entangled qubits $1, 2$, and $3, 4$, the joint state is as shown in Equation 5.93. Particles one and two are mutually entangled and particles three and four are mutually

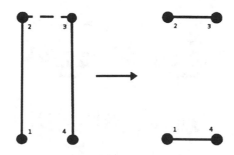

Figure 5.5 A simple illustration of entanglement swapping. Nodes are quantum states, bold edges indicate Bell pairs, and dashed edges indicate quantum measurement. The result is to swap entanglement among the Bell pairs.

entangled, and both entangled pairs are in a Bell state.

$$
\begin{aligned}
|\Psi(1,2)\rangle_+ \otimes |\Psi(3,4)\rangle_+ \; = \;\; & |u_1, u_2, u_3, u_4\rangle + \\
& |u_1^c, u_2^c, u_3^c, u_4^c\rangle + \\
& |u_1, u_2, u_3^c, u_4^c\rangle + \\
& |u_1^c, u_2^c, u_3, u_4\rangle
\end{aligned}
\tag{5.92}
$$

Projecting qubits two and three, one from each pair, to any of the Bell states, yields one of those in Equation 5.93.

$$
\begin{aligned}
|\Phi_1\rangle &= (|u_2, u_3\rangle + |u_2^c, u_3^c\rangle) \otimes (|u_1, u_4\rangle + |u_1^c, u_4^c\rangle) \\
|\Phi_2\rangle &= (|u_2, u_3\rangle - |u_2^c, u_3^c\rangle) \otimes (|u_1, u_4\rangle - |u_1^c, u_4^c\rangle) \\
|\Phi_3\rangle &= (|u_2, u_3^c\rangle + |u_2^c, u_3\rangle) \otimes (|u_1, u_4^c\rangle + |u_1^c, u_4\rangle) \\
|\Phi_4\rangle &= (|u_2, u_3^c\rangle - |u_2^c, u_3\rangle) \otimes (|u_1, u_4^c\rangle - |u_1^c, u_4\rangle)
\end{aligned}
\tag{5.93}
$$

The important point is the orthogonality of $|u_i\rangle$ and $|u_i^c\rangle$. Regardless of the outcome of the measurement, particles one and four will be in one of the Bell states. In Figure 5.5, the bold lines connect particles in Bell states and the dashed lines connect particles on which the Bell state measurement is made as shown. The implementation of a swapping channel is shown in Figure 5.6.

A network of quantum repeaters supports distributed quantum computation by creating high-fidelity end-to-end Bell pairs. Once completed, these pairs can then be used to teleport application data. A quantum repeater (also called a station) is a small, special-purpose quantum computer, holding a few physical qubits that it can couple to a transmission medium. The hardware provides the basic capability for creating short-distance, low-fidelity ("base level") Bell pairs. Each station participates repeatedly in building an end-to-end distributed Bell pair, through purification and the use of teleportation known as entanglement swapping.

Light is the most widely used platform for quantum communication and networking today [161]. Keep in mind, though, that photons are not the only form of quantum transmission; entanglement has been demonstrated among many particles of matter. However, even within the realm of quantum optics, a detailed explanation is beyond the scope of this book; let us simply note that there have been many demonstrations of photon generation from interfaces with single atoms [162–166]. As an example of the advantage of single-atom photon generation, consider that the Heisenberg uncertainty principle comes into play causing uncertainty regarding the photon number. The operators of phase-quadrature and amplitude-quadrature of a

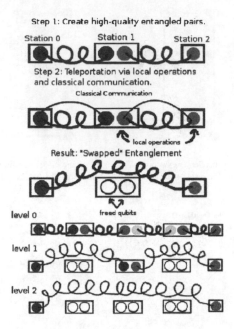

Figure 5.6 Using entanglement swapping to implement a channel. The quantum channel is constructed by swapping Bell pairs over short distances yielding longer channels.

light field are similar to position and momentum of a particle; similar to our inability to precisely measure a particle's position and momentum simultaneously with arbitrary precision, the phase and amplitude of a light wave also cannot be measured simultaneously with perfect accuracy. The more precisely the phase of a light wave is measured, the less determined is its amplitude and vice versa. It turns out that the quantum density matrix, described in detail in Section 6.2, contains the photon number distribution along its diagonal. The diagonal elements of the density matrix contain the intensity distribution of the state while the phase information is in the off-diagonal elements. The photon number follows a Poisson distribution, thus the mean photon number is equal to the spreading of the photon number distribution; that is, the variance. The more enhanced the state's amplitude noise, the more spread out the photon number distribution; that is, the more enhanced its intensity noise. Fortunately, the single-atom generation of photons has been shown to generate "squeezed light" or light in which the uncertainty is reduced to a minimum, making detection of the state more accurate and efficient [167]. In the amplitude-squeezed state the photon number variance is *less* than the mean photon number. Recall that in a Poisson distribution, the variance should be equal to the mean. This is known as sub-Poissonian light. This allows electrical detectors, when struck by photons within a given time interval, to more accurately determine necessary state information. This is just one example of how nanoscale operations may enhance quantum communication.

5.4.5   Quantum wire

What may seem surprising at first is the fact that transmitting qubits; that is, implementing a quantum interconnect system, or quantum wire, is one of the more challenging aspects of quantum computing. As is well known, quantum state cannot be cloned. It must be destroyed at the source and recreated at the destination. This is very different from classical human-scale networking. Decoherence, or entanglement with the environment, quickly destroys the state that one wishes to transmit. This is true even for the relatively short interconnects needed within a quantum computer and within our target nanoscale networking environment. Thus, protection of quantum state has been a critical area of research.

There are two relatively simple ways to implement a quantum wire network. The first is teleportation as we just discussed in the previous section. The second method is to place a chain of devices adjacent to another and let the information flow from one device to the next. This is known as a swapping channel. The usual approach is to place a series of controlled-NOT gates adjacent to one another and allow them to transport the information. The controlled-NOT is shown in Equation 5.94.

$$|\psi\rangle = \alpha|0\rangle + \beta|1\rangle = \begin{bmatrix} \alpha \\ \beta \end{bmatrix} \tag{5.94}$$

Recall that the tensor is used to compose states; Equation 5.95 is a reminder of the notation for tensor products.

$$|\alpha\rangle \otimes |\beta\rangle = |\alpha, \beta\rangle \tag{5.95}$$

Now, we can examine what happens when the tensor state $|0, \psi\rangle$ is formed as shown in Equation 5.96. We can see what happens when the controlled-NOT gate is applied to this state. Since the first qubit is zero, we expect the controlled-NOT to have no effect on the state. Let us see if this is true.

$$|0, \psi\rangle = \alpha|0\rangle|0\rangle + \beta|0\rangle|1\rangle = \begin{bmatrix} \alpha \\ \beta \\ 0 \\ 0 \end{bmatrix} \tag{5.96}$$

Applying the CNOT gate to $|0, \psi\rangle$ yields Equation 5.97. Because α and β are in the top two positions of the tensor product, they are unaffected by the controlled-NOT gate. They, need to be in the lower two positions in order to be swapped. We show how this happens next.

$$CNOT \equiv \begin{bmatrix} 1 & 0 & 0 & 0 \\ 0 & 1 & 0 & 0 \\ 0 & 0 & 0 & 1 \\ 0 & 0 & 1 & 0 \end{bmatrix} \begin{bmatrix} \alpha \\ \beta \\ 0 \\ 0 \end{bmatrix} = \begin{bmatrix} \alpha \\ \beta \\ 0 \\ 0 \end{bmatrix} = |0, \psi\rangle \tag{5.97}$$

Consider the tensor $|1, \psi\rangle$ as shown in Equation 5.98.

$$|1, \psi\rangle = \alpha|1\rangle|0\rangle + \beta|1\rangle|1\rangle = \begin{bmatrix} 0 \\ 0 \\ \alpha \\ \beta \end{bmatrix} \tag{5.98}$$

Applying the CNOT gate to $|1, \psi\rangle$ yields the result in Equation 5.99. Clearly, the amplitudes α and β have been switched.

$$CNOT \equiv \begin{bmatrix} 1 & 0 & 0 & 0 \\ 0 & 1 & 0 & 0 \\ 0 & 0 & 0 & 1 \\ 0 & 0 & 1 & 0 \end{bmatrix} \begin{bmatrix} 0 \\ 0 \\ \alpha \\ \beta \end{bmatrix} = \alpha \begin{bmatrix} 0 \\ 0 \\ 0 \\ 1 \end{bmatrix} + \beta \begin{bmatrix} 0 \\ 0 \\ 1 \\ 0 \end{bmatrix} \tag{5.99}$$

If three controlled-NOT gates are placed in sequence, then the SWAP gate is created. The matrix representation of the SWAP gate is shown in Equation 5.100. Clearly, the middle two amplitudes of a qubit

pair are swapped by the SWAP gate. The concept for nanoscale quantum transport is an adjacent sequence
of such gates in order to transport a quantum state over a spatial distance.

$$SWAP \equiv \begin{bmatrix} 1 & 0 & 0 & 0 \\ 0 & 0 & 1 & 0 \\ 0 & 1 & 0 & 0 \\ 0 & 0 & 0 & 1 \end{bmatrix} \begin{bmatrix} \alpha \\ \beta \\ \gamma \\ \delta \end{bmatrix} = \alpha \left|00\right\rangle + \gamma \left|01\right\rangle + \beta \left|10\right\rangle + \delta \left|11\right\rangle \qquad (5.100)$$

Here we follow the results of [168], in which a simple comparison is made between the bandwidth and
latency of a quantum channel comprised of SWAP gates and one using teleportation. This was done
using a particular quantum technology, known as the Kane quantum computer; however, the results are
generalizable to any technology. In the particular implementation of the proposed quantum computer, a
swap channel, that is, a channel comprised of an adjacent chain of SWAP gates, takes 1 μs to operate and
the swap channel is approximately 60 nm long. Thus, the latency is shown in Equation 5.101 where l is the
latency and d is the total distance of the channel.

$$l = 1 \ \mu s \times \frac{d}{60 \ \text{nm}} \qquad (5.101)$$

The rate of decoherence is a function of $e^{-\lambda \times t}$, where λ depends on the technology and t is the number of
operations. Bits arrive with a fidelity that is proportional to the latency. The more time the qubits spend in
the channel, the lower their fidelity as shown in Equation 5.102 where f is the fidelity and l is the latency.

$$f = e^{-\lambda \times l} \qquad (5.102)$$

Thus, the true bandwidth is a function of fidelity as shown in Equation 5.103 where bw is the true, or
effective bandwidth, and f is the fidelity, and bw_{phy} is the total physical bandwidth.

$$bw = bw_{phy} \times f \qquad (5.103)$$

This can be generalized as follows. Assume a basic gate operation time of T and a channel distance of
D. However, let the distance D be in qubits, the number of qubits in the channel from source to destination.
Then the latency is as shown in Equation 5.104.

$$l = T \times D \qquad (5.104)$$

Following this approach, the bandwidth is shown in Equation 5.105.

$$bw = \frac{1}{T} e^{-\lambda D} \qquad (5.105)$$

We can assume that the fidelity must be above a critical threshold of $C = 10^{-4}$ in order for operation
to be feasible [168]. Then the maximum distance is as shown in Equation 5.106.

$$d_{max} = \log_e \frac{(1 - C)}{-\lambda} \qquad (5.106)$$

Let us compare this swap channel with a teleportation channel. We have explained the operation
of a teleportation channel. In summary, one can begin with initial $\left|0\right\rangle$ states by a variety of means, either
by thermodynamic cooling or applying compression-like techniques that gather entropy into one location

leaving the rest of the states at $|0\rangle$. The necessary entangled Einstein–Podolsky–Rosen (EPR) pair can be created from the $|0\rangle$ states by passing them through a Hadamard gate. This yields the superposition shown in Equation 5.107.

$$\frac{|0\rangle + |1\rangle}{\sqrt{2}} \tag{5.107}$$

The result shown in Equation 5.107 can be used as the control qubit in a controlled-NOT gate, described previously. Using another $|0\rangle$ qubit from the initialized pool of zero qubits, Equation 5.107 is used as the control for the target $|0\rangle$ qubit. The result is the entangled EPR pair shown in Equation 5.108.

$$\frac{|00\rangle + |11\rangle}{\sqrt{2}} \tag{5.108}$$

Finally, because there may be noise and decoherence, the EPR pair may not be perfectly entangled. A purification unit, discussed in more detail in Section 7.7.1, can be used to concentrate the EPR pairs into fewer but better entangled pairs.

The architecture is relatively simple, the EPR generator resides near the middle of the teleportation channel and provides EPR pair qubits to each end point, using the swapping channel just described. Then teleportation, as described earlier is used to transmit information across the channel. Let us see how this compares to the previously described swap channel.

Clearly, the bandwidth of the teleportation channel is directly related to the rate at which the EPR pairs are provided to each end of the channel. As explained regarding teleportation, an EPR qubit will be consumed for each qubit transmitted. The bandwidth depends upon the efficiency of the purification process. The efficiency of purification in turn depends upon the fidelity of the incoming qubits as shown in Equation 5.109, where p is the purification efficiency and f is the fidelity of the incoming EPR pairs.

$$p = f^2 \tag{5.109}$$

If we assume that there are always enough $|0\rangle$ qubits available, the bandwidth of the teleportation channel is shown in Equation 5.110, where d is the channel distance and 10^{-6} is the assumed decoherence rate exponent. Notice that the squared value of the decoherence rate exponent comes from the purification efficiency.

$$bw = 1\ \mu s \times e^{-2 \times 10^{-6} \times d/60}\ \text{nm} \tag{5.110}$$

For a 1 μm distance, the bandwidth of the teleportation channel is smaller than a swap channel. However, the benefit of teleportation is over longer distances. The channel length of the teleportation channel can exceed the maximum distance of the swap channel in Equation 5.106. The latency of the teleportation channel is simply the time for the teleportation operations to take place, assuming the Kane quantum architecture is 20 μs. Similar to the swap channel, a more general characterization of the latency and bandwidth can be derived. The latency is a fixed time period depending upon the number of gates and the time per gate to operate as shown in Equation 5.111, where T is a single gate operation time as in Equation 5.104.

$$l = 10T \tag{5.111}$$

The bandwidth is as shown in Equation 5.112 where D is again the distance in qubits and λ is the decoherence rate exponent similar to Equation 5.105.

$$bw = \frac{1}{T} e^{-2\lambda D} \tag{5.112}$$

A significant point is that, with teleportation, there is no need for a constraint on distance. It is true that the EPR pairs are being supplied by a swap channel; however, the corruption of the EPR pairs due to decoherence is handled by the purification process, which will have to discard more corrupted EPR pairs as distance increases. However, the impact is much less significant than the direct loss over a pure swap channel.

5.5 SUMMARY

As communication networks scale downward in size, the laws of quantum mechanics as discussed in this chapter will become operative. As we have seen, quantum mechanics and quantum computation open up exciting new dimensions for nanoscale networking. It appears that the simple swapping channel will be useful for shorter distance communication and the teleportation channel will be better for longer distance communication. In addition, it appears that there may be the possibility of transporting much more information in new ways. The additional computational power of quantum computation, which inherently operates at the nanoscale, could add significantly more processing power to nodes within a nanoscale network. In fact, many traditional network protocols now have quantum counterparts, for example, quantum network coding [169]. Thus, we can expect that nanoscale networks will be in a unique position to take advantage of both quantum and quantum/classical phenomena. As the relatively new field of quantum biology progresses, we can also expect to see new mechanisms for nanoscale and molecular networking to be discovered.

This is the last chapter dedicated to a particular underlying technology; Chapters 2 and 3 covered biological approaches, Chapter 4 covered nanostructures with a focus on the carbon nanotube, and this chapter focused upon quantum networking. The remaining chapters focus on how these technologies impact communication theory. How is each technology described by information theory? What kind of communication protocols and architectures are best for these technologies? How do they interface with one another and the human-scale world? Finally, what are the next steps for these technologies and where are they headed? These are some of the questions addressed in the remaining chapters.

5.6 EXERCISES

Exercise 33   Unitarity

Recall the discussion of unitary evolution.

1. Precisely why does a unitary matrix preserve normalized probabilities when used as an operator? Sketch a reasonably detailed proof.

Exercise 34   Operators

Recall the discussion of quantum operators. Here are some relatively simple practice questions.

1. Prove that $XY = iZ$

2. Prove that $X^2 = Y^2 = Z^2 = I$

3. A vector $x = [x_1, x_2, \cdots , x_n]$ is normalized if which of the following are true?

- $\|\mathbf{x}\| := \sqrt{x_1^2 + \cdots + x_n^2} = 0$
- $\|\mathbf{x}\| := \sqrt{x_1^2 + \cdots + x_n^2} = 1$
- $x_1^2 + \cdots + x_n^2 = 1$

4. Using Dirac notation, a state $|\psi\rangle$ is normalized if which of the following are true?

   - $\langle \psi \mid \psi \rangle = 1$
   - $|\psi\rangle \, |\psi\rangle = 1$
   - $A^\dagger \, |\psi\rangle = 1$

5. If U is unitary and preserves normalization, then does anything prevent unitary evolution from going backwards, that is, applying the inverse operator to the output state to derive the input state?

6. Which of the following best describes postulate one?

   - Specifies the dynamics of a quantum system
   - Specifies the state space of a quantum system
   - Defines qubits for electron spin

7. Which of the following best describes postulate two?

   - Defines the probabilities of a state space
   - Defines Dirac notation
   - Defines the dynamics of a quantum system through the unitary operator

Exercise 35   Measurement

Recall the discussion of quantum measurement.

1. Which of the following are possible states of a qubit?

   1. $\frac{1}{\sqrt{2}}(|0\rangle + |1\rangle)$
   2. $\frac{\sqrt{3}}{2}|1\rangle - \frac{1}{2}|0\rangle$
   3. $0.7|0\rangle + 0.3|1\rangle$
   4. $0.8|0\rangle + 0.6|1\rangle$
   5. $\cos\theta\,|0\rangle + i\sin\theta\,|1\rangle$
   6. $\cos^2\theta\,|0\rangle - \sin^2\theta\,|1\rangle$

2. What are the probabilities of state $|0\rangle$ and $|1\rangle$ for each valid state above?

Exercise 36   The Tensor Product

1. Which of the following is most true and important about postulate three?

   - Quantum measurement often returns incorrect values
   - We can never know the qubit amplitudes
   - We must project a measurement onto an orthogonal space
   - A unitary operation is required for measurement

2. What is the tensor product of state $|\psi\rangle = \sqrt{2/3}\,|+\rangle + 1/\sqrt{3}\,|-\rangle$ taken with itself twice, namely $|\psi\rangle^{\otimes 2}$ where $|\psi\rangle^{\otimes n}$ indicates $|\psi\rangle$ taken as a tensor product n times with itself?

3. Which of the following is the tensor product of $Y = \begin{vmatrix} 0 & -i \\ i & 0 \end{vmatrix}$ and $Z = \begin{vmatrix} 1 & 0 \\ 0 & -1 \end{vmatrix}$, namely $Y \otimes Z$?

   - $\begin{vmatrix} 1 & -i \\ i & 1 \end{vmatrix}$

   - $\begin{vmatrix} 0 & 0 & -i & 0 \\ 0 & 0 & 0 & i \\ i & 0 & 0 & 0 \\ 0 & -i & 0 & 0 \end{vmatrix}$

   - $\begin{vmatrix} 0 & -1 & 0 & 0 \\ i & 0 & 0 & i \\ i & 0 & 0 & 0 \\ 0 & -i & 0 & 0 \end{vmatrix}$

4. Which of the following is the interpretation of the inner product of A and B in two-dimensional Euclidean space?

   - Length of the projection of vector A on to vector B
   - Length of the projection of vector B on to vector A
   - Length of $A + B$

5. Which of the following states has a relative phase factor?

   - $e^{i\theta}|0\rangle + e^{i\theta}|1\rangle$
   - $e^{i\theta}|0\rangle + 1/\sqrt{2}\,|1\rangle$
   - $e^{i\theta}|0\rangle + e^{i\theta}/\sqrt{2}\,|1\rangle$

6. Which of the following is the best reason for entanglement as discussed in this chapter?

   - There are hidden variables controlling quantum systems that we have not yet discovered
   - Tensor products of systems cannot be factored cleanly into component systems; starts are intermixed

- Quantum states of distinct systems happen to collapse during measurement in a correlated manner for no apparent reason

Exercise 37   The Outer Product

Consider the following examples of outer products.

- $|1\rangle \langle 0| = \begin{bmatrix} 0 \\ 1 \end{bmatrix} \begin{bmatrix} 1 & 0 \end{bmatrix} = \begin{bmatrix} 0 & 0 \\ 1 & 0 \end{bmatrix}$

- $X = \begin{bmatrix} 0 & 1 \\ 1 & 0 \end{bmatrix} = (|0\rangle \langle 1|) - (|1\rangle \langle 0|)$

1. What is an outer product representation for Y?

2. Suppose $|e_1\rangle , ..., |e_d\rangle$ is an orthonormal basis for a state space. Prove that $I = \sum_j |e_j\rangle \langle e_j|$.

3. Prove that $|a\rangle \langle b|^\dagger = |b\rangle \langle a|$.

4. Prove that $tr(|a\rangle \langle b|) = \langle a \,|\, b\rangle$.

Exercise 38   Teleportation

Recall how teleportation works.

1. What role does the Hadamard operator play in teleportation?

2. Can teleportation transfer information faster than the speed of light?

Exercise 39   The Swap Channel

Recall how a swap channel is implemented.

1. Show the construction of a swap gate from three controlled-NOT gates. How does the swap operation occur in detail?

Exercise 40   Quantum Network Latency

Recall the discussion of swap channels versus teleportation.

1. What are the *latencies* of a swap channel and teleportation channel at 1 μm and 6 μm?

Exercise 41   Quantum Network Bandwidth

Recall the discussion of swap channels versus teleportation.

1. What are the *bandwidths* of a swap channel and teleportation channel at 1 μm and 6 μm?

Chapter 6

Information Theory and Nanonetworks

$S = k \log \Omega$

(Carved above his name on his tombstone in the Zentralfriedhopf in Vienna)

Ludwig Boltzmann

Previous chapters discussed specific implementations of nanoscale networks. Chapters 2 and 3 discussed nanoscale networks comprised of subcellular components. Chapter 4 discussed CNT nanoscale networks. Chapter 5 discussed quantum networking and its channel characteristics at the nanoscale. In this chapter, we leave the realm of specific implementations and discuss the more abstract, but ubiquitous, notions of communication and information theory. Nanoscale networking, just as human-scale networking, can be abstracted and distilled into a tractable mathematical language that allows communication and network engineering to take place through information theory. First, this chapter introduces and briefly reviews classical information theory. Then, we seek some of the underlying differences from human-scale networking that may exist at the nanoscale. A prime example of this is quantum information theory, an elegant hybrid of classical and quantum information concepts applied entirely within the quantum domain; thus, we touch on this topic as well. We noted in Chapter 1 that information theory had its roots in physics and Boltzmann's entropy, now nanoscale and molecular networking researchers are relating information theory back to the underlying physics from which it came through characterization of molecular channels.

We review classical information theory, both from the traditional view of Claude Shannon, and an alternative view from Andre Kolmogorov. Then we provide a brief introduction to quantum information theory. The section on quantum information theory ends with a brief discussion of what happens when access to information becomes limited at the most fundamental level, due to size constraints. As we have seen in previous chapters, self-assembly is critical to implementing nanoscale networks. Thus, we discuss self-assembly and self-organization, relating them to information theory. Then we look at the growing field of molecular information theory from the perspective of a nanoscale molecular channel assumed to be implemented by molecular diffusion through a fluid medium.

6.1 INFORMATION THEORY PRIMER

The goal of this section is provide a basic review of information theory and examine what we know about its application within nanoscale networks. Because quantum behavior is one of the unique dimensions of nanoscale networking, this chapter includes an introduction to quantum information theory. We begin by looking at the nature of information, trying to model how it is generated and most efficiently transmitted.

The following assumptions make things tractable. First, each output from a discrete information source comes from a finite set, and we will be particularly focused on an alphabet consisting of only two symbols: 0 and 1. However, there is no loss of generality in supposing that the alphabet is $0, \ldots, n-1$. Furthermore, we will assume that the symbols generated are independent and identically distributed. That is, each output from a source is independent of other outputs from the source, and each output has the same probability distribution describing its rate of occurrence.

We can even assume that we know the source distribution; that is, we know the probability distribution from which the symbols of the alphabet are generated. Most introductory examples begin with coin tosses or the roll of a die. So, as a simple example, assume a sequence of coin tosses of a biased coin with probability p of heads, and $1-p$ of tails. For the more general case of more than two symbols, the distribution on the alphabet of symbols is denoted $p_0, p_1, \ldots, p_n$.

Of course, in actual data, there are dependencies or correlations between symbols. Consider something as mundane as English text. Certain letter combinations appear together very frequently, such as "qu" or "th." These are dependencies that invalidate our independence assumption. Thus, the assumption regarding independent and identically distributed symbols is not truly valid, although it greatly facilitates the analysis by not requiring us to find and model the dependencies. We continue with the development assuming independent and identically distributed symbols.

6.1.1   Compression and the nature of information

Information theory quantifies the amount of information by the rate at which it can be compressed. Let us call the compression rate R. If there are n symbols, then we seek the smallest value of R that allows the symbols to be fully reconstructed. The minimum value of R will then be the information content of the information source. Claude Shannon derived the noiseless channel coding theorem, which provides a concise description of the minimum value of R. As we will see, the minimum value of R is the Shannon entropy, Equation 6.1, of the source distribution, where logarithms are taken to base two, because we have assumed a binary alphabet.

$$H(X) \equiv H(p_x) \equiv -\sum_x p_x \log(p_x) \tag{6.1}$$

Let's examine this a bit more. Suppose we flip coins, getting heads with probability p, and tails with probability $1-p$. For large values of n, it is very likely that we will get roughly np heads, and $n(1-p)$ tails. We can define a typical sequence as one in which the number of heads is between $np(1-\epsilon)$ and $np(1+\epsilon)$. Here ϵ provides a bound around the possible variance in the outcomes.

Now, let's tie this back to a binary sequence, in which there are zeros and ones that are generated with the same probability as a coin flip. We let x be the random variable for a bit *sequence* of length n. Note carefully that x is now a sequence of bits, rather than a single outcome, it is a sequence of outcomes. Let the probability of bit sequence x be $P(x)$. The bounds on $P(x)$ are shown in Equation 6.2. Since there are np heads, and each outcome is independent, there is a p^{np} probability of np heads, or a zero symbol, and correspondingly a $(1-p)^{n(1-p)}$ probability of the remainder being tails, or a one symbol. In Equation 6.2, the result is shown with the ϵ bounds included.

$$p^{np(1+\epsilon)}(1-p)^{n(1-p)(1+\epsilon)} < P(x) < p^{np(1-\epsilon)}(1-p)^{n(1-p)(1-\epsilon)} \tag{6.2}$$

Consider what happens as the variance is reduced, that is, as $\epsilon \to 0$. First, we'll take the logarithm, using base two, of Equation 6.2, which yields $\log P(x) \approx np \log p + n(1-p)\log(1-p)$. To obtain $P(x)$ again, since we are using logarithm base two, make this result the power of two as shown in Equation 6.3. The important thing to notice is that the exponent of two now looks like the previous definition of Shannon entropy from Equation 6.1, except that it is multiplied by n. We have arrived at the notion of entropy in relation to the probability of the sequence x occurring.

$$P(x) \approx 2^{np \log p + n(1-p) \log (1-p)} \approx 2^{-nH(p,1-p)} \tag{6.3}$$

A sequence that comes within the ϵ bound of Equation 6.2 will be called a typical sequence. The number of typical sequences is then $\approx 2^{nH(p,1-p)}$. There are a finite number of typical sequences and they can be enumerated. Thus, it is possible to construct a lookup table containing a list of all $2^{nH(p,1-p)}$ typical sequences. Only an index into the table needs to be transmitted in order to achieve compression. On average, only $H(p, 1-p)$ bits were required to store the compressed string.

Suppose one could compress lower than the rate of Shannon entropy. In other words, suppose that $R < H(p, 1-p)$ could be achieved with a high probability of no loss. Such a probability of no loss would have to be less than or equal to $2^{n(R-H(p,1-p))}$. However, as n increases the now negative exponent decreases causing the probability of no loss to approach zero. In fact, in a binary alphabet, at most 2^{nR} sequences can be correctly compressed and then decompressed by a fixed-length scheme of rate R.

6.1.2 Basic properties of entropy

There are some basic properties and conventions concerning information entropy. First, the probabilities are implicit within the notation $H(X)$ as shown in Equation 6.4.

$$H(X) \equiv H(p_x) \equiv -\sum_x p_x \log(p_x) \tag{6.4}$$

If a letter of the alphabet never occurs, then it has a probability of occurrence of zero. The information content from the source should not be affected by that letter. Thus, in the case where probabilities are zero, we also assume that $0 \log 0 \equiv 0$. Also, if you take the limit of $x \log x$ as x goes to zero, you get zero.

Since probabilities are less than one, the logarithms are negative; however, the negative sign in the definition of entropy makes the final result positive. Thus, entropy is nonnegative having a lower bound of zero. Entropy is at its highest when all letters are equally likely; with d distinct letters the maximum entropy is $\log(d)$. Finally, since the digital world is ubiquitous, the binary entropy is widely used and its entropy is $H(p) \equiv H(p, 1-p)$. The maximum entropy for a binary alphabet is the nice round number of $\log_2(2) = 1$.

6.1.3 Reliable communication in the presence of noise

A binary symmetric channel is one of the simplest channel models used to study a communication channel. If a bit is input to a binary symmetric channel, it is transmitted and received correctly with probability $1 - p$, and flipped to the incorrect value with probability p. This simple channel model allows us to analyze the impact of noise on a binary signal. We can let x be the input to the channel by the source and y be the output that the receiver observes. The impact of noise on the channel is described by conditional probabilities $p(y|x)$. Clearly, we want this probability to be high when x and y are the same; that is, we want the channel to transmit the same data that was input with a high probability. For an error probability of p, the results are shown in Equation 6.5.

$$
\begin{aligned}
p(0|0) &= 1 - p \\
p(1|0) &= p \\
p(0|1) &= p \\
p(1|1) &= 1 - p
\end{aligned}
\tag{6.5}
$$

The channel capacity is the rate at which the channel can send information. As we can see from the binary symmetric channel, the erroneous symbol transmissions will utilize the channel resource but not deliver useful information. Thus, noise reduces the channel capacity. This can be analyzed using mutual information, a relationship between the entropy of two sequences, X and Y. Just as entropy is the amount of information in a sequence, mutual information $I(X;Y)$ is the amount of information shared between two sequences. Similarly, the joint entropy $H(X,Y)$ is the amount of information contained within two sequences. In both cases, mutual and joint entropy, the dependencies, or overlapping information, have to be taken into account.

The joint entropy is shown in Equation 6.6. It is intuitively the sum of the information entropies of each component sequence, but with the common mutual information removed.

$$H(X,Y) \equiv H(X) + H(Y) - I(X;Y) \tag{6.6}$$

The mutual information is shown in Equation 6.7. Here it is also the sum of the information from the component sequences, but with the joint information removed, leaving only the common, or mutual information. This relates to a communication channel with noise, because the higher the mutual information, the greater the chance that we can recover the original transmission.

$$I(X;Y) == \equiv H(X) + H(Y) - H(X,Y) \tag{6.7}$$

Shannon's noisy channel coding theorem is stated in terms of mutual information. It's the maximum mutual information that can be achieved across a noisy channel over all probabilities of sending sequence X, namely, capacity $= \max_{p_x} I(X;Y)$. The maximum is over p_x, which is the marginal distribution of the signals that we choose to send over the channel.

6.1.4 Shannon versus Kolmogorov: Algorithmic information theory

We have seen the use of active networking in Chapter 4 as the network architecture for a nanoscale interconnect network. Traditional networking has a fixed, static architecture; that is, it is designed to transfer static information, information that does not change or execute inside the network itself. It is amenable to analysis as bit sequences using Shannon information entropy. As discussed previously in this book, active networking involves a more dynamic situation, in which executable code is often inserted into packets and injected into the network. Therefore, an information measure that more closely relates to this style of operation has been proposed, known as Kolmogorov complexity [170].

Where Shannon entropy measures information based upon probability of letter occurrence, Kolmogorov complexity measures information based upon the smallest program that can generate the sequence of letters. This may be the an executable program within an active packet for example. More formally, Kolmogorov complexity is described using the following symbols. First, it assumes that the code describing the information content runs on a universal Turing machine ϕ. The program that generates the sequence whose information is of interest is represented by p. Finally, the sequence of interest is x. Kolmogorov complexity $K(x)$ is defined formally as shown in Equation 6.8, where $l(p)$ is the length of program p.

$$K_\phi(x) = \left\{ \min_{\phi(p)=x} l(p) \right\} \tag{6.8}$$

A universal Turing machine is a Turing machine that can simulate any Turing machine on arbitrary input. The universal machine operates by reading both the description of the machine to be simulated as well as the input to be processed from its own tape. The reason a universal Turing machine is used in the definition of Kolmogorov complexity is to provide a consistent (virtual) machine upon which to write the program. Allen Turing designed the first universal Turing machine and Marvin Minsky reduced it to approximately seven states in the early 1960s.

The intuition behind Kolmogorov complexity is that regular sequences, that is, sequences with patterns, should have low complexity, or low information content. As a simple example, an alternating sequence of ones and zeros is easy to understand and predict. It should have a low information content. Yet, given that it has an equal number of ones and zeros, it would have a high Shannon entropy. Other examples are $x^n = 001001001...001$ or $x^n =$ first n digits of π. Relatively concise programs exist to compute π, so that should also have low information, or low complexity. Generally, regular or low information sequences have $K(x^n) \propto O(\log n))$. Random sequences have high complexity, as one would expect. For example, sequences generated by a fair coin have $K(x^n) \propto n + O(1)$. Random sequences have large complexity (e.g., $K(x^n) \propto n + O(1)$). However, we can put an upper bound on the complexity because one can always write a program that simply prints out the sequence itself, no matter how complex it is. In this case, the program is as long as the sequence plus some additional, relatively small, constant length of code to perform the printing.

To summarize, the minimum complexity is obtained when the source produces a single letter, over and over again, with probability one. In this case, there is no need to compress the information, the string contains no information, other than its length. The maximum complexity is obtained when the input distribution is uniform, that is, we know nothing about the potential biases in the source model that is generating the string. Programs can be encoded based upon the probability distribution of their bit-sequence, thus, probability distributions relate to encoded program length.

A comparison between traditional, or legacy, networks viewed in terms of Shannon information and active networks viewed in terms of Kolmogorov complexity is illustrated in Figure 6.1. Just as information entropy is a representation of the amount of information in a sequence of bits, Kolmogorov complexity is also a representation of information in terms of the smallest program that generates a sequence of bits. Variations of complexity, inspired by Kolmogorov complexity, seek to compress information by identifying and compressing patterns within sequences, thus separating "randomness" from structure as illustrated in Table 6.1. Bits are the primary data units for operating with information entropy in traditional networks while active packets contain both static data and executable code. Finally, on the legacy side, mutual information is widely used as an analytical tool for finding maximum channel capacity while an absolute mutual information exists for Kolmogorov complexity.

6.1.5 Minimum description length and sophistication

One of the challenges in using a direct form of Kolmogorov complexity has been the inability to determine the smallest Turing machine program, or even the smallest program for any machine in the general case. The notion behind minimum description length is, in a sense, to ease the exploration of the search space for the smallest program by attempting to explicitly separate the purely random portion of a sequence from the portion of a sequence that is nonrandom and for which a compact program may be more easily discovered. Thus, the nonrandom portion of a sequence can be encoded as a small program while the random portion of the sequence cannot be programmed or encoded and must be included verbatim [171]. This notion of algorithmically compressed information has many variations as shown in Table 6.1.

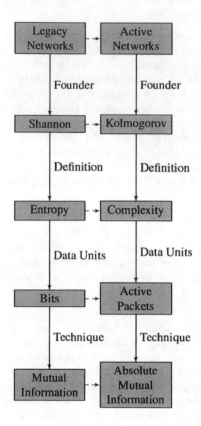

Figure 6.1   Shannon versus Kolmogorov.

Table 6.1

Interpretations of the Random and Nonrandom Parts of a Minimum Description Length Result

| Program | Data | = | Hypothesis | Error | = | Structural Part | Random Part |

6.2 QUANTUM INFORMATION THEORY

Now that traditional information theory has been introduced, the goal of this section is to introduce quantum information theory, since nanoscale networking will be operating near or at the scale where quantum effects become relevant. As we saw in the last chapter, nanoscale networks can be implemented via quantum mechanical principles, in particular, through teleportation channels or swap channels. Quantum information theory, just like classical information theory, can help understand and mitigate the effects of noise in such channels. In order to introduce the topic of information theory, we will need to enhance our concepts and notation from Chapter 5. First, we introduce the density matrix. The density matrix provides a concise representation of a set of quantum states along with the probability of their existence. Then we review the basic postulates of quantum mechanics from Chapter 5, but from the perspective of density matrices. This allows us to introduce the quantum version of information entropy. Then we discuss quantum data compression. There are, of course, many other aspects of quantum information theory that are not covered here, such as quantum error correction. However, the main point is that quantum analogs of classical information theoretic topics exist and can be leveraged for quantum nanoscale networking. Note that it is assumed that the reader is familiar with introductory quantum notation and material from Chapter 5.

The density matrix generalizes the notion of a quantum state, which is ideal for describing a noisy quantum system. The density matrix is primarily a mathematical convenience; it is not required to understand and analyze quantum systems, but it makes it much easier and provides insight and intuition that otherwise might be missed. There are three different views of the density matrix. We will explore each of these views in more detail shortly, but for now, we begin with a summary of each of the views. There is the *ensemble view* in which we imagine a quantum system that could be in any one of a number of different quantum states, $|\psi_j\rangle$, with probability p_j. There is the *subsystem point view*, in which it is possible to represent a subsystem of a larger system as a density matrix. Because a subsystem is entangled with a larger system, one cannot represent the quantum state of the subsystem by itself. However, the density matrix allows the subsystem to be represented as part of the larger system. Finally, there is the *fundamental view* in which all the fundamental postulates of quantum mechanics can be recast in terms of density matrices instead of individual quantum states.

The density matrix can be challenging to understand at first, but it is well worth the invested effort because it makes analysis of including quantum noise, quantum error-correction, quantum entanglement, and quantum communication much easier. Also note that because the density matrix is a unitary matrix, it can be considered an operator. Let's begin slowly with a quick review of the trace operator as shown in Equation 6.9, which is very simply the sum of the diagonal elements of a matrix.

$$tr(A) = \sum_i A_{ii} \qquad (6.9)$$

Also recall that $tr(AB) = \langle A | B \rangle$. In other words, if A and B are vectors, then trace of the matrix that results from their multiplication yields a scalar value that is also the result of the inner product of the vectors.

Suppose that we have a matrix that can be represented in the form $UAU^\dagger$, from the spectral theorem. Then $tr(UAU^\dagger)$ can be determined easily by rearranging the matrices as follows, $tr(U^\dagger U A)$. Because U is unitary, then $U^\dagger U = I$, where I is the identity matrix. The result of $tr(IA)$ is simply $tr(A)$ since the identity matrix multiplied by any matrix is just that matrix. Now, suppose that i are orthonormal bases, then $tr(A |\psi\rangle \langle\psi|) = \sum_i \langle i | A | \psi \rangle \langle \psi | i \rangle$. This is true because the $\bar{i} | i \rangle$ are zero due to their orthogonality. The sum is then taken along the diagonal of the matrix that results after $A |\psi\rangle \langle\psi|$ is evaluated. The result is also the same as $\langle\psi | A | \psi\rangle$, which is the inner product of $|\psi\rangle$ with the matrix A applied as an operator.

Now recall the projection operator. The Gram-Schmidt process is a method for creating a set of orthogonal vectors given an inner product space. In other words, without going into the details of the Gram-Schmidt process, suppose S is a k-dimensional vector subspace of the d-dimensional vector space V. It is possible to construct an orthonormal basis $|1\rangle ... |d\rangle$ for V such that $|1\rangle ... |k\rangle$ is an orthonormal subspace for S. Then the projector $P = \sum_{i=1}^{k} \langle i \,|\, i \rangle$ projects onto the subspace S. This will be important shortly for seeing how the density matrix is composed of individual measurement outcomes.

Now we can look at the ensemble point of view. Start with a quantum system that is in the state $|\psi_j\rangle$ with probability p_j. Notice how the notion of a probabilistic state can be useful for dealing with noise, which may occur probabilistically. We can perform measurements using projectors P_k, which results in associated outcomes k. So now we have multiple probabilistic events going on at once. First, the ensemble of states was in some initial state with probability p_j. Second, the outcome of applying the projector is also probabilistic. So the probability of outcome k, as shown in Equation 6.10, is the probability that we have state $|\psi_j\rangle$, which is p_j and then the probability that we obtain the outcome k, which is $\langle \psi_j \,|\, P_k \,|\, \psi_j \rangle$ as shown in Equation 6.10. Finally, utilizing results from our previous discussion of the trace operator, the last line of Equation 6.10 can be derived.

$$\begin{aligned} P(k) &= \sum_k Pr(k|\text{state } \psi_j)p_j \\ &= \sum_k \langle \psi_j \,|\, P_k \,|\, \psi_j \rangle \, p_j \\ &= \sum_k p_j \, tr(|\psi_j\rangle \langle \psi_j| \, P_k) \end{aligned} \qquad (6.10)$$

Now we are in a position to define the actual density matrix, denoted by ρ as shown in Equation 6.11. So the probability of outcome k, from Equation 6.10 is shown in Equation 6.12. The definition of ρ is an outer product of the possible states and their probabilities. In a sense, ρ is a system level description of the quantum state statistics and ρ completely determines all measurement statistics.

$$\rho = \sum_j p_j \, |\psi_j\rangle \langle \psi_j| \qquad (6.11)$$

$$Pr(k) = tr(\rho P_k) \qquad (6.12)$$

We can look at the simple example where $|\psi\rangle = |0\rangle$ with probability one. The result is as shown in Equation 6.13.

$$\rho = |0\rangle \langle 0| = \begin{bmatrix} 1 \\ 0 \end{bmatrix} \begin{bmatrix} 1 & 0 \end{bmatrix} = \begin{bmatrix} 1 & 0 \\ 0 & 0 \end{bmatrix} \qquad (6.13)$$

Another simple example is $|\psi\rangle = |1\rangle$ with probability one. The result is as shown in Equation 6.14.

$$\rho = |1\rangle \langle 1| = \begin{bmatrix} 0 \\ 1 \end{bmatrix} \begin{bmatrix} 0 & 1 \end{bmatrix} = \begin{bmatrix} 0 & 0 \\ 0 & 1 \end{bmatrix} \qquad (6.14)$$

A slightly more complex example is if $|\psi\rangle = \frac{|0\rangle + i|1\rangle}{\sqrt{2}}$ with probability one. Here the result is as shown in Equation 6.15.

$$\rho = \left(\frac{|0\rangle + i|1\rangle}{\sqrt{2}} \right) \left(\frac{|0\rangle - i|1\rangle}{\sqrt{2}} \right) = 1/2 \begin{bmatrix} 1 \\ i \end{bmatrix} \begin{bmatrix} 1 & -i \end{bmatrix} = 1/2 \begin{bmatrix} 1 & -i \\ i & 1 \end{bmatrix} \qquad (6.15)$$

All the previous examples assumed states existed with probability one. Now let's consider multiple states that exist with other probabilities. Suppose $|\psi\rangle = |0\rangle$ with probability p and $|\psi\rangle = |1\rangle$ with probability $1 - p$. Then the result is shown in Equation 6.16.

$$\rho = p\,|0\rangle\,\langle 0| + (1 - p)\,|1\rangle\,\langle 1| = p\begin{bmatrix} 1 & 0 \\ 0 & 0 \end{bmatrix} + (1 - p)\begin{bmatrix} 0 & 0 \\ 0 & 1 \end{bmatrix} = \begin{bmatrix} p & 0 \\ 0 & 1 - p \end{bmatrix} \tag{6.16}$$

The result of measuring on the computational basis, using Equation 6.12, is shown in Equation 6.17.

$$Pr(0) = tr(\rho\,|0\rangle\,\langle 0|) = \begin{bmatrix} 1 & 0 \end{bmatrix}\begin{bmatrix} p & 0 \\ 0 & 1 - p \end{bmatrix}\begin{bmatrix} 1 \\ 0 \end{bmatrix} \tag{6.17}$$

It is interesting and useful that we can apply operators to the density matrix. This means that we are applying the operator to the entire ensemble of states and their associated probabilities. This is quite a powerful concept. If we go back to our simple example of a quantum system in state $|\psi_j\rangle$ with probability p_j, we can apply some legitimate operation U, which must be unitary, to the system. The entire quantum system is now in the state $U\,|\psi_j\rangle$ with probability p_j. From a density matrix viewpoint, the initial density matrix is $\rho = \sum_j p_j\,|\psi_j\rangle\,\langle\psi_j|$ and the result after applying U is shown in Equation 6.19.

$$\rho' = \sum_j p_j U\,|\psi_j\rangle\,\langle\psi_j|\,U^\dagger \tag{6.18}$$

$$= U\left(\sum_j p_j\,|\psi_j\rangle\,\langle\psi_j|\right)U^\dagger \tag{6.19}$$

Simply put, the next ensemble of states, ρ', after operator U is applied is $\rho' = U\rho U^\dagger$.

There is an interesting phenomenon that occurs when the states in the ensemble are at maximum entropy. First, consider $|\psi\rangle = |0\rangle$ with probability p, and $|\psi\rangle = |1\rangle$ with probability $1 - p$. Then the value of ρ is shown in Equation 6.20.

$$\rho = \begin{bmatrix} p & 0 \\ 0 & 1 - p \end{bmatrix} \tag{6.20}$$

Recall that a Pauli X gate is a NOT operation, so that if an X gate is applied, the next state is as shown in Equation 6.21. The operation simply swapped the probabilities.

$$\rho' = X\rho X = \begin{bmatrix} 1 - p & 0 \\ 0 & p \end{bmatrix} \tag{6.21}$$

Now, suppose $|\psi\rangle = |0\rangle$ and $|\psi\rangle = |1\rangle$ with probability $1/2$. Each state is equally likely; we have the highest entropy of states for our small two-state example. In this case, we are left with some common fraction of the identity matrix. Specifically, $\rho = I/2$. This is known as a *completely mixed state*. Now consider what happens when any legitimate operator U is applied; $\rho' = U(I/2)U^\dagger = I/2$. There is no change to the ensemble regardless of what operator is applied! It would appear that no further useful computation can be done.

Now consider the characteristics of a valid density matrix. Let's start with the definition of the density matrix, namely, $\rho = \sum_j p_j\,|\psi_j\rangle\,\langle\psi_j|$. Consider the trace of ρ shown in Equation 6.22. The trace applies to the matrix outer product $|\psi_j\rangle\,\langle\psi_j|$. In Equation 6.23, we can use the fact that $tr(|\psi\rangle\,\langle\psi|) = \langle\psi\,|\,\psi\rangle = 1$.

Table 6.2

Density Matrix Summary

Definition	The density matrix for a system in state $\lvert\psi_j\rangle$ with probability p_j is $\rho \equiv \sum_j p_j \lvert\psi_j\rangle \langle\psi_j\rvert$.
Dynamics	$\rho \rightarrow \rho' = U\rho U^\dagger$.
Measurement	A measurement described by projectors P_k gives result k with probability $tr(P_k\rho)$, and the post-measurement density matrix is $\rho'_k = \frac{P_k \rho P_k}{tr(P_k \rho P_k)}$.
Characterization	$tr(\rho) = 1$ and ρ is a positive matrix. Conversely, given any matrix satisfying these properties, there exists a set of states $\lvert\psi_j\rangle$ and probabilities p_j such that $\rho = \sum_j p_j \lvert\psi_j\rangle \langle\psi_j\rvert$.

Finally, in Equation 6.24, the sum of the probabilities of each state must be one.

$$tr(\rho) \;=\; \sum_j p_j tr(\lvert\psi_j\rangle \langle\psi_j\rvert) \tag{6.22}$$

$$=\; \sum_j p_j \tag{6.23}$$

$$=\; 1 \tag{6.24}$$

Thus, we know that a valid density matrix must have a trace of one; its diagonal values must sum to one.

For any vector $\lvert a\rangle$, we can treat ρ as an operator in Equation 6.25 and take its inner product. ρ is expanded into its defined probabilistic states ψ_j in Equation 6.26. Finally, in Equation 6.27, we note that there are two inner products $\langle a \vert \psi_j\rangle$ and $\langle \psi_j \vert a\rangle$, which result in a squared value that must be positive. Therefore, the density matrix must be positive. A matrix is positive if $z^T M z \geq 0$ for any real-value vector z.

$$\langle a \vert \rho \vert a \rangle \;=\; \tag{6.25}$$

$$\sum_j p_j \langle a \vert \psi_j\rangle \langle \psi_j \vert a\rangle \;=\; \tag{6.26}$$

$$\sum_j p_j \lvert \langle a \vert \psi_j\rangle \rvert^2 \geq 0 \tag{6.27}$$

A brief summary of the density matrix is shown in Table 6.2.

In quantum information theory, it may be desirable to examine a subset of states from a larger set of states described by a density matrix. Reverting to our anthropomorphic qubits again, suppose that Alice, Bob, Charlie, and Victor each have a qubit with the total system described by a density matrix. If we would like to examine more closely the subsystem of qubits shared by just Alice and Bob we can use the *reduced density operator* and take the *partial trace*, which will return a density matrix expressing only Alice's and Bob's states. This will be explained in more detail next.

Here, we will use a simple example of only two states. Suppose A and B are described by density operator ρ^{AB}, where AB simply indicates that the density matrix describes both systems A and B. The reduced density operator is $\rho^A = tr_B(\rho^{AB})$ where tr_B is the partial trace over system B. Note carefully that the partial trace is taken over system B to return system A. The partial trace operator is defined in

Equation 6.28.

$$tr_B(|a_1\rangle \langle a_2| \otimes |b_1\rangle \langle b_2|) \equiv |a_1\rangle \langle a_2| \, tr(|b_1\rangle \langle b_2|) \quad (6.28)$$

Up to this point, some of the essential tools for analyzing quantum information have been introduced, but we have not yet introduced quantum information theory; that will take place soon. First, we should introduce some terminology. Now that we have worked with the density matrix, two kinds of quantum states become apparent. In the general case, there are mixed states; that is, quantum states whose values are not known exactly, but only probabilistically. In other words, there is some probability, less than one, assigned to each state. On the other hand, there are also pure states. A pure state is a state within a quantum system $|\psi\rangle$ that is known exactly. In general, $tr(\rho^2) \leq 1$. However, $tr(\rho^2) = 1$ if and only if a state is pure. Purification is a method for obtaining a pure state from a larger mixed state. Note that the term "purification" is also used for obtaining maximally entangled states for nanoscale quantum communication. For the general case of purification, given a mixed system A and its density matrix ρ^A, a reference system R can be introduced such that a pure state $|AR\rangle$ exists for the joint system AR and $\rho^A = tr_R(|AR\rangle \langle AR|)$. R is a mathematical convenience and has no direct physical significance.

6.2.1 Quantum information

Now that we have had an introduction to the essential tools of quantum information theory, we can begin to introduce the concept in relation to classical information theory. Let us assume that a quantum information source produces states $|\psi_j\rangle$ with probabilities p_j, similar to the states in the definition of a density matrix. We can follow the definition of information in classical information theory, namely, consider the entropy of states as symbols. However, there is a twist due the unique nature of quantum information and superposition. We can use the computational basis and consider only basis values as letters. For example, $|0\rangle$ occurs with probability $\frac{1}{2}$ and $|1\rangle$ with probability $\frac{1}{2}$. In this case, the entropy $H(1/2)$ is at a maximum of one for a binary-valued source.

Now consider what happens when states other than basis states are used. For example, state $|0\rangle$ with probability $1/2$, just as before, and state $\frac{|0\rangle + |1\rangle}{\sqrt{2}}$ with probability $1/2$. In this case the information entropy is no longer $H(1/2)$ or one. Instead, because the $1/\sqrt{2}$ of state $|0\rangle$ appears with probability 0.5, the classical information entropy is $H(\frac{1+1/\sqrt{2}}{2}) \approx 0.6$. In other words, it would appear, as this example illustrates, that more information can be compressed into a smaller space with quantum superposition than with classical bits. Let us consider a new form of entropy, a quantum version of entropy.

The quantum version of entropy is called von Neumann entropy and was developed as the quantum counterpart of Gibbs entropy and also used as a counterpart of Shannon entropy. Since the density matrix holds a set of quantum states and their probabilities, it is natural to consider expressing quantum information in the form of the density matrix. So we are looking for a form similar to classical Gibbs and Shannon entropy in which there is a probability multiplied by a logarithm of a probability. Consider the definition of the density matrix again, shown in Equation 6.29.

$$\rho = \sum_j p_j |\psi_j\rangle \langle \psi_j| \quad (6.29)$$

The values $p_j |\psi_j\rangle \langle \psi_j|$ are square matrices and their sum is a square matrix. Thus, going back to the spectral theorem, it is possible to use spectral decomposition to represent ρ by a sum of eigenvalues and projectors as shown in Equation 6.30.

$$\rho \equiv \sum_k \lambda_k |e_k\rangle \langle e_k| \quad (6.30)$$

Now consider the logarithm of a matrix. Note that it is possible for matrices to appear as exponents. Specifically, if A and B are matrices and $e^B = A$, then B is equivalent to $\log A$. The matrix exponent is defined in Equation 6.31.

$$e^B = \sum_{k=0}^{\infty} \frac{1}{k!} B^k \tag{6.31}$$

If a matrix is diagonalizable, then it is possible to determine the logarithm by spectral decomposition. If a matrix is written in terms of a diagonal matrix of eigenvalues, namely, as Λ in Equation 6.32, then the logarithm can be found by simply taking the logarithm of the diagonal eigenvalue matrix as shown in Equation 6.33. We know that $UU^{-1} = I$, so it is simply the logarithm of the matrix Λ.

$$B = U\Lambda U^{-1} \tag{6.32}$$

$$\log B = U \log \Lambda U^{-1} \tag{6.33}$$

Quantum entropy is defined as $S(\rho) \equiv -tr(\rho \log \rho)$. Applying the trace operator simplifies the result further. Since the trace is the sum of the diagonal values of a matrix, $tr(\log \rho)$ is simply the sum of the eigenvalues of ρ. So therefore, the von Neumann entropy is shown in Equation 6.34.

$$S(\rho) \equiv H(\lambda_k) = -\sum_k \lambda_k \log \lambda_k \tag{6.34}$$

Notice the pleasing relationship with information entropy H; namely, that $S(\rho) \equiv H(\lambda_k)$, where λ_k are the eigenvalues of ρ. This allows us to make some direct observations about von Neumann, or quantum, entropy. First, quantum entropy will be zero for pure states, that is, states with a probability of one. Second, quantum entropy can have a maximum of the number of orthogonal dimensions of the space. Finally, we briefly mention the relationship with the maximum classical data rate R from earlier in this chapter. Schumacher's noiseless channel coding theorem is the analog of Shannon's noiseless channel coding theorem and states that the minimal achievable value of the rate R is $S(\rho)$.

Quantum information theory has led to the development of quantum analogs to noise and error detection and correction. This is clearly a necessary component of quantum nanoscale networks as quantum state has a tendency to quickly become entangled with its environment. Encoded information will become distorted as it interacts with its environment without some form of error protection. This is a rich and well-explored area that would take us outside the scope of this book to explore in detail. Many of the techniques are close analogs of classical approaches using the density matrix and von Neumann entropy, so you have the tools necessary to understand this area. However, there is one rather unique topic that will end this discussion of quantum information theory at the nanoscale.

6.2.2   The limits of accessible information in a network

As networks scale down in size, the question arises as to the limits of accessible information [172, 173]. This is a situation unique to nanoscale networks. In classical human scale networks, as long as contact with the channel can be made, information is assumed to flow. There may be noise due to the manner in which contact is made, but the noise or attenuation is *external* to the flow of information. In nanoscale networks, a fundamental limit is reached on the ability of accessing the information itself; the inability of quantum state to be fully physically accessible.

Let us approach this by beginning with the analogous classical information theory approach for a communication channel, but using quantum states, in the form of density matrices, instead of classical bits.

As in classical information theory, if X is the information placed on a channel and Y is the information received, then the information gained by the receiver about X is the mutual information $I(X;Y)$. If the channel is perfect, that is, noise-free, then $I(X;Y) = H(X)$. Otherwise, one would like to improve the channel in such a manner as to maximize $I(X;Y)$ to bring it closer to $H(X)$. From a quantum state perspective, one must perform a measurement to obtain a result, thus, the goal is to perform the best possible measurement that maximizes the amount of mutual information. Unfortunately, one can expect that as networks scale down towards the nanoscale, physical constraints limit the ability to measure across a full set of the quantum state or states in which information has been encoded.

If two orthogonal quantum states are used to encode information with probability p and $1 - p$, then the information is the same as the classical information $H(p)$. However, if the quantum states are nonorthogonal, then it is not possible to extract the encoded information through measurement with perfect fidelity. Thus, the information is less than $H(p)$. Recall that the density matrix represents an ensemble of quantum states with various probabilities. The question is how well the states can be distinguished given their initial probabilities and degree of orthogonality.

The Holevo bound places a nice upper limit on the information in this case, as shown in Equation 6.35. Here, ρ is a mixed state that encodes X and the probabilities of each state are $p_0, p_1, \ldots, p_n$. Thus, $\rho = \sum_x p_x \rho_x$, where ρ_x are the density matrices of the individual states within the mixed state. Suppose the information is transmitted over a nanoscale network channel. The receiver measures the results with outcome Y, using the best possible choice of measurements that can be derived. Then the mutual information $I(X;Y)$ is limited by the right-hand side of Equation 6.35.

$$I(X;Y) \leq S(\rho) - \sum_x p_x S(\rho_x) \tag{6.35}$$

In fact, the right-hand side is called the Holevo χ quantity and is so widely used that it is often denoted simply as χ. Volume accessible information has to do with the constraint in which only a subset of the state can be physically accessed and measured, resulting in a loss of information.

6.3 A FEW WORDS ON SELF-ASSEMBLY AND SELF-ORGANIZING SYSTEMS

The need to understand, induce, and control self-assembly in order to create systems at scales too small to manipulate, such as nanoscale networks, should be clear. Moreover, there is a sense that nanoscale networks will operate most efficiently when their own design works with the natural environment, rather than against it. In other words, the nanoscale communication network should fit neatly within the environment within which it resides, ultimately self-assembling as part of the environment. However, the topic of self-assembly and self-organization has been, and continues to be, an unsolved and hotly debated topic. The very definition of order and how to measure it has become a deep and sometimes controversial topic. Self-organization in general is a large topic and would take us outside the scope of nanonetworking. In this section we briefly introduce the topic as it relates to information theory and nanoscale networks. We will discuss this subject in more detail in this chapter as it regards the architecture for nanoscale systems and nanonetworks; here we briefly introduce the relationship between information theory and self-organization. We do this from two perspectives: one that has a strong analog with Kolmogorov complexity and another that utilizes traditional information theoretic techniques, both of which we now discuss. It is important to keep in mind that these are only two of a myriad of definitions of self-organization.

As we have discussed in Chapters 2 and 3, nature provides a very diverse array of biological nanoscale networks and molecular information components, ranging from DNA to microtubules. Kolmogorov complexity was introduced earlier in this chapter as a measure of the complexity of information;

specifically, as the smallest program capable of representing a given target sequence. Suppose we make the analogy that the target sequence is a desired self-assembled device and the program is the set of subcomponents required to self-assemble in order to construct the device. Then the complexity is the smallest number of such distinct components that can self-assemble the target device. This approach has been studied in [174], where tiles represent program code; there are a set of different binding affinities, such that each side of a square tile can have a given affinity for another side of a different tile. The question is what are the minimal number of tiles with any given binding affinity on each side, such that a desired outcome, such as an $N \times N$ square of tiles will self-assemble. Note that this tile-fitting approach has much in common with Markov random fields, which also seeks to find the lowest energy, or most likely configuration, of a connected system. Also recall in Chapter 4 how DNA was utilized to assemble nanotubes into position as interconnects.

Another approach towards measuring self-organization that has been selected for discussion here is that described in [175]. This technique is called observer-based self-organization. It utilizes more traditional information theory rather than Kolmogorov complexity. Also, it explicitly incorporates the subjective observer in the analysis. This is important because there is the question regarding whether "organization" actually exists or is rather a subjective notion within the mind of an observer. The observer acts as a coordinate system within the self-organizing system; the self-organizing system is represented through the observer.

We present this version of self-organization through the language of information theory. An observer of a random variable X is a set of random variables $\{X_1, \ldots, X_n\}$ such that $H(X|X_1, \ldots, X_n)$ goes to zero. Simply put, the information in X and the observer $\{X_1, \ldots, X_n\}$ is equivalent. The *organization information* with respect to the observer is $I(X_1; \ldots; X_n)$, known as multiinformation or interaction information. Thus, a system's self-organization can be quantitatively measured; self-organization occurs as the organization information increases with respect to the observer over time. Note that the multiinformation simply measures the mutual information among all the observer variables. In other words, it measures to what extent they depend upon each other. For example, reaction-diffusion processes, in which a chemical reaction occurs while components of the reaction also diffuse, are fundamental to nanoscale molecular communication, which utilizes differences in concentrations as a communication channel. Conceptually, agents, or communication nodes, act as observers within such reaction-diffusion processes. An increase in the agents' multiinformation is indicative of an increasing degree of coordination within the pattern of a diffusing molecular concentration.

Self-assembly is often simulated within an agent-based framework, similar to the previously discussed self-organizing tiles. The concept in an agent-based approach is to determine the minimal code necessary to implement an individual agent (e.g., molecule) such that the agent interacts with its environment and other agents in such a manner as to self-assemble into a desired system. When approached from the standpoint of implementing code, or algorithms, the size of the program becomes a natural metric to explore. Thus, Kolmogorov complexity, and its derivatives, such as minimum description length and algorithmic information theory, become enticing avenues of exploration. The goal is to examine how the smallest descriptive representation of the system changes over time. The smaller the description, the simpler and more structured the system state. The system should fall into smaller descriptive states as it self-assembles, much like finding a minimal energy configuration or converging to the solution of a Markov random field. A self-assembling system is one that should be compressing its own description. A widely used approach for estimating the complexity is the normalized compression distance [176]. Any convenient compressor can be used to compress the description x, whose compressed form will be $C(x)$. Then the complexity distance between description x and description y is shown in Equation 6.36. If x and y are incompressible when concatenated, represented as $C(x, y)$, then the result is close to one. As x and y

become more compressible, $C(x, y)$ becomes smaller, and the result approaches zero.

$$NCD(x, y) = \frac{C(x, y) - \min(C(x), C(y))}{\max(C(x), C(y))} \tag{6.36}$$

6.3.1 Random nanotube networks, carbon nanotube radios, and information theory

All forms of nanoscale ad hoc networking assume some form of infrastructure, whether it is a microtubule network, a random network of carbon nanotubes, or an aqueous or fluid environment through which molecules can diffuse. The random carbon nanotube network assumes that nanotubes are dispersed or embedded within the environment at a sufficient density to achieve percolation. Once percolation is achieved, autonomous nanoscale devices, in contact with the nanotube network can communicate utilizing the random nanotube network as a shared access channel. Let us assume that on/off current is used. In this case, we would like to know the channel capacity of a shared random carbon nanotube network.

We derived the on/off current for an arbitrary random nanotube network in Chapter 4. We can assume a simple unipolar encoding of bits, with on-current representing one and off-current representing zero. The on/off current levels will vary with the location of the nanomachine and the random topology of the network. Thus, a threshold for a one value θ_1 will have to be chosen carefully. A collision may be detected as an excessive amount of current. This can be detected via a collision threshold value θ_c. Thus, we would like to know the mean and variance of the current values between every pair of nodes in the random nanotube network in order to set the appropriate thresholds. Recall from Chapter 4 that the resistance, and the corresponding on/off current, can be derived from Equation 6.37.

$$R_{\alpha\beta} = \sum_{i=2}^{N} \frac{1}{\lambda_i} |\Psi_{i\alpha} - \Psi_{i\beta}|^2 \tag{6.37}$$

As discussed in [3], the distribution of the eigenvalues and components of the eigenvectors of the graph Laplacian determine $R_{\alpha\beta}$ and thus the variation in on/off current between nodes in the CNT network.

The single carbon nanotube radio follows analysis that is similar to today's techniques for wireless information theory, so the single carbon nanotube radio approach is a somewhat less interesting case to discuss and can be found in the myriad of books on the subject of human-scale wireless networking.

6.4 MOLECULAR COMMUNICATION THEORY

This section applies classical information theory to nanoscale communication in which information is encoded as molecular concentration. We saw the natural biological signaling capability of calcium and pheromone concentrations in Chapter 3. In this section, we examine this form of nanoscale communication as an engineered protocol from an information theoretic point of view. This allows an analysis of the capacity of such a channel. Four types of channels are examined: single channel, multiple-access channel, broadcast channel, and a relay channel. A single channel is simply a sender and receiver using molecular concentration levels to transmit information in half-duplex. A multiple access channel adds the notion of a shared communication medium; namely, multiple transmitters can send to a receiver. A broadcast channel is one in which a single transmitter can simultaneously send information to multiple receivers. Finally, a relay channel is one in which a transmitter uses an intermediate node to relay information to a receiving node. These modes of communication provide the link capability with which to construct a network. We begin with a brief review of information theory.

The starting point for information theory usually begins with the Shannon information associated with an event that occurs with probability p [177]. "Self-information" contained in a probabilistic event depends on the probability of that event; the smaller its probability, the less likely it is to occur and the larger the "surprise" or self-information associated with receiving the information that the event occurred. This is captured in Equation 6.38. Depending upon the base of the logarithm, the resulting self-information can be in bits, nats, or hartleys. A bit is base two, a nat is the natural logarithm, and the hartley is base ten.

$$I(p) = -\ln p \tag{6.38}$$

The information represents the expected value of the information associated with an event that is modeled by a random variable X with a given distribution as shown in Equation 6.39, where x models the occurrence of symbol set A. The entropy can be considered the uncertainty of an outcome measured in number of bits.

$$H(X) = -\sum_{x \in A} p_i \ln p_i \tag{6.39}$$

The conditional probability, $X|Y$ is shown in Equation 6.40, where X models A and Y models B.

$$
\begin{aligned}
H(X|Y) &= -\sum_{x \in A} p(x|y) \sum_{x \in B} \ln p(x|y) \\
&= -\sum_{x \in A} \sum_{x \in B} p(x,y) \ln p(x|y)
\end{aligned}
\tag{6.40}
$$

The mutual information $I(X;Y)$ is the number of bits of information common to both X and Y. Thus, $I(X;X)$ should be the information in X, which is we previously defined to be $H(X)$ in Equation 6.39. More generally, mutual information is shown in Equation 6.41.

$$I(X;Y) = H(X) - H(X|Y) = H(Y) - H(Y|X) \tag{6.41}$$

If X is the input into a channel and Y is the output, then the mutual information $I(X;Y)$ is the uncertainty that X was sent after observing Y as shown in Equation 6.42. If there are no errors in the channel then we have $I(X, X)$ and the capacity is simply the entropy of X. If Y differs from X, then the mutual information, and thus the channel capacity, will decrease.

$$C = \max_{p_i} I(X;Y) \tag{6.42}$$

6.4.1   Brownian motion and order statistics

There are many different ways of encoding information in molecular communication. Among the many possibilities include (1) the order in which *individual* molecules arrive at a receiver, (2) the very nature of a given molecule, and (3) the concentration of molecules. Each of these possibilities can be modulated in order to convey information. The time at which molecules are released can be varied in order to encode information in the arrival times. The structure of a single type of molecule can change in order to convey information. For example the structure of a fluorocarbon can change in order to hold information, which will be described in more detail in Section 8.1.3, and large numbers of molecules can be released at given times in order to change the concentration at the receiver.

A channel model for estimating the information rate for information encoded in individual molecules that propagates across a medium via Brownian motion has been proposed [178, 179]. Several assumptions are made in order to make the analysis tractable. The environment is assumed to be one-dimensional from $(-\infty, d]$, with the transmitter located at the origin and the receiver located at distance d. Distinct molecules of different types can be transmitted and detected. A pattern, determined by the order of arrival of molecules of distinct types, encodes the information. Thus, molecules are assumed to be released from the transmitter in correct order to encode the desired information. However, due to the randomness of Brownian motion, that order may be lost, causing an error. To transmit a message of size $|M|$ bits with T molecules, the rate R is shown in Equation 6.43.

$$R = \frac{\log_2 |M|}{T} \tag{6.43}$$

Since it is assumed that one molecule is transmitted every time unit, R is the bits transmitted per unit time. Unfortunately, due to the random nature of Brownian motion, some molecules may be lost or the order of arrival may change. The model for Brownian motion is assumed to be approximated by a Wiener process, which has a closed form probability density function given by Equation 6.44 in which t_i is the transmission time for the molecule, d is the transmission distance, and σ^2 is determined by the physical nature of the environment.

$$f(t_i) = \frac{d}{\sqrt{2\Pi\sigma^2 t_i^3}} e^{\left(\frac{-d^2}{2\sigma^2 t_i}\right)} \tag{6.44}$$

An approximate model has been developed using Equation 6.44 as the probability density function and assuming independent molecules. If X represents the set of molecules each released at time $t = \{1, 2, 3, \ldots\}$ then using this model, the achievable information rate is proportional to $\log_2(|X|)$. As a specific example, a 1000-bit message can be conveyed by carefully emitting roughly 3000 molecules, which requires little energy or mass. If $d = \sigma^2 = 1$, within 36,000 seconds the 3000 molecules will arrive at the receiver. However, for nanoscale transmission, d will likely be much smaller, which will increase the rate significantly.

6.4.2 Concentration encoding

Now we can move into the realm of molecular reactions and modulation of molecular concentration [180]. Recall the molecular operation of ligand-receptor binding from previous chapters. Suppose that there are receptors R with concentration N. Also assume that ligands A bind with receptors R to create complexes C as shown in Equation 6.45. The unbinding, or release, reaction is shown in Equation 6.46. The important symbols here are the reaction rates k_1 and the release rates k_{-1}. The release rate is in effect, the "washing off" of the previous signal that had bound to receptors so that the receptors are free to operate with respect to the next signal.

$$A + R \xrightarrow{k_1} C \tag{6.45}$$

$$A + R \xleftarrow{k_{-1}} C \tag{6.46}$$

In a traditional human-scale electronic receiver, zero- and one-bit values are encoded as voltage levels. In the simplest form of unipolar line encoding, a one bit is detected when the voltage exceeds some threshold and zero otherwise. In the form of nanoscale media envisioned in this section, molecules are diffusing in a fluid medium. Molecular concentration becomes analogous to electrical potential. Molecules of type A are emitted by the transmitter TN during the interval t_H. No molecules are emitted to indicate a zero. The two bits in this medium are thus A and 0. If the receiver RN senses a concentration of A that is greater than some threshold S, then RN determines that A was transmitted, otherwise RN infers that zero

Table 6.3

Molecular Concentration Encoding Symbols

Name	Symbol	Notes
Transmitter	TN	
Signal	A	$A \rightarrow 1$, $empty \rightarrow 0$
Transmission time slot	t_H	Duration of emission
Amplitude	L	L_{ex} (transmitted) $L(t)$ (received)
Concentration	$C(t)$	Bound receptor complex
Threshold	S	$> S \rightarrow 1, < S \rightarrow 0$
Prob of trans	P_A	Prob of emission during t_H
Receiver	RN	
Receptors	R	On surface of receiver
Receptor concentration	N	Number of receptors
Bind rate	k_1	Binding to receptor
Release rate	K_{-1}	Releasing from receptor
Temperature	T	
Distance	α	Between TN and RN
Molecules delivered	NA	During transmission t_H
Previous complexes	NP	From last transmission interval
Total molecules	S_A	During current transmission + from last one
Successful trans of A	p_1	
Successful trans of 0	p_2	
Delivery distribution	μ_A (mean) σ_A^2 (variance)	Assumed Gaussian

was transmitted. The many symbols used in the analysis are summarized in Table 6.3; the symbols will be explained further as they are used.

Let TN be a transmitting nanomachine and RN be a receiving nanomachine. Suppose that TN releases molecules A with a square wave of amplitude L_{ex}. Further, assume that the transmission of an A occurs with probability P_A as shown in Equation 6.47. The square wave duration is t_H seconds.

$$L(t) = \begin{cases} L_{ex} & \text{with probability} P_A (j \leq t \leq j+1) t_H \\ 0 & \text{with probability} (1 - P_A) \text{ otherwise} \end{cases} \tag{6.47}$$

6.4.3 A single nanoscale molecular channel

The channel operates by the concentration of bound receptors $C(t)$, which is assumed to rise exponentially as signal molecules are released and then decay after time t_H, when the signal molecules are no longer released, as shown in Equation 6.48.

$$C(t) = \begin{cases} C_\infty (1 - e^{-t(k_{-1} + k_1 L_{ex})}) & 0 \leq t \leq t_H \\ C_{t_0} e^{-t k_{-1}(t - t_H)} & t \geq t_H \end{cases} \tag{6.48}$$

Note that, in the second line of the above equation, C_{t_0} is the concentration of bound complexes that existed at time zero. The concentration decreases exponentially over time. The concentration of receptors on the surface of the receiver is N. There are $k_1 L_{ex}$ ligands binding while ligands are being released at the rate k_{-1}. C_∞ is the steady state concentration, which is the rate of binding over the total rate of activity, shown in Equation 6.49.

$$C_\infty = \frac{k_1 L_{ex} N}{k_{-1} + k_1 L_{ex}} \tag{6.49}$$

Since the information is being transmitted in this particular scheme by modulating the molecular concentration, it is important to understand the dynamics of this approach. The binding rate of the signaling ligand to the receptor at the receiver is greatly influenced by the process of molecular diffusion from the transmitter to the receiver, which we have discussed in earlier chapters, particularly Chapters 2 and 3. In particular, the diffusion coefficient, temperature, and distance between the transmitter and receiver all influence the number and rate of molecules that bind at the receiver. In order to simplify the analysis and focus more on the information theoretic aspects of the analysis, many of these details are buried within the binding and release rate constants k_1 and k_{-1}. The binding rate is assumed to be proportional to twice the temperature $k_1 \propto 2T$ and inversely proportional to the distance between the transmitter and receiver $k_1 \propto 1/\alpha$.

The ligand-receptor bond lifetime is related to discussions all the way back in Chapter 2, where we view cellular activity from the standpoint of mechanical forces. Thus, the binding, or adhesion, of a ligand to a receptor can be viewed as a stochastic process that is under the control of applied forces. The release rate as shown in Equation 6.50 is derived from [181] with the following parameters. k_{-1}^0 is the zero-force release rate, that is the rate of release from the binding in the absence of applied force. α has already been mentioned as the distance between the transmitter and receiver. k_B is Boltzmann's constant and T is the absolute temperature. f is an applied force per ligand-receptor bond, which is related to the environmental factors such as energy of the emitted molecules, distance between transmitter and receiver.

$$k_{-1} = k_{-1}^0 e^{\frac{\alpha f}{k_B T}} \tag{6.50}$$

TN transmits A with probability P_A and zero with probability $1 - P_A$. For the channel to work without error, if TN transmits A, then the receiver must detect S or more molecules during the time period t_H in order to correctly detect that A was sent. If TN transmits a zero, then the RN must detect less than S molecules during t_H. Recall from Chapter 1 that the number of molecules and the concentration of molecules are directly related by Avogadro's number. The concentration, in moles, can be converted to the number of molecules by simply multiplying be 6.023×10^{23}. Thus, we can speak interchangeably of "number of molecules" and "concentration." In order to determine the efficiency of this channel, it is necessary to precisely determine how many molecules are bound at the receiver during each potential transmission interval t_H.

Let NA be the number of molecules delivered by the communication channel over the transmission time interval t_H. We define delivery to mean the physical transport and binding of the molecules with the receptor. This can be determined very simply by integrating over the concentration from Equation 6.48 over the time interval of transmission as shown in Equation 6.51. These are the number of new molecules that arrive from the transmitter and bind at the receiver's receptors. This does not include any signal molecules that have already been bound from a previous transmission. We need to account for this next.

$$NA = \int_0^{t_H} \frac{k_1 L_{ex} N}{k_{-1} + k_1 L_{ex} (1 - e^{-t(k_{-1} + k_1 L_{ex})})} dt \tag{6.51}$$

Let NP be the number of bound complexes from the previous transmission. Since the binding release is rapid compared to the transmission time, only the molecules from the last bit transmission would remain. The probability of the last transmission is P_A and the number of molecules that would have been sent we already determined is NA. Thus, all that is required is a multiplier that captures the rate of release as shown in Equation 6.52. The rate of release from binding is an exponential decay from Equation 6.48.

$$NP = P_A NA \int_0^{t_H} e^{-kt} dt \qquad (6.52)$$

The total expected number of delivered molecules is shown in Equation 6.53. This includes both the number of molecules transmitted during the current interval t_H and remaining molecules from the last transmission.

$$E[S_A] = NA + NP \qquad (6.53)$$

Recall that for A to be successfully transmitted, it must meet or exceed the threshold S at the receiver. We let p_1 be the probability of the successful transmission of A. The Markov Inequality can be used to bound the probability that $S_A \geq S$ as shown in Equation 6.54.

$$p_1(S_A \geq S) \leq \frac{E[S_A]}{S} \qquad (6.54)$$

Thus, the probability of success of sending A, p_1, reaches a maximum at $\frac{E[S_A]}{S}$. It also follows that there is a probability of $1 - p_1$ of receiving a zero value in error.

A simplifying assumption is that the number of molecules delivered S_A is normally distributed with mean $E[S_A]$, which we have already determined and variance σ_A^2. The distribution is clearly nonnegative since S_A cannot be negative. So the mean is $\mu_A = E[S_A]$ and the standard deviation is $E[S_A]/3$. Thus, the probability of success is the likelihood of exceeding S, which is shown by the integral of the probability density function shown in Equation 6.55.

$$p_1(S_A \geq S) = \int_S^{\infty} \frac{1}{2\pi\sigma_A} e^{-\frac{(x-\mu_A)^2}{\sigma_A^2}} dx \qquad (6.55)$$

Now consider the transmission of a bit value of zero. In this case, no molecules are emitted during time t_H and the only molecules remaining are ones from the last transmission of an A. In this case, the expected number of molecules transmitted for a zero bit value are shown in Equation 6.56, where the value of NP was previously derived.

$$E[S_0] = NP \qquad (6.56)$$

The probability of a successful transmission of a zero value p_2 can be approximated by the assumption of a normal distribution for S_0 in a manner similar to S_A, as shown in Equation 6.57. Here, $\mu_0 = E[S_0]$ and the standard deviation is $E[S_0]/3$.

$$p_2(S_0 \leq S) = \int_0^S \frac{1}{\sigma_0 2\pi} e^{-\frac{(x-\mu_0)^2}{\sigma_0^2}} dx \qquad (6.57)$$

Thus, again using the Markov inequality, the transmission of a zero bit value is successful with probability $p_2 = S/E[S_0]$ as shown in Equation 6.58.

$$p_2(S_0 \leq S) \leq \frac{S}{E[S_0]} \qquad (6.58)$$

Consider a matrix of the possible channel events that can occur. Let TN transmit X and RN receive Y. A desirable channel would deliver exactly what was transmitted, so we would like the mutual information between X and Y to be maximized. The channel transition matrix shows the probabilities of all the possible events in Equation 6.59.

$$\left(\begin{array}{cc} Pr(A \rightarrow A) & Pr(0 \rightarrow A) \\ Pr(A \rightarrow 0) & Pr(0 \rightarrow 0) \end{array} \right) \tag{6.59}$$

The possible events can be represented in a channel transmission matrix shown in Equation 6.60.

$$P(Y|X) = \left(\begin{array}{cc} p_1 P_A & (1 - p_2)(1 - P_A) \\ (1 - P_1)P_A & p_2(1 - P_A) \end{array} \right) \tag{6.60}$$

The mutual information is the difference between the joint entropy and the sum of the conditional entropies of $X|Y$ and $Y|X$, namely, $I(X;Y) = H(X,Y) - (H(X|Y) + H(Y|X))$. Therefore, the mutual information can be obtained from the channel transition matrix as shown in Equation 6.61.

$$\begin{aligned} I(X;Y) &= (H(p_1 P_A + (1 - p_2)(1 - P_A))) - \\ &\quad (P_A H(p_1) + (1 - P_A)H(p_2)))) \end{aligned} \tag{6.61}$$

The capacity of a single nanoscale molecular channel is the maximum mutual information $(\max I(X;Y))$ over the channel. This derivation is for a single point-to-point channel in which there is only one transmitter and one receiver. The next section generalizes this case to a multiple access channel. A multiple access channel is one in which the channel can be shared among transmitters and receivers while minimizing the impact of collisions during simultaneous transmissions.

6.4.4 A multiple-access nanoscale molecular channel

Following the analysis in [182], consider the case of multiple transmitters from the set $\{TN_1, \ldots, TN_n\}$, which transmit to a single receiver RN. All the transmitter specific parameters will have an additional index i to indicate which transmitter's parameters are being referenced. For example, the transmission probability is P_{Ai} and the binding rates are k_1^i. Note that the release rates are assumed to be the same k_{-1} and the transmission concentrations L_{ex} are assumed to be the same for each transmitter as well. Since the analysis is intended to focus on the information theoretic aspects of the channel only, other architectural components are assumed to be implemented, such as addressing.

We can leverage the previous analysis regarding the expected number of molecules received when A is transmitted as shown in Equation 6.62 and the expected number of molecules received when 0 is transmitted in Equation 6.63.

$$E[S_A^i] = NA + P_{Ai}NA \int_0^{t_H} e^{-k_{-1}t}dt \tag{6.62}$$

$$E[S_0^i] = P_{Ai}NA \int_0^{t_H} e^{-k_{-1}t}dt \tag{6.63}$$

The number of molecules received NA during the transmission interval t_H comes from Equation 6.51, but with k_1^i used instead of k_1 for each respective transmitter. The assumption of a normally distributed number of total received signal molecules S_A and S_0 is also again assumed, using S_A^i and S_0^i for values sent from each respective transmitter. The analysis follows the same pattern as for the single-channel case, namely, there is $\mu_{Ai} = E[S_A^i]$ and $\sigma_{Ai} = E[S_0^i]/3$ for the normal distribution from each respective transmitter.

Now comes the difference due to the underlying shared communication medium. If every transmitter emitted the same number (concentration) of molecules as though it were a single-channel, the total number of molecules in transmission would be higher than expected. Thus, the system must be adjusted to handle the larger concentration caused by multiple transmitters. One approach might be to increase the receiver threshold to detect an A value. Another approach is to limit the transmission rate or amplitude in order to reduce the additional concentration.

The approach limits the signal concentration by introducing a constant K, which is a reducing factor. Thus, the concentration transmitted by TN_i ($j \neq i$) is shown in Equation 6.64 and Equation 6.65. M_A^i and M_0^i are the concentrations to be actually received by the receiver.

$$M_A^i = K S_A^i \tag{6.64}$$

$$M_0^i = K S_0^i \tag{6.65}$$

We are allowing each transmitter to send molecules to the same receiver, so we would like to know how much contribution comes from the other transmitters. The expected contribution from transmitters other than TN_i is shown in Equation 6.66.

$$\sum_{j \neq i}^n \left(P_{A_j} E[S_A^j] + (1 - P_{A_j} E[S_0^j]) \right) \tag{6.66}$$

Recall that N is the number of receptors at the receiver. The proportion of N taken by the other transmitters is shown in Equation 6.67. This is the amount by which the concentration must be reduced in order for the current transmitter's signal concentration to be meaningful.

$$K = \frac{N}{N + \sum_{j \neq i}^n \left(P_{A_j} E[S_A^j] + (1 - P_{A_j} E[S_0^j]) \right)} \tag{6.67}$$

We can again fall back on the assumption of normality. Since the S_A^i is assumed normal and K is constant, then the reduced concentrations M_A^i and M_0^i are also normal with distributions $N(K\mu_{Ai}, (K\sigma_{Ai})^2)$ and $N(K\mu_{0i}, (K\sigma_{0i})^2)$.

The probability of a successful transmission can be found by integrating the normal probability distribution for the number of molecules received. The probability of successful reception at the receiver for transmission of A by transmitter i is shown in Equation 6.68.

$$p_{1i}(M_A^t \geq S) = \int_S^\infty \frac{1}{K 2\pi\sigma_A} e^{-\frac{(x - K\mu_{A_i})^2}{(K\sigma_{A_i})^2}} dx \tag{6.68}$$

Similarly, the probability of successful reception at the receiver for transmission of 0 by transmitter i is shown in Equation 6.69.

$$p_{2i}(M_A^t \leq S) = \int_0^S \frac{1}{K 2\pi\sigma_A} e^{-\frac{(x - K\mu_{A_i})^2}{(K\sigma_{A_i})^2}} dx \tag{6.69}$$

The capacity of the nanoscale molecular multiple-access channel can be found following the same analysis as for the single-channel case. Namely, the mutual information between each transmitter and receiver $I^i(X;Y)$ can be computed using the results just derived for P_{Ai}, p_{1i}, and p_{2i}, with the maximum mutual information shown in Equation 6.70.

$$MC_i = \max I^i(X;Y) \tag{6.70}$$

The total capacity of the multiple-access channel in this case is the sum of the channel capacities as shown in Equation 6.71.

$$MC = \max \left(\sum_{i=1}^{n} I^i(X;Y) \right) \tag{6.71}$$

Thus, we have an analysis for the case of a nanoscale molecular multiple-access channel. There has been an important assumption made in this analysis, namely, that there is no contention among the transmitters. Transmission occurs within the time slot t_H and simultaneous transmissions of an A do not occur. This allowed the single-channel model equations for NA and NP to be directly utilized in the analysis. The next chapter, which discusses higher-level architectural considerations, examines the issue of contention in the finite automata model.

6.4.5 A broadcast nanoscale molecular channel

Now we reverse the situation from the previous section and examine the case of a single transmitter sending to multiple receivers. Similar architectural assumptions are made, namely, that addressing has been implemented in some manner. Also, as in the previous case, we assume that diffusion of molecules occurs uniformly in all directions. Each receiver is assumed to sense the signal independently and without interference in order to make the analysis tractable. If the concentration of receptors on a receiver were on the same order as the concentration of molecules, one could imagine that this assumption could be violated. A receiver positioned behind another receiver, with respect to the gradient of signal flow, may not have the same likelihood of receiving a signal because the blocking receiver may bind a significant portion of the signal concentration. However, this is a complicating factor, which we ignore in this analysis.

There are now a set of receivers $(RN_1, \ldots, RN_n)$ and a single transmitter TN. The transmission to each receiver follows the single-channel analysis summarized by the channel transition matrix in Equation 6.61. Each transmitter to receiver channel capacity in the broadcast is shown in Equation 6.72.

$$BC_i = \max \left(I^i(X;Y) \right) \tag{6.72}$$

Thus, the total nanoscale molecular broadcast channel capacity is shown in Equation 6.73.

$$BC = \sum_{i=1}^{n} \max \left(I^i(X;Y) \right) \tag{6.73}$$

Next, we examine a cooperative technique for information transmission in the nanoscale molecular environment known as the relay channel.

6.4.6 A relay nanoscale molecular channel

Relay channels are a form of cooperative communication, often considered in the traditional human-scale radio frequency domain, in which third-party nodes cooperatively aid a transmitter in communicating with a receiver. From an information theory standpoint, a relay channel is a probability model of the communication between a sender and a receiver in which the transmitter is aided by one or more intermediate relay nodes. It can be thought of as a combination of the broadcast channel, which transfers information from the sender to the relays and receiver, and a multiple-access channel, which also transfers information from the sender and is relayed to the receiver. The same technique can be used in the nanoscale molecular network [182].

In the nanoscale network, the relay node H is assumed to be able to simultaneously receive and transmit information. We assume that addressing is implemented as in previous analyses. A new assumption for the channel is that the relay node H is able to have knowledge of what TN is going to transmit. Clearly, this is a big assumption. The role of the relay H is to transmit the same signal value in time slot t_H with TN in order to increase the likelihood of reception at the receiver.

For this analysis, we can leverage and combine the analyses already derived from the broadcast and multiple-access channels. In the broadcast channel TN transmits to H and RN. The channel capacities BC_h and BC_{rn} were derived previously. In the multiple-access channel, TN and H transmit to RN. The multiple-access capacities from H to RN, which we will call MC_h and TN to RN, which we will call MC_{rn}, have also already been derived from the previous section on the nanoscale multiple-access channel. Figure 6.2 shows the broadcast and multiple-access channels.

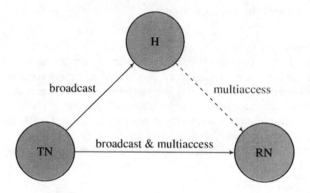

Figure 6.2   The molecular relay channel.

The maximum flow-minimum cut theorem states the maximum flow through a network is the minimum cut. A cut set is the set of edges that if removed, would completely disconnect a graph. In other words, the set of edges that completely divides the network and that also have the minimum capacity limits the maximum flow through the network. Edge (TN, H) and (TN, RN) are cut sets for the broadcast channel. Edges (H, RN) and (TN, RN) are cut sets for the multiple-access channel. Therefore, by the max-flow min-cut theorem, the maximum channel capacity is simply the minimum capacity of the cut sets for both channels. As shown in Equation 6.74, we take the minimum of the maximum broadcast capacity from TN to either H or RN and $\bar{M}C$, which is the multiple-access capacity from TN and H to RN.

$$RC = \min(\max(BC_h, BC_{rn}), \bar{M}C) \tag{6.74}$$

Now we compute $\bar{M}C$. Although TN and H send the same signal, they both bind to the N receptors at the receiver as discussed in the multiple-access analysis. TN is responsible for emitting $E[S_A^{TN}]$ molecules and H is responsible for emitting $E[S_A^H]$ molecules. As discussed previously, the concentration must be reduced on a multiple-access channel by a factor of K as shown Equations 6.64 and 6.65.

The reduced values are shown in Equation 6.75 and Equation 6.76.

$$M_A^{TN} = K_{ATN} S_A^{TN} \tag{6.75}$$

$$M_A^{H} = K_{ATN} S_A^{H} \tag{6.76}$$

The reduction factors are shown in Equation 6.77 and Equation 6.78. Notice that this differs from the general case of multiple-access as shown previously in Equation 6.67 since there are only two nodes assumed to be contending for the channel in this analysis.

$$K_{ATN} = \frac{N}{N + E[S_A^H]} \tag{6.77}$$

$$K_{AH} = \frac{N}{N + E[S_A^{TN}]} \tag{6.78}$$

Now we can compute $\bar{M}C$ as shown in Equation 6.79 where $I^{mc}(X;Y)$ comes from P_A, Equation 6.55, and Equation 6.58. The mean and variance of the random variable $M_A^{TN} + M_A^H$ are $K_{ATN} + K_{AH}\mu_{Ai}$ and $(K_{ATN} + K_{AH}\sigma_{Ai})^2$. A similar approach is done to compute the results for bit zero.

$$\bar{M}C = \max I^{mc}(X;Y) \tag{6.79}$$

All the components have been computed for Equation 6.74 which yields the nanoscale molecular relay channel capacity.

6.5 SUMMARY

We have reviewed classical information theory, both from the traditional view of Claude Shannon, and an alternative view from Andre Kolmogorov. Then we provided a brief tutorial on quantum information theory. The section on quantum information theory ends with a brief discussion of what happens when access to information reaches the extreme limits, due to small size constraints. Self-assembly and self-organization were examined from the perspective of Kolmogorov, namely smallest program to generate a tile sequence, and Shannon, namely, observer-based self-organization and interaction information. Then we assume a molecular transport mechanism and look at information theory from the perspective of a nanoscale molecular channel based upon diffusion through a fluid medium. Information theory has provided analytical tools for estimating self-organization and channel capacity at the nanoscale.

How do these information theoretic requirements and characteristics drive actual implementation? What will a nanoscale or molecular communication network look like? What communication processing components need to exist and where will they reside? In the next chapter, we will look at higher-level, but more tangible, architectural considerations for nanoscale networks.

6.6 EXERCISES

Exercise 42   Information Entropy

Assume a binary transmission channel based upon changes in concentration of a particular molecule (low/high concentration). Assume the probabilities of transmission (p_{low} and p_{high}) and incorrect detection ($p_{low-error}$ and $p_{high-error}$) can be determined for low and high concentration emissions.

1. In general, if knowledge of states becomes less certain as we scale down in size, what happens to the information entropy in such a system from our perspective (explain using the basic equation for information entropy)?

2. What is the relationship between information entropy and mutual information?

3. How is mutual information used to determine channel capacity?

4. How can one use the probabilities above to determine the molecular channel capacity?

Exercise 43   Quantum Density Matrix

Suppose $|\psi\rangle = |0\rangle$ with probability p and $|\psi\rangle = |1\rangle$ with probability $1 - p$.

1. What is the probability of obtaining an outcome of one?

2. If the system is in the state $|a\rangle |b\rangle$, what is ρ_A? (Hint: recall the reduced density operator.)

Exercise 44   Quantum Entropy

Consider the density matrix shown in Equation 6.80.

$$\rho = \begin{bmatrix} 1 & 0 \\ 0 & 0 \end{bmatrix} \tag{6.80}$$

1. What is the von Neumann entropy of ρ in Equation 6.80?

2. What is the entropy of a completely mixed density matrix in d dimensional space? (Hint: state the answer in terms of d.)

Exercise 45   Kolmogorov Complexity

Kolmogorov complexity has its own version of mutual information known as absolute mutual information defined as shown in Equation 6.81.

$$I_K(X;Y) = K(X) - K(X|Y) \tag{6.81}$$

1. Could this be utilized to better determine the capacity of an active network channel? How?

Exercise 46   Self-Organization

Consider the definition of interaction information in Equation 6.82 and conditional mutual information in Equation 6.83.

$$I(X;Y;Z) = I(X;Y|Z) - I(X;Y) \tag{6.82}$$

$$I(X;Y|Z) = H(X|Z) + H(Y|Z) - H(X,Y|Z) \tag{6.83}$$

1. How does interaction information increase (show in terms of information entropy)?

Exercise 47   Single Molecular Channel

Consider the discussion of a single nanoscale molecular channel.

1. How is the detection threshold S determined?

2. What is the impact of distance between the transmitter and receiver on the channel capacity?

3. What is the impact of temperature on the channel capacity?

4. How rapidly can concentration increase and decrease take place? How does this limit the bandwidth?

Exercise 48   Multiple-Access Molecular Channel

Consider the discussion of a multiple-access nanoscale molecular channel.

1. What is the primary difference in the analysis between a single nanoscale molecular channel and a multiple-access relay channel?

2. What happens if a receiving node absorbs a significant amount of the signal molecule and it lies in the path of another receiver?

Exercise 49   Broadcast Molecular Channel

Consider the discussion of a broadcast nanoscale molecular channel.

1. How is the broadcast capacity computed?

2. A nanoscale molecular broadcast signal as proposed in this chapter would need to have a strong enough concentration to reach the farthest node. What impact could this have on nearby nodes?

3. Suggest an alternative approach to broadcast. How does it compare and contrast with the approach discussed here?

Exercise 50   Relay Molecular Channel

Consider the discussion of the relay nanoscale molecular channel.

1. In general, what is a relay channel?

2. How would the nanoscale molecular relay channel described in this chapter need to be modified if the multiple-access portion was shared by n other relays?

Exercise 51   von Neumann entropy

1. Prove that the von Neumann entropy satisfies the inequality $S\left(\sum_j p_j |\psi_j\rangle \langle\psi_j|\right) \leq H(p_j)$.

2. Prove that $S\left(\sum_j p_j \rho_j\right) \leq H(p_j) + \sum_j p_j S(\rho_j)$.

3. von Neumann entropy exhibits subadditivity: $S(\rho_{AB}) \leq S(\rho_A) + S(\rho_B)$. Show that $S(\rho_A \otimes \rho_B) = S(\rho_A) + S(\rho_B)$.

Chapter 7

Architectural Questions

Fools ignore complexity. Pragmatists suffer it. Some can avoid it. Geniuses remove it.

Alan Perlis

This chapter is entitled "Architectural Questions" because as this book is being written, there is no clearly defined communication network architecture for nanoscale and molecular networks. Having read through the prior chapters of this book, the reader should now have a reasonable understanding of the nature of nanoscale channels, ranging from molecular motors and oscillations in chemical concentrations, random interconnected nanostructures and the single carbon nanotube radio, to purely quantum channels. Many elements for a nanoscale network architecture can be found within network architectures in use today. In Chapter 1 we looked at sensor networks and the relationship between sensor networks and nanoscale networks. There should be lessons learned from sensor network architectures that may be applicable to nanoscale networks. The sensor network architecture appears to primarily react to the resource-constrained nature of network nodes and the unique routing situations in sensor networks. In Chapters 2 and 3 we looked at biologically related nanoscale networks. This involved the notion of diffusion-based molecular channels that can be characterized by a variety of models, including Brownian motion, the Langevin equation, and ion concentration waves. This implies an architecture that is optimized to handle information that propagates in such a diffusive manner. In Chapter 4 we looked at carbon nanotube-based nanoscale networks. The architectures in that chapter ranged from chip interconnects and active networks to random nanotube interconnects, to single carbon nanotube radios. The implication is that some architectural features from chip interconnects may be useful in ad hoc nanoscale networks. In Chapter 5 we looked at quantum nanoscale networks. The quantum network architecture is primarily concerned with minimizing the impact of decoherence, whether for the entangled qubits required for teleportation or for the operation of simple swap channels. Finally, the last chapter looked at the information theoretic aspects of nanoscale networking, where we looked briefly at self-organization as well as the channel capacity of the nanoscale approaches from the previous chapters. If aspects of self-assembly and self-organization are required, that will also have a significant impact on the resulting architecture for nanoscale networks.

Thus, in this chapter, we will look at nanoscale architectural questions from several different perspectives, including the need for self-assembly at the nanoscale, architectural concepts that can be drawn from other network architectures, particularly delay-tolerant networks, the quantum network architecture, and the active network architectural framework. First, step back for a minute to consider the scale that we are talking about. Figure 7.1 shows the scales of the components that we have been discussing. It is assumed that a nanomachine will be at the size of a bacterium. The molecular communication structures are at the size of proteins or smaller. Clearly, the nanowire and nanotube are smaller than any structure currently constructed by lithographic techniques.

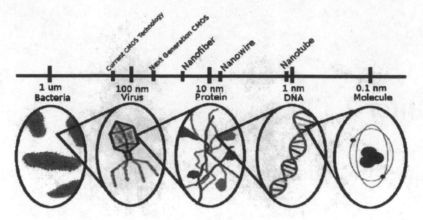

Figure 7.1   Relationship among the small-scale components discussed in previous chapters.

7.1   INTRODUCTION

We began this book with a look at the smallest extant wireless sensor networks as a step towards nanoscale networks. So it is logical to look at wireless sensor network architectures and see how they would extrapolate to the nanoscale. We encountered the benefits of the active network architecture used in a nanoscale processor interconnect system that was ideal for handling the dynamics of stochastic node and link failures with constrained devices. Finally, there is much to be learned from research into handling the stochastic nature of nanoscale chip interconnects. In this chapter, we will introduce in more detail a network architecture alluded to back in Chapter 1 when discussing the vastness of space, namely the interplanetary Internet and delay-tolerant networking. It may seem counterintuitive that such an extreme opposite in scale would have very much in common with a network architecture on the nanoscale. However, as we will see, the delay-tolerant networking architecture has inadvertently solved some of the most interesting nanoscale networking architectural issues. To understand, at least intuitively, why this is the case, step back and look at what both networks are trying to do. Both network architectures must provide a framework and protocols that facilitate operation in which there is an extreme difference in scale between the possible spatial extent of the network and the size of individual nodes and channels. Simply put, human-scale nodes and channels relative to the interplanetary Internet have some useful analogies with nanoscale nodes and channels in a nanoscale network. The interplanetary Internet and delay-tolerant networking provide an architectural interface between the large scale of outer space and the smaller human-scale. Nanoscale networking requires an architecture that interfaces between the relatively large human-scale with the smaller nanoscale.

7.1.1   The definition of an architecture

First, the meaning of "network architecture" needs to be defined. A network architecture is the design of a communications network. It is a framework for the specification of a network's physical components and their functional organization and configuration, its operational principles and procedures, and generally includes the protocol and data formats used in its operation. As mentioned previously, a standard nanoscale architecture does not yet exist, so we cannot go to the level of detail of protocols and data formats. However, we can begin to understand what the architectural framework will look like. Figure 7.2 shows one taxonomy of nanoscale network approaches. This taxonomy is oriented towards the characteristics of the environment in which the network would exist, rather than on channel characteristics, which would be more important for a network architecture. Also, keep in mind that just as today's Internet is a networked collection of many different underlying technologies, all forms of nanoscale networking will ultimately be interconnected into a nanoscale Internet. It is also true that mixing and matching of techniques is currently taking place (e.g., nanotubes and nanorods in an aqueous environment).

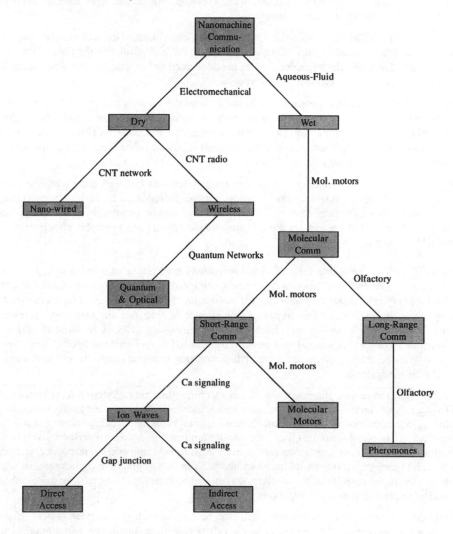

Figure 7.2 Nanoscale communication architecture.

From an architectural standpoint, biological cells and bacterium have served as a template from which we are learning to build nanomachines. There is a simplistic way of dividing up the functional components of what we might anticipate as a general nanomachine architecture. Here are the following architectural components [183]:

- *Control Unit* The control unit executes the instructions to perform the intended tasks controlling all other nanomachine's components. It could include a storage unit, in which the information of the nanomachine is saved. In the cell's case, all the instructions to carry out the intended cell functions are contained in chromosomes within the nucleus.

- *Communication Unit* The main component from our perspective is a transceiver capable of transmitting and receiving messages at the nanoscale. Obvious cellular examples are gap junctions and ligand-receptors, located on the cell membrane.

- *Reproduction Unit* The reproduction unit fabricates each component of the architecture using external elements, and then assembles them to replicate itself. In the biological cell, the code of the nanomachine is stored in molecular sequences, which are duplicated before cell division. Each resulting cell will contain a copy of the original DNA sequence.

- *Power Unit* The goal of the power unit is to supply power to all the other units of the nanomachine by obtaining energy from external sources such as light or temperature, and store it for later distribution and consumption. Cells can include different nanomachines for power generation such as the mitochondria, which generate most of the chemical substances used as energy, and the chloroplast, which converts sunlight into chemical fuel.

- *Sensors and Actuators* Sensors and actuators act as interfaces between the environment and the nanomachine. Several sensors and/or actuators can be included in a single nanomachine (e.g., temperature sensors, chemical sensors, clamps, pumps, motors or locomotion mechanisms). Obvious biological examples of sensors are the olfactory sensor (smell) and examples of actuators include molecular motors and the flagellum.

As we have seen, desirable features of nanomachines are present in a living cell. For example, nanomachines will have a set of instructions to realize specific tasks embedded in their molecular structure. They will most likely exhibit self-assembly and self-replication enabling assembly at the nanoscale as well as maintenance at the nanoscale. This implies nanomachines will contain the necessary corresponding instructions. Communication among nanomachines, just as in living cells, is required to realize more complex tasks in a cooperative manner and enable decentralized and distributed intelligence. There will need to be a "nano-to-macro" interface in the architecture that provides access to nanomachines, their control, and their maintenance.

There are two important characteristics of nanomachines that can be inferred from biological cells that will have an impact on the assumed communication architecture. These are their ability to multitask and their multiple communication interface architecture. As we saw with the amazing learning capabilities of the paramecium in Chapter 4, cells are clearly capable of multitasking. A cell can perform several different actions at the same time. It can ingest nutrients, convert the nutrients into energy, reproduce, and perform respiration. While performing these vital functions, the cell can be sampling the environment or signaling other cells nearby. In the case of the paramecium, the cell can be learning. Thus, nanomachines, as well as cells, should be understood as complex and complete systems.

From the standpoint of communication, cells can be seen as multiple-interface devices. Cells have hundreds, or even thousands, of receivers. A single cell is able to communicate using multiple unique channel access techniques: gap junctions, ligand-receptors, and molecular motors are a few examples. As previously mentioned, cells can be using these communication mechanisms simultaneously. Furthermore, all cells have specific and highly sensitive signal transducing mechanisms. Ligands acting as a signal attach to receptors, which amplify the signal, integrate the signal with input from other receptors, and transmit the resulting signal to the cell. Thus, the communication architecture is capable of four features that characterize signal transduction [184]:

- *Specificity* Specificity is the ability to detect and react to a specific signal as necessary. Each ligand and each receptor are complementary. Different ligands attach to different receptors. Specificity measures the precision with which a signal molecule fits on its molecular complementary receptor, where other signals do not fit.

- *Amplification* Amplification is the ability to increase the strength of the signal. Amplification by enzyme cascades results when an enzyme associated with a signal receptor is activated and, in turn, catalyzes the activation of many molecules of a second enzyme, each of which activates many molecules of a third enzyme, and so on. Such cascades can produce amplifications of several orders of magnitude within milliseconds.

- *Desensitization* Desensitization is the ability to remove the signal once received. The sensitivity of receptor systems is subject to attenuation. When a signal is present continuously, desensitization of the receptor system results; when the stimulus falls below a certain threshold, the system again becomes sensitive.

- *Integration* Integration is the ability to incorporate and interpret the signal properly with other signals. Integration is defined as the ability of the system to receive multiple signals and produce a unified response appropriate to the needs of the cell or organism. For instance, when two signals have opposite effects on a metabolic characteristic, the regulatory outcome results from the integrated input from both receptors by reinforcing the different internal metabolic paths.

As we can see, the cellular environment is extremely complex and rich; cellular network communication is more advanced than any human-scale network in existence today.

The actual mechanism of communication, in terms of a simple communication architecture, can be broken down into five processes among a transmitter, a receiver, and a channel propagation system:

- *Encoding* Encoding is the process of representing information in the form of molecules. A key component of nanoscale networking is that individual molecules are involved in the operation of the channel, whether it involves electrical characteristics of the molecules, changes in molecular concentration, or the utilization of the molecular structure to represent information.

- *Transmission* The transmitter inserts the encoded message into the communication channel medium. This can be the release of current through a chain of interconnected molecules, inducing changes in molecular concentration, or by attaching encoded data to molecular carriers.

- *Propagation* Information, encoded into messages, impacts the receiver by propagation through individual molecules.

- *Reception* Information messages are detected or unloaded from molecular carriers at a receiver.

- *Decoding/Reaction* Upon receiving the message, the receiver decodes the molecular message into useful information such as inducing a biochemical reaction, storing the data, operating an actuator, or reporting a sensed value.

Another architectural consideration is the transmission range of nanoscale communication. The shorter range techniques include molecular motors and diffusion-based techniques. A microtubule is 25 nm in diameter and ranges in length from 200 nm to up to 25 μm. Carbon nanotubes are typically a few micrometers long. Thus, one may somewhat arbitrarily define short-range communication as from a few nanometers to up to a few micrometers. Pheromones are an example of long-range nanoscale molecular communication, they may travel up to several kilometers. Thus, we can define long-range nanoscale communication from a few micrometers up to a few kilometers.

The relatively slow rate and short distance of nanoscale communication raises interesting questions about the impact on congestion and queuing. The fact that information transfer at the nanoscale requires the physical displacement of molecules will add a new dimension to traditional network metrics. Here we need to explicitly mention another architectural difference, that between *concentration encoding* and *molecular encoding*. Concentration encoding involves modulating the concentration of molecules in order to transmit information. Molecular encoding uses the structure of a molecule in order to encode information. In either case, matter must travel to the receiver in order for information to be transferred. Queuing could literally become a queue, that is, a line of molecules waiting to proceed along their path.

An architectural feature of nanoscale networks relative to existing human-scale networks is their low, self-powered operation. Chemical and microscopic physical processes drive nanoscale communication. This is ideal for in vivo applications in which the human body's own natural sources of energy can drive communication.

The long-range pheromone communication technique has some architectural disadvantages. All nanoscale molecular communication techniques that rely upon diffusion are subject to turbulence that adds noise to the signal. This turbulence can overwhelm whatever diffusive processes are assumed, whether simple Brownian motion, a Langevin process, or other variations of drift-diffusion processes. Even simple changes in temperature can have a large impact. We can expect that molecular encoding will be less susceptible to noise, as the information is embedded within the structure of a molecule and not its concentration. This is another simple example of how the relationship between physics and information becomes important in nanoscale networks; choosing natural flows in which molecular or concentration coded molecules can flow will yield a significant boost in performance. Self-assembly and self-organization may play a role in determining optimal channels. An advantage that counteracts the noise problem in long-range pheromone communication is the fact that, in nature, only small amounts of the messenger molecule are required. Natural receivers are highly sensitive.

All these techniques, and others, could be integrated in various ways. In other words, not simply having subnetworks comprised of each technology interface to one another, but combine the technologies in fundamental ways. Molecular motors can carry quantum dots that represent information, carbon nanotubes can hold quantum information and perhaps provide a platform for quantum communication.

Given the potential for long delays when approaching human-scale distances and the density at which information can be packed into a nanoscale molecular channel, we can anticipate the potential for a large bandwidth-delay product for such channels. The bandwidth-delay product of a channel is a significant architectural parameter.

7.2 ARCHITECTURAL PROPERTIES DERIVED FROM A FINITE AUTOMATA MODEL

A protocol has been developed for the reliable transfer of information using a molecular channel and, in the process, lays out the groundwork for what an appropriate architecture looks like [185]. Each node is abstracted to the level of a probabilistic timed automaton with multiple input ports, or receptors, and output ports, or emitters, for the reception and transmission of molecular information. Each node may operate in two modes, a receiving mode, in which the node can read all symbols entering its receptor ports and a sending mode in which the node can transmit information by emitting molecules from its ports. The node cannot be in both a sending and receiving mode simultaneously; molecules that arrive during a transmission cannot be received. If a node receives transmissions form multiple senders simultaneously, it will not detect either one. Information is considered to be received correctly only if all information read on all input ports are identical. If different information is read on different input ports, then a collision is assumed to have occurred.

There is an assumption that molecules can travel a maximum distance known as the communication radius, before becoming ineffective. As we have seen in Chapter 3, a maximum distance is useful in clearing molecular concentrations from the system so as not to interfere with newer transmissions. All nodes operate asynchronously. Although there is no global clock, each node is assumed to have its own notion of time and the ability to maintain multiple timers. Local clocks are assumed to tick at approximately the same rate. Being asynchronous is a severe restriction since it prevents simple global timing to coordinate sending and receiving times between nodes. Also, the automata are assumed to have no identifiers; that is, no unique addresses are required.

As in almost all ad hoc network analysis, the communication radius of each node enables connections to other nodes such that a graph results in which the automata are vertices and the nodes within communication radius are edges. The graph diameter D is the length of the longest communication path. The node degree, which is the number of adjacent nodes in the graph, is denoted by Q. The number of adjacent, that is, connected, nodes indicates the likelihood of a collision on the communication channel.

As an architecture, this model is useful in that it minimizes the number of assumptions required on the capabilities of nodes. A well-behaved nanoscale network is one that has a reasonable degree of connectivity, a bounded graph diameter D, and a bounded maximal number of neighbors Q.

Let T be the time to transmit over the distance of the communication radius. Each node has a timer that measures $2T$ time units, the time needed for a round trip. Each node is assumed to either listen for input for the duration of the $2T$ time interval, or transmit and then listen for that interval. If a message needs to be sent, the following send algorithm is assumed to take place. The node waits for time $2T$ and then with probability p sends the message. This wait period and probabilistic send sequence is repeated k times in order to increase the reliability of the transmission. The message is comprised of a signal s with probability of failure ϵ and the message will reach any node within the communication radius of the sending node. All nodes are assumed to work concurrently, asynchronously, and all using the same protocol.

Thus, each node is transmitting with probability p, which includes both the sender and its adjacent neighbors. As mentioned there are up to Q adjacent neighbors. Thus, there is probability $(1-p)^Q$ that none of the adjacent neighbors is transmitting. The joint probability p of the sender transmitting and $(1-p)^Q$ that none of the neighbors is transmitting is $p(1-p)^Q$. This value is convex and has a maximum at $p = \frac{1}{Q+1}$. After k attempts, the probability of failure is shown in Equation 7.1.

$$\left(1 - p(1-p)^Q\right)^k = \epsilon \tag{7.1}$$

Solving for k, the result is shown in Equation 7.2.

$$k = O\left(Q \log\left(\frac{1}{\epsilon}\right)\right) \tag{7.2}$$

Thus, a node transmits s with probability $\frac{1}{Q+1}$ and, for reliability, repeats the transmission $k = O(Q \log \frac{1}{\epsilon})$ times with an overall probability of error in transmission of ϵ. The send protocol takes the amount of time to send as shown in Equation 7.2. This indicates what one could expect in terms of channel capacity.

A broadcast protocol could use a flooding algorithm, namely, using the send algorithm to transmit a signal that is forwarded by adjacent neighbors. This process is repeated by each neighbor node until the signal is received by all connected nodes. Once the broadcast is forwarded, nodes would no longer respond to the same signal in order to avoid an infinite propagation of the signal. The time for the broadcast to complete is of the order shown in Equation 7.3. Recall that D is the graph diameter, which is the longest of all shortest paths between all node pairs, and n is the number of nodes in the nanonetwork. The probability of broadcast success is $1 - \epsilon$.

$$O\left(DQ\log\left(\frac{n}{\epsilon}\right)\right) \tag{7.3}$$

Note that for a broadcast algorithm, in order to achieve the failure probability ϵ, the send protocol must be performed with error probability $\frac{\epsilon}{n}$. This calls for repeating each send operation in a node $k = O(Q\log(\frac{\epsilon}{n}))$ times; k should grow both with a decreasing value of ϵ and with an increasing size of the nanonetwork.

Let us pause a minute to look at the bigger picture. Recall that in Chapter 2 we discussed DNA computing for nanoscale communication using the Benenson model for DNA finite automata. That provided an operational mechanism for nanoscale computation. Now we have a protocol and architecture to handle communication using such automata. Next, let us consider lessons that may be useful from existing and developing network architectures when applied to the nanoscale.

7.3 APPLYING LESSONS FROM OUTER SPACE TO INNER SPACE

There is a somewhat unexpected relationship between a developing network architecture, used for the interplanetary Internet, and communication in inner space. As we have seen in the introductory exercises in Chapter 1, there are interesting parallels between communication at the nanoscale relative to the human-scale and communication at the human-scale relative to communication at the interplanetary scale. Many of the design decisions may well be similar, as we will discuss. Delay-tolerant networking (DTN) is the specific interplanetary Internet communication architecture we will be examining in detail shortly. One of the primary goals of DTN was to handle communications in which information must be transferred from a transmitter to a receiver when there is no fully connected route between them. For example, deep-space probes may have to relay information to artificial satellites around other planets before the information is relayed to Earth. The satellites may be on the opposite of the planet, blocking communication. There may never exist a full path from the deep-space transmitter to Earth. Yet, the deep-space probe can relay the information to a satellite, which can store the information. When the satellite later moves into position that is accessible to Earth, it can then relay the information to Earth. The point is that the information is eventually transferred to its destination without a complete route in place. Consider nanoscale relays, such as molecular motors, handing off information in the form of attached cargo from one to the other. In fact, the motion models used to describe molecular motors that we discussed in Chapter 2 are being examined today in human-scale networks for examining the performance of DTN routing. Although it's a bit of an exaggeration, one could *almost* use the same simulation models from DTN routing and change the name to nanoscale routing.

A few other similarities between DTN and nanoscale networking include the goal of handling heterogeneous gateways and the blurring between naming and addressing. DTN is concerned about the wide variety of possible interfaces on Earth that could potentially have to be supported in communicating with the interplanetary Internet. The same will be true of the nanoscale Internet. As mentioned previously, we expect to see many different underlying nanoscale technologies seamlessly connecting with one another. Nanoscale information may flow over molecular motors, then interface with a cell signaling protocol such as an ion wave propagation signaling system, then interface with a random nanotube sensing mechanism, as a hypothetical example. The point is that there would be a diverse set of gateways between nanoscale technologies all required to work together seamlessly.

The delay-tolerant networking protocol has also begun addressing the address space issue given such large differences in scale. Information from a deep-space probe that makes it to Earth does not necessary need to be concerned with reaching a specific port on a specific host. Any host within a region capable of receiving the message is all that is required. The application to nanoscale networking is similar; consider the extreme case of the number of molecules in a mole of a substance. Avogadro's constant, the number of molecules in a mole, is 6.022×10^{23}. As nanomachines approach this level of size and density, does one need to individually assign addresses to each and every nanomachine? Or would a similar notion of region-host-application be more sensible?

Another consideration of DTN, given the long distances and lack of a complete routing path, is the use of bundles as a data unit to minimize transmissions. Rather than rely upon many quick exchanges, bundles attempt to anticipate any protocol exchanges and include as much as possible in each transmission. The latency will, by the very nature of interplanetary communication, be extremely high, thus, the best way to utilize the bandwidth-delay product of each channel is to use very large messages. DTN also enables the destination address to be specified as a function of the payload contents, similar to content-based routing.

A relatively unique feature of the DTN protocol is the fact that routing takes place opportunistically and there is no complete route. Intermediate nodes that relay the information may come and go randomly or be destroyed soon after transmission. In fact, the initial transmitter may be designed to survive only long enough to make a single transmission before succumbing to its harsh environment. The challenges of debugging such a dynamic situation, particularly one that can never be duplicated again, have been anticipated. DTN has been endowed with a unique reporting mechanism in which information regarding the status of messages as they pass through the network can be generated. An analogous feature for nanoscale networks may be of interest as well given the difficulties in manipulating objects at that scale.

Delay-tolerant networking also points to some fundamental requirements that are needed in order for the architecture to operate properly. The first is the requirement for time synchronization. DTN keeps the option open for scheduled, or predictable, routes to be utilized; this requires knowledge of the current time and the time when routes will be available. Another requirement is the ability to perform localization. Transmitters and intermediate relay nodes in DTN need to have some notion of their relative location to one another. A discussion of space navigation would take us outside the scope of this book; however, localization in both inner and outer space is a significant challenge. Nanoscale localization will be useful not only for communication, but also for extremely precise primary operation, whether it is used for precision treatment of diseases or extreme resolution material inspection.

Table 7.1 summarizes the relationship between delay-tolerant networking and nanoscale networking. Time synchronization is required for DTN in order to take advantage of predictable link formation. Most nanoscale molecular techniques that have been proposed to date assume synchronized send times. Localization, the ability to estimate a node's current location is required for DTN, again for routing purposes. Localization is not required for the nanoscale techniques proposed so far; however, it would be very useful for any application to know where the nanomachine is located. Both content-based addressing and addressing to nodes in a region rather than individual nodes, as previously discussed, is supported by DTN and useful for nanoscale networking. Routing, using contact-based approaches, is specified by many DTN routing protocols and will be very useful for many nanoscale networking approaches. Very minimal error correction is supported by DTN; the same is expected to be true for nanoscale networks. The unique "report-to" endpoint identifier (EID) feature of DTN is anticipated to be very useful for debugging nanoscale contact-based routing protocols. This feature allows direct diagnostic output from a message during its journey to arbitrary entities other than the source or destination. Finally, security will be an interesting issue. DTN takes the traditional view of protecting the data with its security architecture. Nanoscale networks will most likely be physically secure. The security compromises most likely to occur may be by contamination with adverse nanomaterials, either accidentally or maliciously.

There is much to learn regarding architecture from some of the novel nanoscale chip interconnect technologies that have been explored [51]. We have described chip interconnects in Chapters 1 and 4. In Chapter 4, the carbon nanotube field effect transistor was introduced as well as the method of using DNA

Table 7.1
Delay-Tolerant Network Architecture Versus Nanonetwork Architecture

Component	Interplanetary Internet (DTN)	Nanoscale Network
Time synchronization	Required	Required
Localization	Required	Useful
Addressing	Very flexible	Flexible
Routing	Contact-based	Contact/diffusion
Error correction	Limited	Limited
"Report-to" EID	Critical for debugging	Critical for debugging
Security	Encryption/data protection	Anticontamination

to control the nanoscale assembly of interconnects. An architecture comprised of a grid of carbon nanotube field effect transistors has been interconnected with conducting carbon nanotubes. The problem of partially matching DNA strands was noted; namely, that if interconnects with enough complementary DNA bases are in close proximity, they may hybridize, causing an invalid connection. The analogy has been made between the DNA partial matching problem and the Hamming distance in error correction. Specifically, if one wants to protect more bits with the same overhead, the Hamming distance decreases and errors are more likely to occur. This is similar to trying to add more DNA sequences to make a more complex interconnection network.

7.3.1 Routing with Brownian motion

As we mentioned, perhaps one of the best architectures to learn from for nanoscale networking is at the extreme opposite in scale, namely the interplanetary Internet, or delay-tolerant networking. Delay-tolerant networking does not require a complete path from the transmitter to the receiver for routing to take place. The notion is that communication in deep space will require relaying information from one node to another, but nodes may be temporarily hidden behind a planet, for example. In fact, in some cases, expendable transmitters may be designed to transmit their information to the next relay just before being destroyed by harsh conditions. Thus, there is a very low expectation that a complete path from a transmitter to receiver will exist over interplanetary links, which is required by most of today's routing protocols.

      Also, given that communication is limited, even using quantum entanglement, by the speed of light, the notion of interplanetary links means that the delay portion of the delay-bandwidth product will be extremely large. Many short interactive protocol exchanges would simply take too long for any reasonable transmission. The delay-bandwidth product is the amount of information that can be stored in channel, similar to the volume of water that can be held in a pipe. Thus, delay-tolerant networks utilize the full volume of the "pipe" by sending bundles. Bundles are larger amounts of information that are anticipated to be needed than the typical network packet of information, with the goal of reducing the need for multiple round-trip exchanges of information.

      While routing algorithms are still an area of research for delay-tolerant networking, commonly used routing protocols are based upon variants of epidemic routing. Epidemic routing uses the simple notion of contact among organisms in which a disease is spread. A transmitter infects mobile nodes with the message that needs to be transmitted in such a manner that the message is passed by node "contact" until the message reaches its destination. The goal is to maximize message delivery rate to the destination nodes and minimize latency, while simultaneously reducing the load on the network. These goals are trade-offs because spreading the message to more nodes increases the likelihood that it will reach the destination, however, it can also utilize more network resources; copies of the message have to be maintained by more nodes within the network.

The routing operation within the basic epidemic routing protocol is very simple [186]. Each node maintains a list of messages that it has sent as well as messages that it is currently holding on behalf of other transmitters. When hosts come into contact with one another they exchange messages in what is sometimes called an antientropy session. The term antientropy comes from our discussion of information entropy in Chapter 6; namely, contact among nodes and message exchange decreases the disorder among the different data stores throughout the network as follows. This poorly chosen term has been used in the past and can be confusing depending upon the perspective in which it is used. One could say that because information is spreading throughout the network its entropy is increasing, much like heat spreading through a system and the corresponding second law of thermodynamics. However, the term is being applied from the perspective of the distribution of data across all node data stores, which will tend to converge to a common set of data. At this point, even though the protocol is fundamental to the interplanetary Internet, the notion of random motion, intermittent contact, and information or "molecular" exchange should begin to look familiar, for example, from a large number of molecular motors and their information cargoes and cell signaling concentration gradients to very low power single-carbon nanotube radios.

In the basic epidemic routing protocol, when nodes come into contact, each node determines which messages the other node has that it does not have. An exchange takes place such that each node obtains the set of messages that it is missing relative to what the other node has. As nodes move, whether interplanetary or nanoscale, and given a sufficient amount of time and data storage on each node, messages will eventually arrive at their destination. Of course, the basic epidemic routing protocol can be enhanced when any knowledge about message routes and node motion can be taken into account. The theoretical limits of predictive routing, including "flash routes," which utilizes predictive knowledge to anticipate and prepare for contact of very short duration, are discussed in [57]. While networking in outer space has been conquered with delay-tolerant networking, networking in inner space with nanoscale networking is only now starting to be seriously considered. In the nanoscale environment, it would seem natural to leverage as much of the natural physics of the environment as possible so as to minimize the limited on-board processing that would be required. Thus, we see the importance of the prior chapters of this book, namely that a thorough understanding of the underlying physics is required in order to leverage as much as possible within the guiding principles of a nanoscale architecture.

Continuing the fundamentals of basic epidemic routing, each message contains an identifier, hop count, and an acknowledgment request. The identifier is the intended final destination node. The hop count limits the maximum number of antientropy exchanges allowed by a message. This allows some level of control over the resources versus delay. A large hop count value will propagate messages more quickly, reducing latency. But more network resources are consumed with a larger hop count as the message may spread in unnecessary directions due to the stochastic nature of node motion and contact. As we will discuss in more detail later, recall the Langevin model of molecular motor motion in which there is both drift and diffusion occurring. One question is how the physics of motion at the nanoscale affects node contact and the spread and information when such a routing protocol is to be used. The acknowledgment request indicates whether a message should be acknowledged. This can serve a dual purpose; first it can enable the implementation of reliability by allowing the sender to retransmit if necessary, that is, after a specified timeout period. In addition, because the acknowledgment message returns to the sender using the same epidemic routing process, it can serve as an antimessage to cancel any pending original messages that are still in route to the receiver [85]. This could help to more quickly clear the network of unnecessary overhead. Clearly, the data storage size on all nodes must be of sufficient size to hold all messages that are in transit. Generally, when the data storage capacity is exceeded, older messages are dropped, under the assumption that they are most likely to have already reached their destination.

Let us consider the impact of node motion in a nanoscale network and its impact on a simple epidemic routing protocol. Equation 7.4 shows the mean square displacement, which is the second moment of the displacement from the initial position at Z_0 over time t.

$$M(t) \equiv E(|Z_t - Z_0|^2) \tag{7.4}$$

Consider random walks and Brownian motion, which characterize many nanoscale particles due to the reasons discussed in Chapter 2. If the step length variance $\sigma_L^2 < \infty$, the mean square displacement will be linear with t, that is $M(t) \propto t$. The particles *are* the channel and their contact rate, the rate at which they become close enough to one another to exchange information in a manner that propagates the information in the desired direction, is proportional to the efficiency of the channel. When $\sigma_L^2 < \infty$, this is known as normal diffusion.

Studies have been done on delay-tolerant networking information spread when the step length variance approaches infinity $\sigma_L^2 = \infty$ [187]. This is known as superdiffusion; since longer step lengths occur more often, the nodes tend to spread much more quickly. The relationship between mean square displacement and time can be represented by examining the slope of the function $M(t)$ on a log-log scale. The slope is γ in Equation 7.5.

$$M(t) \propto t^\gamma \tag{7.5}$$

For the case of normal diffusion, γ is one. $\gamma > 1$ for the case of superdiffusion. The case of $\gamma = 2$ is known as ballistic movement. It's important to note that in all cases, we are talking about nodes with a constant velocity and operating over the same duration of time. So it is not the case that a node happens to travel further because its velocity is greater or it has more time to travel. The effect is due strictly to the greater variance in step length. The log-log plot of Equation 7.5 is characteristic of many scale-free phenomena that we have seen in early chapters, such as the structure of the current Internet connectivity. Data collected from human-scale movement patterns in [187] shows γ values that range between one and two and are thus superdiffusive.

The Levy random walk model allows easy representation and control of the diffusion properties in this case be allowing explicit representation of the step length distribution as shown in Equation 7.6, where l is the step length and μ is the distribution shape parameter. The distribution is valid when μ is greater than one.

$$f_L(l) \propto l^{-\mu} \tag{7.6}$$

Looking at Equation 7.6, it is apparent that for a large value of μ, the distribution decreases rapidly, resulting in a smaller tail, or less chance of large step lengths occurring. As μ decreases in value, the tail becomes fatter and there is a greater chance for a potentially rare, but sudden large step length to occur. Thus, there is an inverse relationship between γ and μ. In fact, it turns out that $M(t) \propto t$ when $\mu > 3$ and $M(t) \propto t^2$ when $1 < \mu \leq 2$. Thus, with $\mu = 3$, normal Brownian motion occurs.

An important point is that the efficiency of delay-tolerant routing appears to be dependent upon the rate of *new* contacts between nodes. Intuitively, the so-called antientropy exchange has little effect when nodes in the same spatial location meet one another; there is little information to be exchanged once information has saturated a local area. It is the scale-free step length that allows nodes to drift within a local area in order to spread, and collect, information, then take a much longer step in order to carry the information to a new spatial area and repeat the process.

7.3.2 Localization in outer space

Just as we have drawn inspiration for nanoscale network protocols from the interplanetary Internet and DTN, we might also address the problem of localization by looking first to outer space before peering into inner space. Consider how a nanomachine, whether autonomous or not, can determine its position within inner space. The nanomachine is too small to be directly seen or manipulated; its position in space, which may be important for routing purposes, is unknown without some means of localization. Clearly, nanomachines may need to be cognizant of their position in order to successfully carry out their primary tasks; for example, to accomplish navigation, to map the location of defects, or to map other observations. Such localization capability can be useful even in a static environment, such as individual tubes within a carbon nanotube network.

In a similar manner, spacecraft traveling into deep space are outside the ring of satellites comprising the Global Positioning System (GPS). So our goal is to consider how spacecraft, particularly deep space probes, determine their position and what we can learn to apply at the nanoscale. Spacecraft still require precise navigation for trajectory calculations, instrument operations, and communication. The current primary source of navigation information for deep space probes is range and range rate obtained as part of the uplink and downlink communication with ground stations such as NASA's Deep Space Network (DSN) [188]. The problems with this approach are that the DSN is not only expensive, but it is having to manage a growing number of concurrent missions with limited resources and time for each new mission. A more fundamental problem is the lack of accuracy in the localization itself. While the distance from Earth can be considered accurate, the precise position of the spacecraft in three-dimensional space is much less accurate. For these reasons, better, more accurate, and autonomous localization techniques are being researched. All of these characteristics would be ideal at the nanoscale as well.

The ancient art of navigation by the light of the stars is an approach that can be used. Star cameras using light visible to the human eye may be mounted aboard spacecraft and the images used to estimate location. However, a more interesting and potentially more accurate approach is patented that provides not only position in space, but also position in time; namely, clock synchronization. Remember that synchronized clocks are an assumption for delay-tolerant networking. X-ray pulsar sources from deep space provide natural beacons and oscillators for both localization and timing. Cosmic x-ray sources have regular time signatures that can be predicted far in advance. A spacecraft need only record the arrival time of x-rays from multiple sources from different directions simultaneously. The result is a system similar to GPS, but with natural cosmic information sources. Thus, one could look for analogous naturally occurring sources of radiation at the nanoscale, which, of course, there are.

In fact, it is suggested that a full suite of on-board navigation solutions can be engineered from the extrinsic x-ray and pulsar sources. From variable cosmic x-ray sources there are opportunities to create accurate three-dimensional solutions, including vehicle attitude determination, position and velocity determination, as well as clock correction for maintaining accurate time. Pulsars are particularly good sources for time synchronization. Pulsars are rotating neutron stars that emit a beam of electromagnetic radiation. The observed period of their pulses range from 1.4 milliseconds to 8.5 seconds. The radiation can only be observed when the emission beam sweeps past the observer, which is known as the "lighthouse effect," further emphasizing their potential for navigational use. Neutron stars are extremely dense, therefore the rotation period and the interval between observed pulses is very regular. For some pulsars, the regularity of pulsation is as precise as an atomic clock [189]. As an aside, pulsars are known to have planets orbiting them. Now let us consider localization in inner space.

7.3.3 Localization in inner space

As mentioned earlier, we are nearly as helpless in manipulating a nanoscale object in inner space as in outer space. Recall that, from the perspective of this book, the reason for localization is as an aid to routing; if nanomachines have knowledge of their relative locations, the network layer can route packets only in the required direction, thus reducing the overhead of spreading messages to all nanomachines. Localization and navigation of nanomachines in general is a large and growing field on its own, so we only briefly introduce the topic here as it might relate to nanoscale network routing.

Reasonably accurate nanoscale positioning exists today for manufacturing purposes [190]. The manufacturing approaches assume that the target object is held firmly in place on a human-scale platform. The entire platform then makes carefully controlled movements on the order of tens to thousands of nanometers along x and y axes. A common mechanism for accomplishing this is the piezoelectric actuator. It is well known that squeezing piezoelectric materials yields an electric charge. Thus, they are commonly used in strain sensors. However, this process can be reversed; an electric voltage can be applied to a piezoelectric material that causes mechanical strain. This converse process of inducing strain in piezoelectric materials has also been known for some time. However, more recently, thin-film actuators based on this process have been developed. The piezoelectric material lattice structure can have its dipoles aligned, known as the poling direction. When an applied electric field has the same polarity as the poling direction, the strain will be positive, which causes an elongation of the material. If the applied electric field is opposite in polarity to the poling alignment, the strain will be negative. This causes a contraction of the material. The important point is the dipole alignment constrains the resulting expansion and contraction to only one direction. Thus, very precise and minute changes can be made to the platform holding a target object, which allows manufacturing activities to take place with nanoscale changes in position. Certainly there are many other mechanisms for achieving similar nanoscale translation of position in spatial coordinates. However, the solutions to manufacturing related positioning differs from the type of localization and navigation that an autonomous nanoscale machine would require.

Results that are closer to solving the problem of localization and navigation of autonomous nanomachines take the form of in vivo control and tracking using magnetic resonance imaging (MRI) [191]. Before discussing the localization aspect, we need to discuss the architecture of the small-scale autonomous machines used in this approach and how they communicate. This approach uses a hybrid architecture, one that many researchers are using to try to combine the "mature" technology of electronics (which is unfortunately still primitive compared to biological cells) with organic systems, such as bacteria and other single-celled organisms [192]. In some respects, it is similar to the dawning of the age when humans first learned to ride the horse, except we are doing it on the cellular scale. These bioelectronic architectures are a reasonable step in the direction towards true nanoscale robotics and communications.

The particular approach used in [192] is to embed wireless transmission into a microrobot, whose size is slightly larger than the thickness of human hair. The actual size of the machine is 300 square μm. The electronics within the machine are powered by an on-board photo-voltaic source. The machine also contains an on-board electronic sensor capable of detecting medically significant targets such as pH or oxygen. The machine can generate a small electric current whose field is strong enough to be detected when close enough to sensitive magnetic receptors. The machine is designed to alternately switch the photo-voltaic current from the sensor to the transmission pad, which creates perturbations in the electric field to implement communication. Essentially, the transmission pad is an area on the machine where the conductor that carries the communication current is exposed from the magnetic shielding around the remainder of the device. This multiplexing of input power between providing sensor energy and communication saves power. The strength of the signal from the sensor is used to control the modulation of the current through the transmission pad. Thus, the stronger the signal, the higher the transmission frequency. This is similar to power-line carrier approaches to communication in which modulation of the power source is used to implement communication. Experiments have been conducted examining the trade-off between the transmission pad area and the area required for the photo-voltaic input, both of which need to be minimized in order continue to scale down the size of the overall machine. The hardware architecture of the machine avoids the large antenna and power consumption of traditional RF designs.

As pointed out in [193], a key concept in understanding communication, steering, and localization in this approach is the Biot-Savart law. At any given point, a magnetic field B is specified by two components:

a magnitude, or strength, and a direction. Thus, a magnetic field is a vector field. The magnetic field B is defined in terms of force on a moving charge. The current flowing through a wire on the transmission pad in the above machine, contributes to the magnetic field that surrounds the wire. The wire through which the current flows can be broken down into infinitesimal lengths dL. Each infinitesimal length contributes to the magnetic field B. Consider some point P in space near the wire through which the current is flowing. The contribution that each infinitesimal length of wire carrying the current IdL makes to the magnetic field at point P is perpendicular to the current element and perpendicular to the radius vector from the current element to the field point P. The infinitesimal value of the magnetic field $d\vec{B}$ is shown in Equation 7.7. In this equation, $\vec{1_r}$ is a unit vector that specifies the direction of the vector distance r from the current to the field point P. μ_0 is the magnetic permeability of free space with a value $4\pi \times 10^{-7} N/A^2$, where N is newtons and A is amperes.

$$d\vec{B} = \frac{\mu_0 IdL \times \vec{1_r}}{4\pi r^2} \tag{7.7}$$

As pointed out in [193], a nice simplification occurs when the transmission for the machine is implemented as a loop. In this case, the magnetic field at the center of the loop is equidistant from each infinitesimal length of wire in the loop dL. Using the vector product $\vec{A} \times \vec{B} = AB\sin\theta$, where A and B are the magnitudes of the vectors $\vec{A}$ and $\vec{B}$, we can assume that $\theta = \pi/2$, a constant value. The resulting loop form for B is shown in Equation 7.8.

$$d\vec{B} = \frac{\mu_0 I \sin\theta}{4\pi r^2} \tag{7.8}$$

The solution for B can be found by integration as shown in Equation 7.9.

$$B = \frac{\mu_0 I}{4\pi r^2} \oint dL = \frac{\mu_0 I}{4\pi r^2} 2\pi r = \frac{\mu_0 I}{2r} \tag{7.9}$$

The transmission distance can then be found by determining the z component perpendicular to the transmission pad as shown in Equation 7.10.

$$dB_z = \frac{\mu_0 IdL}{4\pi} \frac{r}{(z^2 + r^2)^{3/2}} \tag{7.10}$$

The circular integral over the current loop simply yields the circumference of the loop, while all other terms are constant. Thus, Equation 7.11 shows the magnetic field strength at distance z from the transmitter. This indicates the sensitivity of the receiver that would be required in order to detect the magnetic field channel.

$$B_z = \frac{\mu_0}{4\pi} \frac{2\pi r^2 I}{(z^2 + r^2)^{3/2}} \tag{7.11}$$

Magnetic resonance propulsion (MRP) is a patented technique for using an MRI magnetic field to induce controlled movement of a nanomachine by adjusting the induced magnetic fields. It is natural then, to attempt to use the MRI to also track the location of the nanomachine as will be discussed next. The propulsion system for the communicating sensor device just discussed is a set of bacteria known as magnetotactic bacteria (MTB) [193]. These are a fascinating class of one-celled organisms that have the ability to orient themselves along the magnetic field lines of Earth's magnetic field (recall early in Chapter 1 the discussion of magnetic storms). These bacteria have organelles called magnetosomes that contain chains of magnetic crystals, usually aligned along the main axis of the bacterial cell body. This class of bacteria was discovered in 1963. However, fossils of these bacteria, with their chains of crystals preserved, have been discovered in fossils (magnetofossils) that are 1.9 billion years old. These bacteria are usually found in a special layer between oxygen-rich and oxygen-starved water or sediment. It is thought that the phenomenon of magnetotaxis, the ability to move, detect, and respond to magnetic fields, evolved out of the need for the bacteria to navigate within this special layer with very steep chemical gradients.

Since the communicating machine just discussed has no power remaining for propulsion, the use of bacteria with a good sense of direction seems like a promising alternative. An obvious steering approach is to leverage the bacterias' tendency to align along a magnetic pole and apply the appropriate external magnetic field to create an artificial pole to which the bacteria will align. A particular MTB bacteria, MC-1, is discussed in [193] for a propulsion system. The diameter of the MC-1 bacteria is $\approx 2~\mu$m, which is only half the size of the smallest human capillary. An individual MC-1 bacteria can provide 4 pN of thrust and achieve an average velocity of 200 μm/s. The bacteria can be induced to move under computer control by a conductive electrical network positioned near the bacteria. Recall that the machine dimensions are 300 square μm, which is large relative to the size of individual bacteria. Thus, a swarm of bacteria are used to propel the device. However, the bacteria have a lifetime of only a few hours, which would render the device immobile after that time. Thus, the bacteria are not firmly attached, but rather induced to push the device as needed. New bacteria can be injected as needed as older ones expire.

Having introduced the architecture of such a nanorobotic system, we can now proceed to a brief discussion regarding the notion of localization in this proposed architecture, which is based upon nuclear magnetic resonance imaging (NMRI) or more euphemistically termed magnetic resonance imaging (MRI) in order to remove the word "nuclear." The concept is to use the magnetic field from MRI to both navigate by inducing movement of the MTB bacteria as well as detect the location of the bacteria. Simply put, nuclear magnetic resonance (NMR) is a property that magnetic nuclei have in a magnetic field. An applied electromagnetic (EM) pulse causes the nuclei to absorb energy from the EM pulse and then radiate the energy. Any nucleus that contains an odd number of protons and/or neutrons has an intrinsic magnetic moment (spin). The principle of NMR usually involves two sequential steps: (1) alignment (polarization) of the magnetic nuclear spins in an applied, constant magnetic field and (2) perturbation of the nuclear spin alignment using an electromagnetic, usually a radio frequency (RF), pulse. MRI uses a powerful magnetic field to align the nuclear magnetization of hydrogen atoms of water within the body. RF fields are used to systematically alter the alignment of this magnetization, causing the hydrogen nuclei to produce a rotating magnetic field detectable by a scanner.

The body is largely composed of water molecules that each contain two hydrogen nuclei or protons. The magnetic moments of these protons align with the direction of the magnetic field. A radio frequency electromagnetic field is then briefly turned on, causing the protons to alter their alignment relative to the magnetic field. When this field is turned off the protons return to the original magnetic alignment. These alignment changes create a signal that can be detected by the scanner. The frequency with which the protons resonate depends upon the strength of the magnetic field. The position of protons in the body can be determined by applying additional magnetic fields during the scan, which allows an image of the body to be built up. These are created by turning gradient coils on and off, which creates the knocking sounds heard during an MR scan. Protons in different tissues return to their equilibrium state at different rates. By changing parameters of the scan, this effect is used to create a contrast between different types of body tissues.

The magnetosomes within the magnetotactic bacteria significantly disturb the local magnetic field to the extent that they are detectable within commercial MRI machines. These perturbations mostly affect the spin-spin (T2) relaxation times [193]. The spin-spin relaxation time is the time it takes for the coherence in the spin precession to decay; that is, for the elementary particle spin to lose alignment. This allows tracking of position and concentration of swarms of MTB. Unfortunately, the naive approach of taking a sequence of MRI images is far too slow to apply to real-time navigation and control.

While these are necessary and impressive steps towards nanoscale robotic control in vivo, it still assumes a large external structure, namely, an MRI machine, in order to accomplish control and localization. The nanorobots are not yet making independent attempts at coordinated behavior to accomplish tasks. However, they provide an ideal platform for experimenting with nanoscale communication protocols. In particular, they enable experimental verification of communication networking and routing techniques for inter-nanorobotic communication and ad hoc routing.

7.4 ARCHITECTURE OF EXTANT IN VIVO WIRELESS SYSTEMS

One of the most beneficial and lucrative uses of nanoscale networking will be medical applications. Many of today's most challenging medical problems will be greatly aided or solved by having precise ubiquitous knowledge, at the cellular level, of internal processes within a patient's body. Unfortunately, cellular-level observation still requires invasive techniques such as biopsies. Approaches to place sensors within the body attempt to use wireless techniques; there is rapidly growing research and standards on implant communications, however, it is currently very crude, invasive, and dangerous. Most of these problems are due to the relatively large transceiver size, significant energy requirements to penetrate the tissue, and wireless radiation damage caused to the surrounding tissues caused by the high power requirements. Thus, it is instructive to take a brief look at the current crude implant communications architecture in order to learn what challenges are being faced in legacy implant systems and how they are being addressed.

The specific absorption rate (SAR) is a measure of the rate at which energy is absorbed by the body when exposed to RF energy. An estimate of this absorption is given in Equation 7.12. The units of SAR are watts per kilogram. E is the induced electric field and $|E|^2$ is the root mean square of the field vectors. ρ is the density of the tissue, and σ is the electrical conductivity of the tissue. Note that SAR is a measure used for mobile cell phone and MRI scans as well as implant radiation.

$$SAR = \frac{\sigma |E|^2}{\rho} \tag{7.12}$$

There have been approaches to implant network routing that focus on minimizing thermal effects. Temperature increase comes from heating when RF is used to recharge the sensor (assuming this technique is used for power), radiation from the antenna, and temperature increase from the implanted sensor circuitry. All of these potentially damaging effects are combined into a single equation that estimates the increase in temperature of tissue with blood flow [22]. Network routing metrics can be used such that the value of this equation is minimized. The basic idea here is to route through ex vivo body coordinator. To use SAR as a routing metric such that SAR is minimized during routing. In other words, routing is based upon minimizing radiation exposure to the patient.

The Medical Implant Communication Service (MICS) is a specification for in vivo wireless communication. The MICS standard is defined by the Federal Communications Commission (FCC) and the European Telecommunications Standards Institute (ETSI). The frequency band is between 402 and 405 MHz. The maximum transmit power is 25 microwatts. The maximum bandwidth is 300 kHz, making it a relatively low bandwidth architecture. The main advantage of MICS over previously used inductive technologies is a small amount of additional convenience, because the inductive technologies required the external transceiver to touch the skin of the patient. However, MICS gives a range of only a few meters. While this may sound like a very constrained system to work with, there are good reasons for the constraints. The primary reason is both the signal attenuation due to biological tissue and the requirement for safety in order to prevent damage to the tissue. The very same component of tissue that allows MRI to work so well for imaging—high water content also impedes wireless communication.

The IEEE 802.15 Task Group, Body Area Network (TG-BAN) is also looking at in vivo wireless communications. The goal for this group is the development of standards for both ex vivo and in vivo body area networks that would be used for waveform sampling of biological signals. A primary concern for any communication standard that is integrated with the human body is safety. The specific absorption ratio (SAR) is a critical safety measure for such systems in order to avoid radiation damage to tissue. Another safety concern is, of course, energy consumption. Battery power is not viable for numerous safety reasons, including danger of chemical leakage, the requirement for painful and risky surgery to replace, and large size. Thus, energy scavenging is a preferred approach by current technologies.

Examples of implants that would benefit from in vivo communication are [194]:

- *Pacemaker*, including an implantable defibrillator;
- *Prosthetics*, including artificial retina, cochlear implants, and brain pacemakers for patients with Parkinson's disease;
- *Capsular endoscopes*, for diagnosis of the gastrointestinal tract;
- *In vivo sensors*, for health monitoring, including glucose monitoring;
- *In vivo actuators*, including implanted insulin pumps and bladder controllers.
- *In vivo bone stress sensors*
- *Nanoscale devices* to maintain homeostasis in injured or sick patients
- *Therapeutic nanoparticles* released when tumor cells are detected to eliminate malignant tumors
- *Anticoagulant* release when a stroke is detected
- *T cell (white blood cell) recruitment* when bacteria are detected on implanted prosthetic devices
- *Medical nanorobotic device* communication

Communication should be bidirectional in almost all cases. The data rates can vary from a few kilobits per second to several megabytes per second for live video from an endoscope. The network should support communication from one implant device to another, for example, from an implanted glucose monitor to an implanted insulin pump ultimately, from any location within the body to any other location. At this point we can see that today's wireless communication is foreign to the human body in many ways. The implanted devices themselves may be rejected by the body, which views the material as an invading object. The promise of safer, more benign communication network mechanisms that naturally use biological communication mechanisms are much more appealing.

Because of the high path loss through tissue and the safety requirement to maintain low power radiation, carrier sense multiple access (CSMA) protocols do not work well. Many "hidden nodes" within the body may also contend with one another for a channel. The alternative is a simple time division multiple access (TDMA) approach. Another architectural constraint due to the high path loss through tissue is that direct implant-to-implant communication is avoided. Instead, implants first communicate with an ex vivo body "coordinator" node, which then forwards the messages back into the body if necessary. It turns out that path loss is reduced by transmitting out of the body first and then back in, rather than going directly through from implant to implant. This also has advantages in terms of in vivo power savings; it is much easier to change a battery on the ex vivo relay device.

Having looked at human-scale in vivo challenges, we examine another existing human-scale network architecture from which we may learn much that will apply to the self-assembly of nanoscale networks. The active network architecture enables the high degree of flexibility necessary to accomplish self-assembly.

7.5 ACTIVE NETWORK ARCHITECTURE

The goal of active networking is to create an architectural network framework that, in stark contrast to current communication networks, is easy to evolve and that also enables efficient application-specific customization. The mechanism to achieve these goals is to make the network as programmable as possible. This architectural concept for networking has reappeared periodically since the early days of packet radio and is likely to reappear again for nanoscale networks. One of the reasons is that active networking provides a mechanism for a network to evolve; in other words, to self-assemble [1]. Research into early forms of active networking has focused on network management and self-healing capabilities as well as providing highly dynamic capabilities for wireless ad hoc networks. For example, many creative techniques for evolutionary and genetic algorithms to "grow" the network were developed [54, 195–197]. In fact, fear of the extreme flexibility of active networks has been one of the main reasons it has not been fully adopted in traditional network architectures.

This section provides an overview of the basic architecture for human-scale active networks. It describes the components and introduces common terminology. It also discusses a few execution environments that implement various forms of active networks. We discuss the active networking framework developed in early 2000, which provides a programmable infrastructure using a well-defined structure for packets that contain general-purpose program code and a uniform standardized execution platform within the nodes of the network.

The traditional networking protocol stack on a network node is fixed; network nodes manipulate protocols only up to the network layer of the protocol stack and no higher. The narrow waist of the Internet Protocol is a bottleneck that separates and isolates intelligence at the higher layers from reaching lower layers where it can be most effective. New types of information may have unique communication needs in order to meet their quality-of-service requirements. Unfortunately, with legacy protocols such as the current Internet Protocol, any new type of information must be transformed into a format suitable to meet the common lower layers of the protocol stack. This transformation onto the "one-size-fits-all" network and lower layers prevents custom actions from taking place to meet unique communication needs. It can be argued that when the need becomes desperate enough, the standards slowly change to accommodate the necessary change, but this takes years and sometimes decades, as in the case of IPv6. As a simple example, data, audio, and video must reside over the same network, data link, and physical layers from source to destination. With respect to these lower layers, each type of information is treated equally. However, if the I, P, and B frames within video packets could be recognized and acted upon in a custom manner, more intelligent decisions could be made to avoid congestion with minimal impact to video quality [198–200]. For example, critical I frames could receive priority while other frame types may be delayed or dropped if necessary. The point is that this kind of custom application handling by the network is very limited, if not impossible, using today's passive networks. An active network allows code to be injected into the network that enables well-defined targeted changes to the operation of the network that are customized for particular application traffic.

The traditional architectural layered approach was reexamined in the innovation of active networking. The purpose of layering in an architecture is to isolate one protocol from another so that development, debugging, and maintenance of the network are tractable. However, as changes in the implementation of a protocol accumulate over time, there may be redundancies between layers. Layers may perform the same functionality as the layers beneath or above it. Also, protocol layers may operate more efficiently if they could access information directly from a layer below or layer above. As previously mentioned, the inability to dynamically change layers to better support differences in applications is an inefficiency in current network design; it is impossible to determine a priori what functions are needed in a network for all possible applications [196]. Standardization is a time-consuming process. The time from conceptualization of a protocol to its actual deployment in a network can be an extraordinarily long process. As an example, work on the design on the Internet Protocol version 6 (IPv6) was initially started in 1995 but the protocol took at least a decade to develop and deploy. To further complicate matters, variations in implementation by different network hardware suppliers cause problems for interoperability. Suppliers may not completely implement a standard feature or may add extra features that better exploit their unique hardware. This complicates the network architecture and protocols, which have to be cognizant of all these possible variations. A similar issue that suppliers have to deal with is backward compatibility. A revision of a protocol may need to change the position of a field in the header of a protocol to accommodate more information. However, network devices upgraded with the new protocol still have to support data that conforms to an earlier revision.

Active networking provides a flexible, programmable model of networking that addresses the concerns and limitations that we have just described at the human scale. In an active networking paradigm, the nodes of the network provide execution environments that allow execution of code that is loaded and installed while the network remains in operation. In fact, the dynamic loading and installation of code is the main part of the active network architecture; it is part of normal operation. In a sense, the network self-assembles just as we would like a nanoscale network to do.

Instead of standardizing individual protocols, the nodes of the network present a standardized execution platform for code-carrying packets. This approach eliminates the need for network-wide standardization of individual protocols. New protocols and services can be rapidly integrated into the network. Standardizing the execution platform implies that the *format* of the code inside the packets is also agreed upon. But users and developers can code their own custom protocols within active packets. The active packet code may range from a new protocol for transporting video packets or it may implement a custom routing algorithm for a specific application. The ability to introduce custom protocols breaks down barriers to innovation and enables developers to customize network resources to effectively meet their application's needs.

7.5.1   The active network framework

There is much to learn from self-organization and the active network architecture that may benefit nanoscale networks. Thus, we examine the human-scale architecture in more detail. Human-scale custom network code is injected in the network in one of two ways. One approach is to download active network code to network nodes separately from data packets. The downloaded code carries computation for some or all of the data packets flowing through the network node. The code is invoked when the data packets that it has been designed to handle reach the node. This is known as the discrete approach [201]. The other approach is the capsule or integrated approach [202]. In this approach, packets carry both computation and data. As packets flow through network nodes, custom code is installed at some or all nodes in the network. These packets are called capsules or SmartPackets, which are the basic unit of communication in an active network. Applications inject SmartPackets in the network and active nodes process the code inside the SmartPackets. A hybrid of the discrete and integrated approaches is Packet Language for Active Networks (PLAN) [203].

The nodes of an active network, called active nodes, provide the platform for SmartPackets to execute their custom code. Active nodes participate in the customization of the network instead of passively switching or routing data. The active node provides the platform for execution that is called an execution environment (EE). An execution environment defines a virtual machine and a programming interface for users of the active network. The EE exports an application program interface (API) that users can program or control by directing packets to it. An active node can have multiple EEs. This enables the creation of a backbone of active nodes spanning different research institutions and yet enables researchers to independently investigate different execution models and features.

There have been a number of active network execution environments developed for active nodes to study various issues such as resource allocation, service composition, and security. The following

Table 7.2

Table of Active Network Execution Environments

EE Name	Institution	VM platform	Approach
Magician	University of Kansas	Java	Capsule
ANTS	Massachusetts Inst. of Technology	Java	Capsule
Switchware	University of Pennsylvania	Caml	Hybrid
Netscript	Columbia University	Netscript	Discrete
SmartPackets	BBN	Assembly code	Capsule
CANeS	Georgia Tech.	UNITY	Capsule

section discusses features of execution environments that have been explored. A few popular execution environments are listed in Table 7.2. Caml, which stands for Categorical Abstract Machine Language, is a strongly typed functional programming language from the ML, which stands for metalanguage, family of languages, designed for program safety and reliability [204]. UNITY is a programming language that is rather theoretical and attempts to focus on the what, rather than the where, when, or how of implementation. The language has no flow control; statements in the program run in a random order, until none of the statements causes change. A correct program converges into a fixed point [205]. Finally, NetScript is a programming language geared towards building networked systems. Netscript programs are organized as mobile agents that are executed remotely under local or remote control. The goal is to ease the development of new code within intermediate network nodes [206]. In all of these approaches, there is a clear attempt to understand the fundamental relationship between computation and communication, and its attendant aspects such as safety and security. At the nanoscale and molecular level, we can anticipate computation will take on a different form, no longer relying upon the traditional, well-defined registers and central processing units to which most people are accustomed, but rather new forms of computation that are more compatible with the underlying physics, for example, DNA and enzyme computing, quantum computing, and hybrids of these approaches with more traditional processing.

7.5.2 Properties of execution environments

As mentioned in the previous section, researchers have developed many different types of execution environments that focus upon exercising particular facets of active networking mechanisms. This provides an opportunity to enumerate and classify properties that execution environments possess to make code transport and execution possible. This section discusses the following principal features of execution environments:

- Code download;
- Implementation platform and language;
- Composition;
- Security and authentication;
- Resource allocation and monitoring;
- Interpacket communication.

The vision of active networks is a framework from which services are (self-)composed to provide users with a flexible and customizable networking infrastructure. Active networks allow the composition of user-injected services with those that are preinstalled on network nodes, thus enabling users to develop highly customized solutions within the network. The execution environment must provide modular service components for users and developers to utilize as building blocks. Users compose custom network services by using the software interfaces and connections provided within the execution environment in which their code executes. These connections can be in the form of library modules, a scripting language, or an API. The programming language can be a high-level language such as Java, C++ and C, or it can be a scripting language like Perl or Tcl. The expressiveness of the language has been found to be a trade-off among efficiency, security, and ease of network coding. A fully implemented and highly expressive language such as Java may provide ease in programming, but may also be inefficient inside a network and expose security vulnerabilities. On the other hand a uniquely customized language oriented towards a particular piece of network hardware may be efficient, but also tedious to develop.

Active network architectural components have a formal, well-defined interface that abstracts the inner workings, yet presents the user with a rich representation of its behavior. This is known as introspection, and enables a user to access an implementation state and work with abstractions to it. This increases reusability because the component can be reused in different protocol frameworks. The property of intercession enables a user to modify the behavior of an abstraction. The user can refine existing behavior or adjust the implementation strategy to improve performance. This is usually achieved through the mechanism of inheritance in object modeling. This enables users to customize previously developed active network components tailoring them to the needs of the application without having to write new components from scratch.

Given that the active network architecture allows for such a high degree of dynamic change, the notion of a self-healing and self-organizing network follows naturally. The self-assembly of active networks will couple with the self-assembly of the molecular environment. The next section considers the role of complexity theory in a self-healing communication network.

7.5.3 Active networks and self-healing

Fault-tolerant and self-healing systems should have the ability to self-compose solutions to faults. Such self-healing should be an inherent part of system operation, rather than a structure imposed from "outside" the system as it is today. Genetic algorithms are one form of self-composing solution, however, genetic algorithms as currently implemented require external control to manipulate the genetic material. In other words, the genetic algorithm itself must be programmed into the system. If the genetic algorithm code fails, then the self-healing capability fails. While this situation is not ideal, it has been explored as a possible step towards a truly self-healing network.

Many active network components and services have been designed, implemented, and have undergone experimentation. However, the fundamental science required to understand and take full advantage of active networking lags behind the ability to engineer and build such networks. In fact, the current Internet, whose protocols were built upon the ill-defined goal of simplicity, is only slowly being understood. An outcry from the Internet community, with its carefully crafted, static protocols and layering, with massive documentation ($O\,(6,000)$ Request for Comments) of passive (nonexecutable) packets is that the Internet is already "too" complex.

In fact, the fundamental notions of complexity, order, and self-assembly are intimately related. How can such systems, which require complexity to be adaptive, at the same time, appear simple to understand and manage? Are active networks really more complex than the current Internet? Are adaptive applications built upon active networks any more or less complex than the same applications built upon the legacy Internet? Does a measure of complexity exist that would allow an objective comparison to be made? What are the benefits of an active network with respect to passive networks? While these are extremely difficult questions to be answered, we lay the groundwork to answer these questions by proposing a complexity measure inspired by Kolmogorov complexity, and proposing an adaptation mechanism, genetic programming, based upon an analogy with biological systems. Kolmogorov complexity was applied to optimize the combined use of communication and computation within an active network in order to determine the optimal amount of code versus data [170]. It was shown that if an estimate of the complexity of the information related to the prediction of the future state of the network is estimated to be high, then the ability to develop code representing the nonrandom or algorithmic portion of that information is low. In other words, the harder it is to predict future events, the longer and more complex the code had to be in order to represent that event. This results in a relatively low efficiency for algorithmic coding of the information. The benefit of having code within an active packet would appear to be minimal in such cases.

Conversely, if the complexity estimate is low, then there is great potential benefit in representing information in algorithmic form within an active packet. It was suggested that if the algorithmic portion of information changes often and impacts the operation of network devices, then active networking provides the best framework for implementing solutions. This is precisely the case in genetically programmed network services, a new class of services that are not predefined but that evolve in response to the state of the network. In this section, we restrict this class to those services that are programmatic solutions for perceived faults that occur in a network.

Frameworks for protocol and service composition have been developed for human-scale active networks [207]. Demonstrations have shown that it is possible for the network to generate code rapidly and in a manner that can never be known a priori for every possible condition. The inspiration for genetic algorithm-based approaches to active network service composition comes from nature in the form of molecular biology's docking problem. The docking problem addresses precisely how one molecule lines up with another in a preferred orientation, for example in the ligand-receptor binding discussed in Chapter 3. In the case of a self-healing active network, solutions that efficiently match a particular fault should be able to "dock" with that fault. Prediction for successful docking in biology can be attempted by searching for minimal energy or minimal geometric construction combinations. It will be important to study the relationship between complexity and solution composition. In particular, it has been hypothesized that the complexity of the fault coupled with the potential solution will decrease as the optimal solution is composed. Examples of specific network faults are: (1) network misconfiguration, (2) bandwidth and processor misallocation, (3) faults caused by distributed denial-of-service and virus attacks, (4) poor traffic shaping routing problems, (5) nonoptimal fused data within the network, (6) poor link quality in wireless and mobile environments, (7) mal-composed protocol framework models in the network, and (8) poorly tuned components of network services.

We will show an example of a simple self-composing solution to a fault, namely misallocation of bandwidth and processing capability resulting in packet jitter. A fitness function defines a metric for "goodness" of the solution as it self-assembles, or evolves. In this case, "goodness" is the reduction in the variance of packet interarrival times. The fault is represented by the difference between the actual system and a minimum required fitness. Genetic material will self-assemble to minimize the effect of the fault. The complexity of the combined fault-solution pair should be at a minimum when the fitness is optimal. Note that by applying the active network architecture to the extremely dynamic approach of self-assembly, we have come full circle in emulating the nature-inspired technique of genetic programming on a nanoscale network that could actually be implemented with real biological components, including actual ligand-receptors.

7.5.4 Complexity and evolutionary control

Complexity and evolution are intimately linked and descriptive complexity, specifically Kolmogorov complexity, is one way to quantify this linkage. Kolmogorov complexity ($K(x)$) is the optimal compression of string x. Active networks form an ideal environment in which to study the effects of trade-offs in algorithmic and static information representation, because an active packet is concerned with the efficient transport of both code and data. We have already seen active networks applied in Chapter 4 [51], that demonstrated code flowing from one nanoscale processing element to another over a potentially faulty nanoscale network.

As noted in Figure 7.3, there is a striking similarity between an active packet and DNA. This is no coincidence. Active packet code can modify the operation of the network and effect "transcription" control on succeeding active packets. Both active packets and DNA carry information having algorithmic and nonalgorithmic portions. The algorithmic portion of DNA has transcription control elements as well as the codons, which is the actual code for proteins. The active packet has control code and may contain data as well.

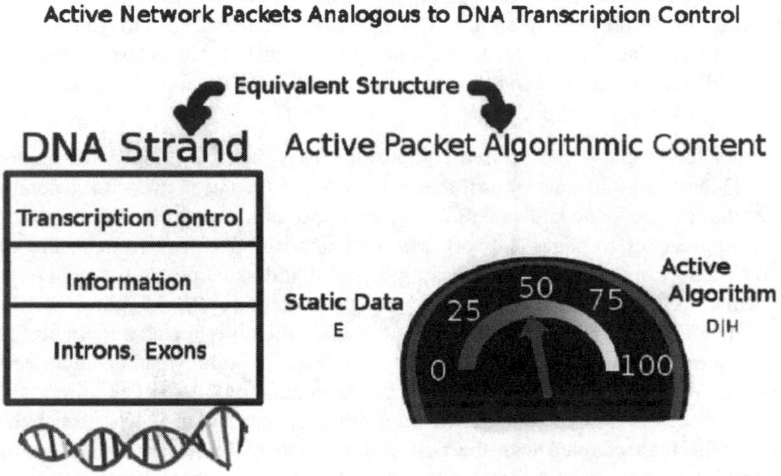

Active Network Packets Analogous to DNA Transcription Control

Figure 7.3 Active packet as DNA.

7.5.5 The application of a complexity measure in a communication network

The active network architecture should enable the active network to automatically generate solutions that bring the network back into line with a healthy model of the system. A fitness function can be used to describe the distance of the network from its desired state. The concept of molecular docking, mentioned previously, requires a more precise measurement of the degree of "fit" in the docking of a fault with its best solution. Kolmogorov complexity, estimated via the minimum description length algorithm, can be a means to measure the fit between the fault and the desired state. The next paragraph describes the minimum description length complexity estimator and its relationship to active networking.

Direct application of minimum description length (MDL) can be applied to an active packet. Let D_x be a binary string representing an arbitrary sequence x. Let H_x be a hypothesis or model, in algorithmic form, which attempts to explain how x is formed. MDL states that the sum of the length of the shortest encoding of a hypothesis of two components will yield an estimate of the Kolmogorov complexity. The two components are the length of a model generating sequence x and the length of the shortest encoding of x using the given hypothesis. This can be represented mathematically as shown in Equation 7.13, where $K(H_x)$ is the complexity of the hypothesis and $K(D_x|H_x)$ is the complexity of the data given the hypothesis. The intuitive notion is that the hypothesis or algorithm used to "compress" the data must be included in the measure of the complexity of x. A complex algorithm used to compress the data to a smaller size essentially shifts the complexity from the resulting data size to the algorithm. There is a trade-off between the algorithm and the data; this is a fundamental concept that was shown in Figure 7.3 by the algorithm versus data dial.

$$K(x) = K(H_x) + K(D_x|H_x) \tag{7.13}$$

Now we have to consider the impact of imperfect, or lossy, compression as a representation of complexity. We will simply call this error. The error in the hypothesis or model must be compensated within the encoding. A small hypothesis with a large amount of error does not yield the smallest encoding, nor does an excessively large hypothesis with little or with no error. A method for determining $K(x)$ can be viewed as separating randomness from nonrandomness in x by "squeezing out" nonrandomness, which is computable, and representing the nonrandomness algorithmically. The random part of the string, that is the part remaining after all patterns have been removed, represents pure randomness, unpredictability, or simply, error. Thus, the goal is to minimize Equation 7.14 where $l(x)$ is the length of string x, H_e is the estimated hypothesis used to encode the string (D_x) and E is the error due to an imperfect hypothesis.

$$l(H_e) + l(D_x|H_e) + l(E) \tag{7.14}$$

The more accurately the hypothesis describes string x and the shorter the hypothesis, the shorter the encoding of the string. Choosing an optimal proportion of code and data minimizes the active packet length. The Kolmogorov complexity of a combined fault and solution description is minimized when the optimal solution to mitigate the fault is composed.

The goal is to utilize the active network architecture along with genetic programming to self-assemble nanoscale networks. Genetic algorithms are widely known for their ability to find optimal solutions while avoiding local extremes by using evolutionary processes dependent on "random" mutation. The Kolmogorov complexity of the genetic material during the evolution of a genetic algorithm can be estimated and yields interesting clues about the underlying physics of the information during its evolution towards maximizing fitness. As the evolution proceeds and the fitness level of the genetic material rises, the complexity decreases. This result yields an interesting insight that supports the hypothesis that "solutions" that self-compose to mitigate a fault will tend to decrease in complexity. It should also be noted that if the fitness function is designed to compress a sequence, then the genetic algorithm can also return an estimate of the complexity.

7.5.6 Genetic network programming architecture

The goal in this section is to provide a brief example of how an active network architecture might enable self-assembly at the nanoscale. Thus, only a very brief overview of genetic programming is given. Genetic material begins in a random state and converges to the complexity of the optimal value guided by a fitness function. This enables solution composition from a wide range of possible solutions. One problem with this approach is the time required in evolving towards a feasible solution. Another problem is the fitness function itself may need to be self-generated.

An active network was used to test the feasibility of a genetically programmed network service [54]. An active packet representing the nucleus (assuming network nodes are like eukaroytes, that is, cells containing nuclei) was injected into all network nodes. The nucleus contained a population of chromosomes, which are concatenations of functional units. Operation of genetic network programming begins with injecting basic building blocks, known as functional units, into the network. This "genetic material" is

Table 7.3
An Adaptive In-Line Genetic Communication Network Programming Algorithm

Step	Operation
1)	Inject functional units
1.a.	Propagate to all nodes
2)	Inject nucleus
2.a.	Propagate to all nodes
3)	Inject fitness function
3.a.	Propagate to all nodes
3.b.	Choose three functional units to form three independent chromosomes
4)	Clone "normal" packet traffic through each chromosome while simultaneously performing Step 5.
5)	Begin operation
5.a.	If mutation event is to occur
5.a.i.	Generate a random change in one of the functional units to another available functional unit
5.b.	If recombination event is to occur
5.b.i.	Select a crossover point and swap portions of best parent chromosomes
5.c.	Otherwise, for each chromosome, select an available functional unit at random and append to each chromosome
5.d.	Compute fitness
5.d.i.	"Normal" packets flow through each chromosome
5.d.ii.	Compute a fitness value based upon monitoring desired chromosome objectives, such as latency or jitter
5.d.iii.	Return fitness value to the nucleus
5.e.	Assign fitness to each chromosome
5.f.	Repeat from Step 5.a until a desired fitness threshold is reached

flooded into each active node. This material remains inactive within each node until a fitness function is injected into the active network. Receipt of a fitness function causes genetic evolution to proceed. From the algorithm description in Table 7.3, Step 5 is similar to known genetic programming techniques. The primary difference from known genetic programming techniques is that it is occurring in real-time, during normal operation along with other network traffic, within a communication network. Step 4 indicates that normal communication traffic flow occurs while Step 5 is repeatedly executed.

Functional units are very short (minimum complexity) pieces of active packet code that perform simple, well-defined operations upon other active packets. Examples of functional units are Delay, Split, Join, Clone, and Forward. Chromosomes are strings of functional units as shown in Figure 7.4. In biological cells, chromosomes package DNA sequences and the DNA codons are translated from amino acids into proteins at the ribosome. In the active network implementation of genetic programming, the string of functional units operates upon active packets from other applications (or other functional units) that traverse through the node. The chromosome is represented in the code in a form similar to a Lisp symbolic expression, for example: ((Null Join Split) (Delay Split Join Delay)). Mutation and recombination occur among a population of genes. Mutation is a probabilistic change of a functional unit to another functional unit. Recombination is the exchange of chromosome sections from two different chromosomes.

Fitness functions can be designed to measure quality of service (QoS) at different layers of the traditional protocol stack. As a particular example, jitter control might have a fitness function that minimizes per frame variance at the link layer. The network layer uses its capabilities to maximize successful packet reception probability, in other words, perform the function of routing packets. The transport layer would have a fitness function that attempts to minimize end-to-end packet variance. The key is that each of these fitness functions must work together towards reaching the stated goal.

In keeping with the active network architectural philosophy of customized application handling within the network, fitness functions are "user" defined and injected into the network to control the evolution. For example, in our initial tests, minimization of variance in transmission time was used as a simple fitness function to reduce jitter. Initial experiments quickly demonstrated that the proper design of

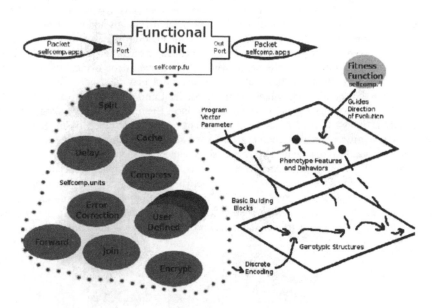

Figure 7.4   Active network self-healing.

the fitness function is crucial. It reminds one of the saying, "Be careful of what you pray for because you might get it." Often the fitness was achieved, but in ways that were unexpected and sometimes detrimental to the intended operation of the network. As a trivial example, slowing the traffic to a near halt can minimize the variance. Thus, a low latency term had to be added to the jitter control fitness function indicating that reducing jitter is good, but is also important to minimize latency.

While a priori techniques have obviously been developed for jitter control in legacy networks, this simple jitter control experiment demonstrates the adaptive in-line network genetic programming technique. The functional units injected into the network should allow evolution of a variety of interesting solutions to reduce variance, including adding delays, forwarding along different paths, or perhaps *new approaches that have not been considered.* In this particular experimental validation, the active network, genetically programmed, jitter control experiment evolved a solution that reduced packet transit variance by a factor of over 100.

7.6   CARBON NANOTUBE NETWORK ARCHITECTURES

In this section we look at carbon nanotube related architectural aspects, namely, ad hoc random carbon nanonetworks and the single carbon nanotube radio architecture.

7.6.1   Random carbon nanotube network architecture

Figure 7.5 compares a random carbon nanotube network architecture with that of a traditional network. In the case of random carbon nanotube networks, the traditional networking stack is conceptually merged because, rather than the network layer being positioned above the physical and link layers, network routing of information is an integral part of the physical carbon nanotube layer. It would be ideal to avoid or minimize network and routing as it is traditionally implemented through additional processing on each node in today's network nodes. At the bottom level, communication links in random carbon nanotube networks may be carbon nanotubes overlapping at points that will be identified as nodes, instead of the links between hosts and routers found in traditional networks. Data transmission occurs via modulated current flow through the random carbon nanotube network. Gate control may be used to induce routes through the random carbon nanotube network, which is initially divided into different gate areas [3]. Information is routed through nodes by varying the nanotubes' resistance. Even though the topology is random, the network resistance is easily analyzed as per the discussion on network resistance in Chapter 4. For example, when an external gate is turned on, the nanotubes within that area become conducting. Memristance will also likely play a part in routing at the nanoscale. The sensing elements, which sense by variation in resistance, may act simultaneously as routing elements. In other words, the very act of sensing changes the resistance, which changes how information is routed through the network. This is enabled by changing state, similar to the manner of a routing table on a router; a gate-controlled electromagnetic field controls the resistance within the specific area of the random carbon nanotube network. It remains to be seen in what manner this architecture is generalizable to random memristor networks, in which resistance has a memory based upon past current flow.

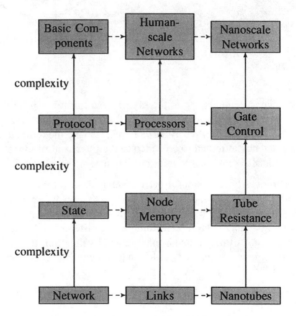

Figure 7.5   Nanoscale network architecture.

7.6.2   Single carbon nanotube radio architecture

In this section, we look at the architectural implications of the single carbon nanotube radio network. In Chapter 4, we discussed the receiver portion of the single carbon nanotube radio. However, work on a single carbon nanotube transmitter is ongoing [208]. It is understandable that the mindset of many researchers will be to develop an architecture as close as possible to human-scale wireless transmission at the nanoscale; we proceed along these lines.

The hardware for a single carbon nanotube radio system can be divided into five fundamental components: transceiver, power, processor, memory, and sensing units all at the nanoscale. The nanoscale transceiver is the carbon nanotube radio, discussed in detail in Chapter 4. The nanotube radio transmitter consists of an oscillator, modulator, amplifier and antenna, all within a single nanotube. The nanotube is mechanically oscillating at the frequency of the carrier signal. When a DC voltage is applied to the nanotube, charge is concentrated at the tip of the nanotube. Therefore, when the nanotube oscillates it radiates an oscillating electromagnetic field. Self-oscillation of a nanotube occurs with only an applied DC voltage [209]. A counterelectrode controls the oscillation frequency. An appropriate DC voltage can be applied that will cause simultaneous field emission and self-oscillation. Frequency modulation (FM) in the nanotube transmitter can be accomplished by modulating the mechanical resonant frequency of the nanotube by changing the tension on the nanotube with a voltage on a stationary electrode beneath the nanotube. The mechanically oscillating charge on the tip of the nanotube is the analog of a small antenna. The difference is that the movement of the electrons creating the electromagnetic wave in the nanomechanical transmitter is driven by mechanical motion as opposed to electrical current in a traditional wireless antenna. Power control might be implemented by changing the magnitude of the mechanical self-oscillation. Note that this is a mechanical modification of power control, rather than a traditional electronic approach. Certainly, increasing the charge on the tip of the nanotube will also increase the radiated power. Amplitude modulation may be implemented with these techniques as well. An analogy to beamforming might be used with an array of nanotransmitters [208].

Just as in the discussion of MRI-induced navigation of magnetotactic bacteria discussed earlier, the nanotube volume for a transmitter is 3.9×10^4 nm^3, which is not only small enough to travel through capillaries, but able to fit within a single cell. However, this size does not include any auxiliary components such as power supply and control circuitry. Also, path loss through aqueous media such as blood and tissue would likely rapidly attenuate such weak signals. Molecular communication would be a more suitable choice for biological environments. Our last topic in this chapter regards the architectural aspects of quantum nanoscale networking, in particular, the need for purification of entangled states and how that impacts the architecture.

7.7 THE QUANTUM NETWORK ARCHITECTURE

This section assumes that the reader has a grasp of the material on quantum computation from Chapter 5 and quantum information theory from Chapter 6. Here we discuss quantum entanglement purification and a proposed quantum network architecture. Entanglement purification is required in order to increase the fidelity of the entangled pairs to the degree that they are "pure" enough to enable teleportation. Without purification, the entangled pairs suffer entanglement with the environment, losing their entanglement with one another. Entanglement purification along with entanglement swapping significantly impacts the architecture of a quantum network. There are several algorithms for purification, also called "distillation," since some entangled pairs are sacrificed in order to increase the concentration of higher fidelity, or more pure, entangled pairs. Then we discuss a proposed quantum network architecture.

7.7.1 Quantum entanglement purification

First, it should be made clear that experimental metropolitan-area quantum key distribution networks exist and have been deployed around the world including Boston [210], Europe, [211], and Japan [212]. These networks are human-scale and, of course, intended for distribution of quantum keys. The underlying mechanism of quantum transport is teleportation. So, a good starting point for nanoscale architectures that leverage quantum phenomena is to consider existing quantum architectures.

As we have discussed in Chapter 5, we require entangled Bell pair states in order to perform teleportation. Unfortunately, transmission of quantum states results in corruption of the quantum state as the states become entangled with their environment. Fidelity of quantum states is a metric used to describe how near the states are to being uncorrupted. Fidelity is defined as the probability that a measurement of two qubits would show them to be in the correct state, a Bell state in this case.

In this section we will discuss purification, a necessary step in providing proper entanglement in order to enable teleportation, and then review a corresponding architecture. Note that, in general, the term "purification" can be used to obtain a pure state from a mixed state. Here we are discussing the more specific case of "entanglement purification."

Recall that a mixed state cannot be described as a ket vector. It must instead be described by its associated density matrix. Density matrices can describe both mixed and pure states. A simple criterion for checking whether a density matrix is describing a pure or mixed state is that the trace is equal to one if the state is pure, and less than one if the state is mixed. Equivalently the von Neumann entropy is zero for a pure state, and strictly positive for a mixed state. Assume an entangled state shown in Equation 7.15.

$$\psi^- = \frac{|10\rangle - |01\rangle}{\sqrt{2}} \tag{7.15}$$

The fidelity F is shown in Equation 7.16, where ψ^- is the perfect entangled state and M is a general mixed state. F will also be known as the purity of the entangled state, which should approach one if M is pure.

$$F = \langle \psi^- \,|\, M \,|\, \psi^- \rangle \tag{7.16}$$

Let M be a pure state of two particles $|\phi\rangle \langle\phi|$. The entanglement as mentioned earlier is $E(\phi) = S(\rho_A) = S(\rho_B)$. Recall from the previous chapter that $\rho_A = tr_B(\langle\phi\,|\,\phi\rangle)$.

The algorithm for entanglement purification is comprised of four steps as follows. In the first step, Alice and Bob perform a random but equivalent $SU(2)$ rotation. The special unitary group of degree n, denoted $SU(n)$, is the group of $n \times n$ unitary matrices with a determinant of one. The group operation is that of matrix multiplication. The operator for the rotation is shown in Equation 7.17.

$$U = \begin{pmatrix} \alpha & -\bar{\beta} \\ \beta & \bar{\alpha} \end{pmatrix} \tag{7.17}$$

This yields Equation 7.18.

$$W_F = F|\phi^-\rangle \langle\phi^-| + \frac{1-F}{3}|\phi^+\rangle \langle\phi^+| + \frac{1-F}{3}|\psi^+\rangle \langle\psi^+| + \frac{1-F}{3}|\phi^-\rangle \langle\phi^-| \tag{7.18}$$

In the second step, each of the pairs is acted upon by σ_y which effectively converts the result from containing a majority of ϕ^- states to a majority of ψ^+ states. The component of ψ^+ states has $F > 1/2$ with the remaining states equal.

In the third step, an exclusive-or XOR is applied on both sides impacting the two ϕ^+ states. Then, both Alice and Bob measure the target pair along the z axis. The unmeasured source pair is kept if the target pair's spins come out equal, which is the case of both inputs being ψ^+ states, and is discarded otherwise. It is in this step that classical communication is required in order to compare the results of the XOR operations. Finally, if the source pair is not discarded, it is converted back by a unilateral σ_y rotation. Then a random bilateral rotation is applied. This entire four-step process can be repeated, potentially increasing the entanglement of the remaining pairs.

It is possible, after each state from the entangled pair is distributed to our two favorite people, Alice and Bob, that they can each measure the resulting states locally on the same prearranged arbitrary basis. Then, the probability P of obtaining the same outcome can be determined and the fidelity can be computed based upon these results. Namely, $F = 1 - \frac{3P}{2}$. If they always obtain opposite outcomes, $P = 0$ and the fidelity is perfect. If the outcomes that Alice and Bob obtain are the same up to 2/3 of the time, then the fidelity of the entangled state is near zero.

A Bell state is shown in Equation 7.19. Suppose the first qubit in each ket is Alice's and the second is Bob's. It is clear immediately that there are only two possible outcomes, either both are one or both are zero. In other words, the resulting measurements between Alice and Bob are correlated. In addition, there is a 50% chance for each of the correlated outcomes, either one or zero. Thus, there is a high (maximum) entropy for such a sequence of outcomes.

$$\frac{|00\rangle + |11\rangle}{\sqrt{2}} \tag{7.19}$$

If m is the number of high-fidelity Bell states and n is the total number then n/m is called the distillable entanglement, which provides a simple measure of the amount of entanglement in a system. The goal of entanglement distillation is to reach a value of unity for this measure.

The number of copies of a pure state that can be converted to maximally entangled states is simply equivalent to the von Neumann entropy $S(\rho)$ as discussed in the last chapter. The von Neumann entropy ranges from zero for a simple product-form state to one for a maximally entangled state. The entanglement measure E is shown in Equation 7.20, where A and B refer to Alice and Bob, respectively, whose outcomes are correlated.

$$E = tr(\rho_A \ln \rho_A) = tr(\rho_B \ln \rho_B) \tag{7.20}$$

Entanglement distillation converts N copies of an arbitrary entangled state ρ into $S(\rho)N$ maximally entangled Bell pairs. The goal is to do this using only local operations and classical communication (LOCC). The result is that both spatially separated parties will share maximally entangled states enabling teleportation and quantum cryptography.

For a pure state, we can use the following method, known as the Procrustean technique. The method is named after Procrustes, a mythical Greek giant who stretched or shortened his captives in order to fit them into his beds. Suppose only maximally entangled or product states fit into his "beds." Then, he can make any state fit into a "bed" using the method we will describe, but he may have to chop off all the entanglement to do it. We describe the technique using Alice and Bob again, who share the state shown in Equation 7.21, where $1/2 < p < 1$.

$$\sqrt{p}|00\rangle + \sqrt{1-p}|11\rangle \tag{7.21}$$

Alice performs a measurement, with outcomes corresponding to the operators shown in Equations 7.22 and 7.23.

$$M_1 = \sqrt{\frac{1-p}{p}}|0\rangle\langle 0| + |1\rangle\langle 1| \tag{7.22}$$

$$M_2 = \sqrt{\frac{2p-1}{p}}|0\rangle\langle 0| \tag{7.23}$$

When outcome 1 is obtained, the state is left in a maximally entangled state; on obtaining outcome 2, the state is left in a product state. The scheme is successful with probability $2(1 - p)$ and this is actually the optimal success probability. It is relatively easy to check that these are valid measurement operators, that is, $M_1^{\dagger} M_1 + M_2^{\dagger} M_2 = I$, where I is the identity operator. Hence, it can be implemented by bringing in an ancillary qubit, performing a unitary interaction between Alice's qubit and the ancilla, and then measuring the ancilla.

Now that we have introduced entanglement purification, the next section introduces a quantum network architecture. The quantum architecture is presented a manner similar to traditional network architectures, including protocols, a notion of fields in packets, and protocol layering.

7.7.2   Quantum network architecture

A quantum architectural diagram is show in Figure 7.6 [213]. As is traditional in quantum drawings, wavy lines indicate quantum interactions and straight lines indicate classical communication. The first aspect to notice about the architecture is the difference in the quantum stations in the figure. The stations at the extreme left and extreme right are the sender and receiver nodes. Recall from Chapter 5 that in order to increase the distance of a quantum channel and avoid decoherence, entanglement swapping can be used. Thus, the stations between the sender and receiver in the architectural diagram of Figure 7.6 are intermediate nodes known as quantum repeaters. Their function is to act as "bridges" over which entangled pairs can be transported. Recall from Chapter 5 that entanglement swapping involves an intermediate node to share two sets of entangled pairs with two different nodes, one on each side of the quantum repeater. The repeater essentially links the two sets of pairs together, through the process of entanglement swapping. The result is a direct connection between the nodes on either side of the quantum repeater. The swapping operation can continue indefinitely, in a pairwise manner, through chains of quantum repeaters. For example, in Figure 7.6, stations one and two swapped with stations two and three to connect stations one and three, and similarly, stations three and four swapped with stations four and five to directly connect stations three and five. Then stations one and three swapped with stations three and five to connect station one directly with station five, thus completing the end-to-end channel from the sender to the receiver.

At the lowest level, the physical layer, physical entanglement (PE) occurs. At the sending node, an optical pulse is entangled with physical qubits, then transmitted over a long-distance optical fiber. At the receiver, the optical pulses are entangled with a free qubit. Properties of the pulses can be measured to determine whether entanglement was successful [214]. The receiver responds with "keep" or "discard" values for each qubit. Thus, the sender will be aware of which entangled pairs were successfully created. This is the entanglement control (EC) layer. Entanglement purification can take place between the lowest layer channels at the repeaters, before any entanglement swapping occurs. This is the purification control (PC) layer. Then entanglement swapping occurs at the entanglement swapping control (ESC) layer. Since, as has been mentioned, there may be an indefinitely long chain of quantum repeaters, the process of entanglement swapping may take place more than once. In fact, as should be clear from the explanation, each entanglement swap combines two n hop paths into one $2n$ hop path. Thus, the height of the protocol stacks as shown in Figure 7.6 will grow taller as more quantum repeaters are used. After all the entanglement swapping is completed and an end-to-end channel is formed, entanglement purification is again applied in a final purification control (PC) layer.

The example shown in the previous figure is a simple linear path. However, if there is a more complex network of repeaters, then the architecture must support routing through the repeaters. Such a routing protocol would need to be able to determine the center repeater for each iteration of swapping that is to take place. Also, both the stations and the qubits must be addressable so that the intermediate stations can keep track of which qubits are being swapped and the outcome of the swap.

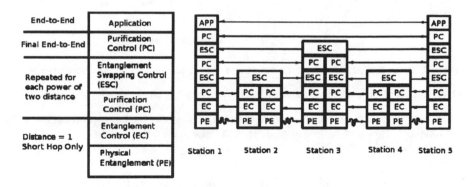

Figure 7.6 A quantum network architecture.

7.8 SUMMARY

This chapter has reviewed several existing network and hardware architectures for use in nanoscale networking. We began by looking at a finite automata model of a nanoscale network, which helps to put architectural considerations into an analytical form. Then we introduced the relationship between the network architecture required for deep space to earth communications, as specified in delay-tolerant networking, and the requirements for a nanoscale network architecture. We looked for similarities between interplanetary Internet and nanoscale networking. It turns out that both have similar problems that are already being addressed by the delay-tolerant networking architecture. Then we looked at what we might learn from research into network architectures for implant communications. The primary goal of this architecture is safety as well as trying to overcome the unnatural fit between wireless and biological systems. Then we looked at the active network architecture. In particular, we considered how active networks have features that may contribute to a self-assembling nanoscale network. We examined recent work on the carbon nanotube radio transmitter and how it may lead to a wireless carbon nanotube radio architecture. Finally, we looked at aspects that drive quantum network architectures.

How are nanoscale and molecular networks likely to evolve? What new nanoscale technologies are on the horizon? What applications will benefit and drive nanoscale networking? Finally, what is the future of nanoscale and molecular communications? All these and other questions are addressed in the next chapter.

7.9 EXERCISES

Exercise 52   Superdiffusion

Recall the discussion of superdiffusive node movement and the impact on delay-tolerant networking.

1. How does superdiffusive node movement relate to scale-free phenomena?

2. Suppose molecular motors are network nodes and that they obey the Langevin equation as discussed in Chapter 1. How does molecular motor motion relate to the Levy walk?

Exercise 53   Timed Automata Communication

Recall the discussion of timed automata for nanoscale communication.

1. Given a timed automaton, what is the impact of increasing the communication radius on the underlying graph diameter?

2. Given a timed automaton, what is the impact of increasing the communication radius on the optimal retransmission rate k for a *nonbroadcast* transmission?

3. Given a timed automaton, what is the impact of increasing the communication radius on the optimal retransmission rate k for a *broadcast* transmission?

Exercise 54   Epidemic Routing

Recall the discussion of epidemic routing as a mechanism for delay-tolerant network routing.

1. Use the definition of information entropy from Chapter 6 to explain how the notion of "antientropy" applies in epidemic routing.

Exercise 55   Quantum Network Architecture

Entanglement distillation and purification are critical aspects of the quantum network architecture.

1. What is a measure for the amount of entanglement in an ensemble of quantum states?

2. Explain two methods for achieving entanglement distillation.

3. What overhead does the need for high-quality entanglement add to a quantum network architecture?

Exercise 56   Nanotube Radio Network Architecture

Recall the discussion of a single carbon nanotube radio network.

1. Why is an amplifier not required to drive the antenna in the nanomechanical radio transmitter?

2. What technique(s) may be used to achieve frequency modulation?

3. What technique(s) may be used to achieve amplitude modulation?

4. What challenge will the nanoradio need to overcome to transmit through biological tissue?

5. Is collision detection possible with the nanoradio? What are other alternatives to requiring collision detection?

Exercise 57   Impact of Self-Assembly

The previous chapter considered two approaches to self-organization: Wang tiles and interaction information. How do these approaches impact the nanoscale network architecture?

1. How does the notion of Hamming distance apply to the complexity of a network that self-organizes by means of complementary DNA base pairs?

2. Is there a relationship between nanoscale assembly with complementary DNA base pairs and Shannon's graph capacity [215]? If so, what is the relationship?

3. Is there a relationship between nanoscale assembly with complementary DNA base pairs and Korner's graph entropy [216]? If so, explain the relationship.

Exercise 58   Active Network Architecture

Consider a vesicle released that contains a molecular computation reaction as one form of active nanoscale networking. Also consider a sender choosing to emit from a set of chemicals that can interact within the channel.

1. What are the benefits of active networking in the NANA architecture?

2. What benefits might be achieved by molecular interaction within a channel? Consider that the sender and receiver can both emit interacting chemicals. What is the analogy with beamforming and smart antennae?

3. Suppose a transmitter can release a chemical that improves channel performance before sending an actual message. What other ways can *active* network approaches be applied to nanoscale networking?

Exercise 59   Legacy In Vivo Network Architecture

Recall the discussion of today's in vivo network architectures, for example, MICS and IEEE TG-BAN.

1. Why do current implant architectures have such a short range?

2. How do implant network routing techniques enforce user safety?

3. Is implant routing that minimizes energy consumption equivalent to routing that minimizes thermal radiation? Explain how they differ or why they lead to the same result.

Exercise 60   Delay-Tolerant Networks

Recall the discussion on the relationships between delay-tolerant networking (DTN) and nanoscale networks.

1. What are all the analogous properties between DTN and nanoscale networks?

2. How can DTN routing techniques be applied in nanoscale networks?

Exercise 61   Nanoscale Localization

Recall the discussion on nanomachine localization.

1. Why is localization important for nanoscale networking?

2. How could nanomachine steering and localization be implemented? Explain at least one approach.

3. Consider a wire loop of radius 50 nm carrying 100 mA of current. What is the magnetic field strength at a distance of 1 micron away from the loop? Plot the rate at which the magnetic field strength decreases with distance.

Chapter 8

Conclusion

Prediction is very difficult, especially if it's about the future.

Niels Bohr

In this final chapter, the future of nanoscale networks is posited. The previous chapters have presented topics that have been vetted through research. In this chapter, the topics look towards the future of nanoscale networks and thus may tend to be a bit more fanciful and speculative, although we will continue to cite research backing up predictions whenever possible. We begin with a discussion of a hypothetical olfactory communication network, in which molecules transmitted through the air are used to communicate. Components of a such a communication system already exist; until the notion of nanoscale and molecular communications was developed, few researchers had thought of using these components as a digital communication channel. This could be a relatively near-term advance. Applications will directly impact the human scale, for example, communication through harsh environments, in which today's electronic devices disintegrate, such as monitoring the internal environment of running turbines, engines, or factory processes. High temperatures, pressures, electromagnetic noise, and corrosive chemicals make traditional communication, in any form, impossible. Such an environment may actually be ideal for some types of nanoscale and molecular communication that rely upon molecular diffusion for transmission of information.

A longer-term advance will be the interconnection of different underlying nanoscale network technologies. Today's communication networks, including the Internet, are seamless connections of networks. Each network may be comprised of a different underlying technology, yet the networks are seamlessly connected. Thus, we consider an Internet of nanoscale networks, namely, connecting together a variety of different nanoscale technologies in a seamless manner as well connecting such networks to the macroscopic world.

Next, we take the potentially foolhardy step of predicting the future of nanoscale network technology. In particular, we look at what technologies and applications are likely next steps for advances in nanoscale networking, standards that are needed, conferences that are likely to advance the state of the art, and finally some thoughts on new theory that is either needed or may be developed. Finally, we end with a speculative look at applications that will benefit from nanoscale network technology. These include all aspects of our daily life from energy to health care.

8.1   OLFACTORY COMMUNICATION

A potential future application of nanonetworks may be olfactory communication. It has long been known that the sense of smell is one of the most primitive senses; olfactory information connects directly to an area deep within the limbic system of the human brain that also includes the area where emotion, behavior, and long-term memory are located. Thus, it is postulated that the sense of smell has a deep impact upon humans of which they are largely unaware. Scent marketing [217], or the branding of products by companies by means of smell, is an example of attempting to exploit this deep, but little understood, connection within the human brain. Odors are closely tied to memory and emotions in long-term memory with the brain. The Proust phenomenon is the term for odor-combined memory recall. Smell and taste are closely linked; both rely on the olfactory nerves. Color-odor combinations are very strong; the food industry relies heavily on the ability of color to influence what we taste. There is some evidence that humans utilize pheromones as well as animals, including mothers who can accurately identify their infants by smell [218–220]. Given our sensory overload, commercial marketing research would like to find new ways to reach into and manipulate our minds and emotions; given the direct connection between the olfactory nerve and the emotional part of our brain, odors are definitely being explored as a communication channel. There are many interesting examples of products from hotels to automobiles that use odors and scents in a carefully controlled manner in order to increase our pleasure and impulsiveness. Because the olfactory nerve connects to our memory and emotions, the interpretation of odors becomes subjective, based upon the user's past experiences. Thus, researchers are experimenting with dynamically controlling odors in order to adapt them to each individual. Olfactory technology is becoming more dynamic and intelligent. Digitally controlled odors are now moving beyond marketing to more direct communication applications. For example, olfactory displays are being developed as a human-communication interface. There is an olfactory reminder system [221]. Olfactory email has been developed and patented as well as cell phone smell emitters and smell ring tones [222]. Thus, it appears that olfactory communication in various forms, is a small, but rapidly growing research area. The psychological impact of smell communication is a fascinating topic in itself, however, our goal is to explore the more mundane engineering task of simply communicating digital information via the olfactory system. The psychological impact of smell upon emotions and behavior hints at the possibility that our commonly accepted definitions of information from an engineering and information theory standpoint may be lacking in some deep aspects.

Information flow has often been modeled as fluid flow; here we want to explore whether we can literally transport information over a fluid. Given the complex pattern recognition techniques required to identify gases, gases contain a considerable amount of information; the flow of such gases could provide significant communication bandwidth.

8.1.1   Towards odor communication

Clearly, our focus is on molecular-level olfactory communication by the most efficient means possible, not constrained to odors that humans can detect. It should be noted that there is a difference between an odor and a pheromone. Both may be detected by olfactory organs. However, an odor is a smell detectable by humans while a pheromone may not be detectable as a smell. Electronic components for an olfactory communication system exist today. Electronic noses can detect odors and gases with some level of specificity. Mechanisms for the controlled release of odors are being developed, although not necessarily for an olfactory communication channel, but rather for digital transmission of smell over cell phone and Internet. However, our exploration involves inverting this approach, that is, rather than using traditional communication technology to transmit a sense of smell or taste, we seek to understand the capabilities of a communication channel comprised of odor molecules.

It is interesting to note that there have been odor communication systems considered in the past [223]. However, these are communication systems designed to transmit odors over conventional communication channels. The problems involved in the transmission of odors over conventional communication channels suggest the challenges that will need to be overcome in constructing an olfactory communication channel.

First, the olfactory sense has been one of our neglected senses; less intense research has focused on it than on the more lucrative senses of sight, where consumers will pay for quality video, and sound, where consumers will pay for high-quality music. The underlying physics of the sense of smell is complex and less straightforward than dealing with well-known electromagnetic and sound waves. The sense of smell of an odorant is determined by complex, and less well-known, interactions between ligand molecules and olfactory receptor molecules. The olfactory receptors within the nose are sensitive to hundreds of different kinds of odorants, interacting with each in different ways. The dimensionality of the sense of smell appears to be orders of magnitude larger than that of vision. Odor emission technology appears to be immature. Clearly, artificial generation of visual and auditory stimuli is done in high speed and with high quality; smells appear to be less well understood and harder to reproduce. Current techniques rely upon the interactive release of extracts that are prepared in advance, much like filling an ink cartridge in a printer.

While electronic noses are reasonably well advanced in order to provide the basis for an olfactory communication receiver, an automated process serving as a transmitter appears to be a challenge. Such a transmitter should be autonomous and flexible enough to emit any specific odor by accurately mixing and releasing its store of odors.

8.1.2 The odor receiver: The electronic nose

Electronic noses have been developed since the early 1980s. The challenge has been to improve their specificity, that is, it is not an easy task to determine precisely what odor or gas is being detected. Most accurate electronic noses today are known as quartz crystal microbalance sensors (QCM). This device measures mass per unit area by sampling the change in frequency of a quartz crystal resonator. The resonance frequency is disturbed by the addition or removal of a small mass due to oxide growth or decay at the surface of the acoustic resonator. They have a constant frequency of vibration at a constant temperature. The quartz crystals are usually coated with a chemical that tends to reduce the frequency. When exposed to certain gases and depending upon the nature of the gases, the frequency will change during exposure until it reaches a steady state. This new steady state, or resonant frequency, helps to identify the gas. The change in resonant frequency is based upon the change in mass of the gas. Let $f_k - f_{k+1}$ be the QCM change in resonant frequency and let the change in mass of the gas be $m_k - m_{k+1}$. Also, C_f is the mass sensitivity constant of the quartz crystal and A is the sensitive area of the device. The fundamental resonance of the quartz crystal is f_0, measured in hertz. The relationship between frequency and mass of the gas is shown in Equation 8.1 [224]. As A scales down in size, the sensitivity increases.

$$f_k - f_{k+1} = -\frac{C_f f_0^2}{A}(m_k - m_{k+1}) \tag{8.1}$$

The change in frequency of the QCM detector is not permanent. Once the target gas is removed and the air returns to its original state, the frequency also returns to its original steady state value. The application of the original air, usually room air, to the device is know as purging or cleaning. Multiple QCM sensors can be coated with different chemicals in order to be sensitive to different gases. For an odor communication channel, this could serve a few purposes. First, it can provide diversity, so that an odor transmitter could send multiple gases simultaneously in order to increase bandwidth. Second, multiple sensors may be able to better filter out noise, reducing the probability of error in detection. Finally, it may also improve the sensitivity to lower volumes of gas, reducing the need for the transmitter to consume its store of gases.

However, simply adding more QCM sensors with different chemical coatings does not necessarily result in better performance, in fact the opposite is possible. The choice of sensor becomes a feature classification problem, that is, determining which set of sensors best distinguishes the target gas of interest. Principle components analysis (PCA) is used to accomplish this task. Although PCA is well-known, we take the time here to provide a quick review. The ability to correctly distinguish a signal is a critical aspect of communication networking.

Principle components analysis provides a mathematical mechanism for finding the best basis to use when viewing a noisy set of data [225]. Intuitively, we can rotate the data into a better position, a position in which the strongest underlying pattern is clearly visible and the noise is minimized. The simplest basis that one could choose would be the identity matrix shown in Equation 8.2. Notice that the basis is an $m \times m$ matrix in this example.

$$B = \begin{bmatrix} b_1 \\ b_2 \\ \vdots \\ b_m \end{bmatrix} = \begin{bmatrix} 1 & 0\ldots & 0 \\ 0 & 1\ldots & 0 \\ \ldots & \ldots\ldots & \ldots \\ 0 & 0\ldots & 1 \end{bmatrix} = I \qquad (8.2)$$

We can let X in Equation 8.3 be the original data.

$$\vec{X} = \begin{bmatrix} x_A \\ y_A \\ x_B \\ y_B \\ x_C \\ y_C \end{bmatrix} \qquad (8.3)$$

We are assuming that the data somehow characterizes the dynamics of the system that it represents. The superposition principal of linearity allows us to interpolate between the data points. Thus, PCA may legitimately represent the data by any linear combination of its basis vectors. Let X be the original data, let Y be the transformed data, and let P be a matrix operator that transforms the data as shown in Equation 8.4.

$$PX = Y \qquad (8.4)$$

P can be thought of as a rotation and stretch that transforms X into Y. Also, the rows of P are new basis vectors for representing the columns of X, as shown in Equation 8.5.

$$PX = \begin{bmatrix} p_1 \\ \ldots \\ p_m \end{bmatrix} \begin{bmatrix} x_1 & \ldots & x_n \end{bmatrix} \qquad (8.5)$$

The result is shown more explicitly in Equation 8.6.

$$Y = \begin{bmatrix} p_1 \cdot x_1 & \ldots & p_1 \cdot X_n \\ \vdots & \ddots & \vdots \\ p_m \cdot x_1 & \ldots & p_m \cdot x_n \end{bmatrix} \qquad (8.6)$$

In Equation 8.7, each column of Y is shown. Thus, each coefficient of y_i is a dot product of x_i with the respective row in P. The jth coefficient of y_i is a projection onto the jth row of P, by definition of the dot product.

$$y_i = \begin{bmatrix} p_1 \cdot x_i \\ \ldots \\ p_m \cdot x_i \end{bmatrix} \tag{8.7}$$

A signal-to-noise ratio (SNR) falls out of the analysis and is defined in Equation 8.8, where the σ^2 are the variances. Let us determine where these variances originate.

$$SNR = \frac{\sigma^2_{signal}}{\sigma^2_{noise}} \tag{8.8}$$

Suppose we have two sets of data A and B shown in Equation 8.9.

$$\begin{aligned} A &= \{a_1, a_2, \ldots, a_3\} \\ B &= \{b_1, b_2, \ldots, b_3\} \end{aligned} \tag{8.9}$$

If we assume the mean has been subtracted, resulting in a zero mean, and we use the notation $\langle \cdot \rangle_i$ for the average value over i, then the variances are shown in Equation 8.10.

$$\begin{aligned} \sigma^2_A &= \langle a_i a_i \rangle \\ \sigma^2_B &= \langle b_i b_i \rangle \end{aligned} \tag{8.10}$$

Now, we can define the covariance as shown in Equation 8.11.

$$\text{covariance of A and B} \equiv \sigma^2_{AB} = \langle a_i b_i \rangle_i \tag{8.11}$$

Now, consider the data in row vector form as shown in Equation 8.12.

$$\begin{aligned} a &= [a_1, a_2, \ldots, a_n] \\ b &= [b_1, b_2, \ldots, b_n] \end{aligned} \tag{8.12}$$

The covariance can be expressed as the matrix computation shown in Equation 8.13. Note that the result is normalized by $n - 1$ rather than n to provide an unbiased estimate, which is a technicality that would take us off topic to explain.

$$\sigma^2_{ab} = \frac{1}{n-1} ab^T \tag{8.13}$$

It is possible to expand the covariance matrix operation to an arbitrary number of dimensions, beyond the two shown in the previous equation. This is done in Equation 8.14, which is an $m \times n$ matrix, where the rows of X can be thought of as all measurements of a particular type and the columns of X are all measurements for a particular trial.

$$X = \begin{bmatrix} x_1 \\ x_2 \\ \vdots \\ x_m \end{bmatrix} \tag{8.14}$$

The covariance matrix S_X for our new multidimensional X is shown in Equation 8.15. The diagonal elements of the matrix are the variances of measurement types and the off-diagonal elements are the covariances between measurement types.

$$S_X = \frac{1}{n-1} X X^T \tag{8.15}$$

Recalling the SNR from Equation 8.8, an underlying assumption is that the signal of interest has a large variance and the noise has a low variance. We are trying to linearly transform the data in such as manner as to best distinguish between the signal and the noise; the assumption is that the noise is smaller than the signal. The goal is to find a matrix P, such that $Y = PX$ and $S_Y = \frac{1}{n-1} Y Y^T$ is diagonalized. The rows of P are known as the principle components. Eigenvalue decomposition can be used to find the principal components as shown in Equation 8.16 where the matrix $A = X X^T$.

$$
\begin{aligned}
S_Y &= \frac{1}{n-1} Y Y^T \\
&\quad \frac{1}{n-1} (PX)(PX)^T \\
&\quad \frac{1}{n-1} P X X^T P^T \\
&\quad \frac{1}{n-1} P (X X)^T P^T \\
&\quad \frac{1}{n-1} P A P^T
\end{aligned}
\tag{8.16}
$$

Given that A is symmetric, it can be diagonalized in the form shown in Equation 8.17. Here, D is a diagonal matrix of eigenvalues and the columns of E are the eigenvectors.

$$A = E D E^T \tag{8.17}$$

By selecting the rows of P to be eigenvectors of $X X^T$, $P = E^T$. Using Equation 8.18, we see that $A = P^T A P$. Now, continue from the last line of Equation 8.16 as shown in Equation 8.18.

$$
\begin{aligned}
S_Y &= \frac{1}{n-1} P A P^T \\
&\quad \frac{1}{n-1} P (P^T D P) P^T \\
&\quad \frac{1}{n-1} (P P^T) D (P P^T) \\
&\quad \frac{1}{n-1} P P^{-1} D (P P^{-1}) \\
&\quad \frac{1}{n-1} D
\end{aligned}
\tag{8.18}
$$

Since the diagonal of D contains the eigenvalues of $X X^T$, the largest eigenvalues yield the most interesting principle components. In other words, the signal is in the largest eigenvalues and the noise in the lower ones. For our odor communication system, this provides an indication of our signal-to-noise ratio, the mutual information, and ultimately, the channel capacity.

Work on high resolution sensing of gases, for example, via carbon nanotube networks [154], will enable not only identification, but also very accurate measures of concentration and even localization of molecules, all of which hold communication information. Nanotubes allow high resolution detection and localization of particles. This could enable an electronic nose composed of nanotube sensors to detect vast amounts of information from fluids, that is, not just the type of fluid molecule, but their relative locations to one another. Carbon nanotube networks were discussed in detail in Chapter 4, here we consider them as sensors for an electronic nose. Note that, as was just discussed, human-scale electronic noses depend upon change in frequency of quartz crystals for detection; the random nanotube networks are based upon a change in electrical conductance for detection.

As an example, a carbon nanotube electronic nose can detect as low as 44 parts per billion of NO_2 and 262 parts per billion of nitrotoluene [226]. Nitrotoluene has several uses including as a taggant for explosives, allowing pre- and postdetonation determination of where an explosive was manufactured. The release and detection of a taggant is itself a form of information transmission. The random carbon nanotube network discussed in Chapter 4 can be used as a detector. In one specific implementation, a random single-walled carbon nanotube network was deposited on gold contacts in the shape of a fence. A random nanotube network was exposed between the openings in the fence, allowing greater contact with target chemicals. The mechanism of detection is a change in conductance as described in Chapter 4. During operation, there is both a detection time and a recovery time. In the specific detector of [226] detection time is on the order of seconds and recovery time is on the order of minutes. Adsorption, or the collection of target molecules on the nanotube network, results in the donation or acceptance of an electron which changes the Fermi level in semiconducting nanotubes and changes the conductance. An interesting hypothesis is made that there are two types of changes in conductance, one in which target molecules simply adsorb to a nanotube and change its conductance, and another in which target molecules actually form a nanotube-molecule-nanotube junction, essentially changing the topology of the random resistor network. This particular nanotube network was also able to detect large organic molecules such as benzene, acetone, and nitrotoluene. Since these molecules are larger, it is assumed that they tend to increase conductance by forming intertube junctions.

8.1.3 Pheromone impulse response

While our interest is in nanoscale and molecular communication, that is, communication by means of information encoded into individual molecules, it will be necessary to step back and consider the bulk behavior of gases in general. Let us take a short historical interlude into the nature of gases and diffusion of gases in particular, since this is important for understanding the nature of how such information would travel. We have already discussed Brownian motion in Chapter 2. Diffusion can be considered from the point of view of individual atoms or molecules in such random motion. Diffusion can also be considered from a higher-level view using Fick's laws, which will be described shortly.

Just as Brownian motion was recognized since ancient times, so too diffusion was not only recognized in ancient times, but it was harnessed for the making of steel. The cementation process involved the diffusion of carbon through iron to create early forms of steel [227]. Gaius Plinius Secundus who lived from AD 23 to August 25, 79, and was also better known as Pliny the Elder, mentioned the process of cementation. He was a natural philosopher as well as a naval and army commander of the early Roman Empire. As a side note, the date of his death is well-known because it corresponds to the eruption of Mount Vesuvius; he died after quickly sailing in to examine the effects of the destruction caused by the eruption, which was an example of diffusion on a grand scale. The process of diffusion appears to have been largely ignored until the Scottish chemist Thomas Graham noticed that gases don't appear to separate into layers based upon their density with the heaviest on the bottom, but rather, mix together. He was the first to experimentally measure this mixing and publish the results in 1833. The notion of the coefficient of diffusion had not yet been deduced, however, Maxwell, in 1867, used the results of Graham to compute the coefficient of diffusion for CO_2 in air; he was accurate to within $\pm 5\%$. However, it was Adolf Fick who, in 1855, developed the laws of diffusion that later allowed Maxwell to compute the coefficient of diffusion. Fick was neither a chemist nor a physicist, but a physiologist. His interests were in medicine and the roles diffusion played in the human body. Fick knew of Graham's results and also of work by Fourier on heat flow as well as Ohm's law for electric current. Fick had the deep insight that diffusion must operate in a similar manner. Fick theorized that the flux, or flow through unit area per unit time, of a substance is proportional to the gradient of its concentration. He called the constant of proportionality "a constant dependent upon the nature of the substances," which we will call the diffusion coefficient. The modern form for diffusion along one dimension is shown in Equation 8.19. Here J is the flux in $\frac{mol}{m^2 s}$, D is the diffusion coefficient $\frac{m^2}{s}$, ϕ is the concentration in $\frac{mol}{m^3}$, and x is the distance in m.

$$J = -D \frac{\partial \phi}{\partial x} \tag{8.19}$$

To relate this back to individual particles, D is proportional to the squared velocity of the diffusing particles. This is dependent upon factors such as temperature, viscosity, and the size of the particles. In the more general case of multiple dimensions Fick's first law takes the form shown in Equation 8.20.

$$J = -D \nabla \phi \tag{8.20}$$

Fick's second law quantifies how the concentration of a substance changes with time and can be derived from the first law. If the mass of the substance remains constant, then the concentration will decrease by the rate at which the flux area expands as shown in Equation 8.21.

$$\frac{\partial \phi}{\partial t} = -\frac{\partial}{\partial x} J \tag{8.21}$$

Inserting Fick's first law for J, Equation 8.22 is obtained.

$$-\frac{\partial}{\partial x} J = -\frac{\partial}{\partial x} - D \frac{\partial \phi}{\partial x} \tag{8.22}$$

Finally, after some simplification, the final form shown in Equation 8.23 is obtained.

$$\frac{\partial \phi}{\partial t} = D \frac{\partial^2 \phi}{\partial x^2} \tag{8.23}$$

The general form for multiple dimensions is shown in Equation 8.24.

$$\frac{\partial \phi}{\partial t} = D \nabla^2 \phi \tag{8.24}$$

A classic work that examined olfactory communication among animals is [228]. This work examined the transmission characteristics of a pheromone transmitted from one animal to another of the same species, thus, considering a pheromone in the strict sense of its definition as olfactory communication among the same specifies. The assumption is that the pheromone becomes active once it exceeds a given threshold. Thus, there is no notion of explicitly encoding multiple signals. The focus is upon the diffusion process, rate of emission, threshold concentration, and rate of diffusion. Four cases are considered: (1) pheromone released as instantaneously in still air, (2) pheromone released continuously in still air, (3) pheromone released continuously from a moving source, and (4) pheromone released continuously in wind.

These results are applicable with some adjustment to any fluid media, both air, which was the media originally focused upon, as well as aqueous and other fluid media. Of course, we are making very idealized assumptions. Any notion of wind, or flow, introduced into the transport system could also introduce turbulence, which would greatly complicate the analysis. Thus, the simplest initial cases to consider are instantaneous and continuous emissions in still air. The instantaneous case is actually slightly more complex, since the initial "puff" of pheromone will fade out, while in the continuous case, we can consider an approximate constant concentration.

The second case, continuous emission in still air, can be thought of as an expanding sphere of pheromone centered on the emitting source. We assume a three-dimensional (x, y, z) Cartesian coordinate system with the transmitter located at position $(0, 0, 0)$. There is also an assumption that the transmitter is at ground level, so that there is a ground plane intersecting at $z = 0$; any pheromone hitting the ground is reflected back into the air. Let r be the distance from the transmitter and equal to $z^2 + y^2 + z^2$. D is the diffusion coefficient in $\frac{cm^2}{s}$. The original equation for this case was put forth in [229] and is shown in Equation 8.25. This equation can be derived from Fick's laws discussed earlier in this chapter [230]. Here $Q_{message}(t)$ is the rate of pheromone emission, or message molecules, at time t in molecules per second. $U(r, t)$ is the concentration of molecules at distance r from the transmitter.

$$U(r, t) = \int_0^t \frac{Q_{message}(t^*)}{4\pi D(t - t^*)^{\frac{3}{2}}} e^{\frac{r^2}{4D(t - t^*)}} dt^* \tag{8.25}$$

Since this is a continuous release of a pheromone, the release begins at time t^*. Also, the coefficient of diffusion, D, is shown in Equation 8.26. Here k is Boltzmann's constant, T is temperature, η is the viscosity, and R is the radius of the size of the particle. It is important to realize that D is only an approximation, since it can vary with the concentration and shape of the particles.

$$D = \frac{kT}{6\pi\eta R} \tag{8.26}$$

In the simplest case, the continuous release can be held constant, rather than modulated. In this case, $Q_{message}$ is not a function of t as shown in Equation 8.27.

$$U(r, t) = \frac{Q_{message}}{4D\pi r} \operatorname{erfc}(\frac{r}{\sqrt{4Dt}}) \tag{8.27}$$

As $t \to \infty$, the complementary error function disappears and Equation 8.28 results. This is the steady-state concentration, as it is no longer a function of time. The function simply decreases at a linear rate as one moves away from the source.

$$U(r) = \frac{Q_{message}}{4D\pi r} \tag{8.28}$$

Now let us consider the threshold required for activation of a receptor that will lead to Equation 8.32. Assume that a nanorobot releases messenger molecules at rate $Q_{message}$ carrying $I_{message}$ bits per molecule. The density, or concentration, of message molecules, as a function of time t and distance r from the source is shown in Equation 8.29.

$$U_{message} = \frac{Q_{message}}{4\pi Dt}^{3/2} e^{\frac{r^2}{4Dt}} \tag{8.29}$$

The units of $U_{message}$ are $molecules/m^3$. D is the diffusion coefficient for message molecules.

$$D = \left(\frac{kT}{\eta}\right) \left(\frac{D_{message}}{162\pi^2 I_{message}}\right) \tag{8.30}$$

Fluorocarbons are a potential example of a type of messenger molecule that could hold relatively large amounts of information [231]. These organic molecules can utilize one carbon atom C to store a bit. An H atom connected to the carbon chain backbone represents a 0 and an F atom in a similar location represents a 1. Given this form of molecular encoding, the size of message given the molecule is shown in Equation 8.31. $r_{message}$ is the spherical radius.

$$I_{message} = \frac{4}{3}\pi r_{message}^3 D_{message} \tag{8.31}$$

We can assume that reception requires a concentration-dependent minimum sensor cycle time $t_{sensor} = t_{EQ}$ and that there is a minimum threshold concentration required for detection as shown in Equation 8.32. Here $N_{encounters}$ is the number of receptor-ligand encounters, N_A is Avogadro's number, $r_{message}$ is the radius of the message size, and MW is the target ligand molecular weight.

$$C_{min} = \frac{N_{encounters}}{r_{message}^2 t_{sensor}} \sqrt{\frac{I_{message} MW_{unit}}{4\pi k T N_A}} \tag{8.32}$$

So the minimum concentration required C_{min} clearly increases when the number of encounters required increases and it also increases with molecular weight and the size of the message in bits. It decreases with temperature and it also decreases with a longer cycle time, as there would be more time for an encounter between ligand-receptor to occur. Finally, it also decreases with the radius of the message molecule, a larger radius having better chance of an encounter as well.

Recall that we are assuming a continuous emission at a fixed rate, so that signal concentration is an expanding sphere with a fixed value assumed to be above the detection threshold. Therefore, we need only know the radius of this sphere to find the range of receptors that will detect the signal. The range is shown in Equation 8.33. Again, $Q_{message}$ is the message release rate and D is the coefficient of diffusion, and C_{min} was just derived in Equation 8.32.

$$R_{max} = \frac{Q_{message}}{4\pi C_{min} D} \tag{8.33}$$

The time that it takes for the expanding message sphere to reach a proportion p of its maximum radius is approximated by Equation 8.34. Here, f_R is the fraction $0 < f_R < 1$ of the expansion of the message sphere. Also, $\text{erfc}(p^*) = p$, where erfc is the complementary error function, which was discussed following Equation 3.1 in Chapter 3.

$$t_R = \left(\frac{pQ_{message}}{p^* 4\pi C_{th} D^{\frac{3}{2}}}\right)^2 \approx \left(\frac{1.1 f_R Q_{message}}{8\pi C_{th}(1 - f_R)} D^{\frac{3}{2}}\right)^2 \tag{8.34}$$

8.1.4 Towards an olfactory transmitter

In this section, we consider the more challenging aspect of developing a nanoscale olfactory transmitter. Olfactory receivers, or sensors, can be constructed at the nanoscale. However, odor transmitters are only now being developed at the human scale; development at the nanoscale is likely to occur in the future. Therefore, we will take some time in this section to consider some of the challenges in odor transmission. Pheromone transmission that tries to use complex mixtures of pheromones to increase bandwidth, is likely to have similar challenges. First, the human olfactory process generally receives little attention, yet is the most complex sense with which we are endowed [232]. There are many more types of odor receptors than there are photo or audio receptors. Unlike vision and hearing, which have a huge commercial market driving ever-increasing improvements in video and audio, our olfactory sense is little understood. In [232] and [223], the concept of an odor recording and playback system is explored. The question is whether a recorded odor can actually be played from a minimal mixture of different odors. In essence, the question is analogous to a printer in which only a few primary colors are required in order to print any transmitted color. Unfortunately, as mentioned previously, odor is much more complex and there does not appear to be a simple set of "primary" odors. However, there has been some progress in this direction. First, researchers would like to increase the range of types of odors that can be created while at the same time reduce the number of components required to create the odors. In [232], an artificial citrus flavor was recorded and "played back." A typical orange odor contains 14 different liquid components. This orange odor was analyzed by an electronic nose. Next, only three of the 14 different components were used. The relative ratios of each of the components was dynamically changed and the result analyzed by the electronic nose. The goal was to adjust the values of the three components until they matched the analysis of the original orange flavor comprised of the 14 components. A rigorous test by a panel of humans indicated that they were unable to tell the difference between the original and synthetically generated flavor. This is the first known attempt at automated odor recording and playback in which the playback synthetically generated a result using a mixture with fewer odorants than the recording. It should be noted again that this is a human-scale exercise designed for sensing by the human olfactory system. However, we can imagine a nanoscale transmitter releasing a controlled mixture of pheromones or odors in a combined concentration and molecular encoded signal.

The automated human-scale odor transmission concept is taken a step further by developing a general algorithm for mixing odorants to generate a target odor [233]. Keep in mind that while they are concerned with recording and replaying human detectable odors, we are thinking in terms of a communication transmitter and receiver. They assume a quartz-microbalance sensor (QMB) electric nose, which is analogous to our receiver. Thus, the results are specific to this type of receiver. Active odor mixing systems already exist as discussed in the previous section. An active system simply automates the task of experimentally mixing odorants and checking how closely they match a target odor. In some sense, this is a brute-force way of finding the correct mixture. The work we discuss in this section is a step towards developing an algorithm for predetermining the odorant mixture that matches a given target odor. There are some interesting results that may be applicable to a molecular encoded odor communication channel.

Let us proceed using the notation in [233]. A "pure" chemical or odorant will be denoted o and its concentration c. The pair $(o; c)$ stands for an odorant at a given concentration. When an electronic nose "sniffs," or receives the molecular signal, it registers the result by recording a list of features, depending upon the type of electronic nose and its sensing elements. Assume that there are m such features. These features are represented by an m-dimensional *response vector* $r(o; c)$. Each feature can change in a concentration-dependent manner, called the response curve. Thus, there are m possible response curves for odorant o that will fully characterize the behavior of an odorant in an electronic nose, or our receiver.

So far, we have been discussing only a single odorant; it may activate m sensors, and activate them in a different manner. In fact, it is of interest that human olfactory receptors are nonspecific as well. In other words, there is not a one-to-one mapping of receptors to odorants, but rather, a single odorant may activate multiple receptors. Artificial electromechanical noses work in a similar manner. Therefore, let $r(o_1, \ldots, o_n; c_1, \ldots, c_n)$ be the electronic nose response to the mixture of pure chemicals $o_1, \ldots, o_n$ in concentrations $c_1, \ldots, c_n$, respectively. Now comes the most complicating factor: Knowing the response curves of the individual pure chemicals, or odorants, does not necessarily allow us to determine the response curve of a mixture of such chemicals. Since the odorant detectors are nonspecific the response form of one odorant can be affected by the presence of another odorant. In other words, they interfere with one another in a potentially nonlinear manner.

Recall that there is a difference between the response vector of m-dimensions and a response curve, the change in response over a range of concentrations. With regard to the response vector, the linear law of mixtures was demonstrated as shown in Equation 8.35 where $\alpha_1, \alpha_2, \ldots, \alpha_n$ are constants called the mixing coefficients. This result is not perfectly accurate; there are still small variances of one or two percent from this linear approximation, but it appears to be a valid and useful result.

$$r(o_1, \ldots, o_n; c_1, \ldots, c_n) = \alpha_1 r(o_1; c_1) + \ldots + \alpha_n r(o_n; c_n) \tag{8.35}$$

Suppose we have a "palette" of a transmitter consisting of n odorants $t_1, \ldots, t_n$. Assume the target odorant is $(o; c)$; we are interested in finding the mixing vector $v = (v_1, \ldots, v_n)^T$ that is the solution of the objective function in Equation 8.36. The L_2 norm is assumed for the vector norm.

$$v = \left\{ \begin{array}{l} \text{argmin}_v \, \|r(t_1, \ldots, t_n; v_1, \ldots, v_n) - r(o; c)\| \\ \text{such that } v_i \geq 0 \; \forall i \end{array} \right. \tag{8.36}$$

Again, keep in mind that response curves are created by variations in concentration whereas the response vector is comprised of m features due to the electronic noise detector responses. As we mentioned above, we are dealing with n odorants. Assume the response curve of the jth feature of the ith odorant to be $r_j(t_i; v_i) = \beta_{ji} v_i$. Thus, β_{ji} is a coefficient that adjusts the concentration v_i of odorant i due to the presence of odorant j, where A is an $m \times n$ matrix $A_{ij} = \alpha_i \beta_{ij}$, as shown in Equation 8.37.

$$v = \left\{ \begin{array}{l} \text{argmin}_v \, \|A_v - r(o; c)\| \\ \text{such that } v_i \geq 0 \; \forall i \end{array} \right. \tag{8.37}$$

Again, if all relationships were linear, this could be easily solved. Unfortunately, the response curves are actually rarely linear. An approach is to iteratively solve this nonlinear problem [233] based upon the linear approximation we have just derived. Thus, let $r_j^p(t_i; v_i)$ represent the response curve at the pth iteration of the algorithm we are about to explain. As is often done in situations that require linearity, a Taylor expansion around the point of interest can be used to approximate a linear portion of the function. The Taylor expansion is an approximation that is accurate near the point of interest. Our interest is the expansion of $r_j(t_i; v_i)$ around the point v_i^p which is shown in Equation 8.38.

$$r_j^p(t_i, v_i^p) = r_j(t_j, v_i) + \left(\frac{dr_j(t_i; v_i)}{dv_i} \right)_{v_i^p} (v_i - v_i^p) \tag{8.38}$$

In order to keep the notation from becoming too large and unwieldy, let β_{ji}^p be substituted in the expansion of Equation 8.38, shown in Equation 8.39.

$$\beta_{ji}^p = \left(\frac{dr_j(t_i; v_i)}{dv_i} \right)_{v_i^p} \tag{8.39}$$

In Equation 8.40, we simply apply the first order Taylor expansion, shown in Equation 8.40, to each of the odorants i that affect j's response.

$$\begin{aligned} r_j(t_1 \ldots t_n; v_1 \ldots v_n) &\approx \alpha_1 r_j^p(t_1; v_1) + \ldots + \alpha_n r_j^p(t_n; v_n) \\ &= \alpha_1(r_j(t_1; v_1^p) + \beta_{j1}^p(v_1 - v_1^p)) + \ldots + \\ &\quad \alpha_n(r_j(t_n; v_n^p) + \beta_{jn}^p(v_n - v_n^p)) \end{aligned} \tag{8.40}$$

The target odor vector is now shown in Equation 8.41, where again, p is the iteration through the algorithm and v_i^p are the concentrations required to create the target odor at iteration p. Equation 8.41 is the error in creating target odor τ using mixing vector v^p.

$$\tau^p = r(o; c) - [\alpha_1 r(t_1; v_1^p) + \ldots + \alpha_n r(t_n; v_n^p)] \tag{8.41}$$

The optimization problem can be stated as $dv^p = \text{argmin}\,_{dv} |A^p dv - \tau^p|$ where $A_{ij}^p = \alpha_{ij}^p = \alpha_j \beta_{ij}^p$ and $dv = v - v^p$. The goal is to adjust v in order to minimize the error. A small correction is made in each iteration. Letting $0 \le \gamma \le 1$ be a parameter used by the algorithm, the error adjustment is shown in Equation 8.42.

$$dv^p = \text{argmin}\,_d v |A^p dv - \gamma \tau^p| \tag{8.42}$$

Once Equation 8.42 is solved to find dv^p, then $v_i^{p+1} = \max(v_i^p + dv_i^p, 0)$. This value is then used to update the target as shown in Equation 8.43.

$$\tau^{p+1} = r(o; c) - \left[\alpha_1 r(t_1; v_1^{p+1}) + \ldots + \alpha_n r(t_n; v_n^{p+1})\right] \tag{8.43}$$

If the error is not reduced, that is, $|\tau^{p+1}| > |\tau^p|$, then γ is halved and the iteration is recomputed. There is no guarantee of converging to the global minimum error; however, the algorithm has been reported to work well in actual trials.

8.2 AN INTERNET OF NANOSCALE NETWORKS

A future consideration for nanonetworks will be how nanoscale networks will interface with one another. As usual, nature has already accomplished this via signal transduction in cells. Nature has provided interesting ways of assembling and self-assembling necessary resources at the nanoscale. These are known as nonequilibrium approaches, since such assembly is not the direct result of material "falling into place" or reaching physical equilibrium, but rather directed control of material movement. As we have already seen in Chapter 2, molecular motors can carry cargo from one location to another. In [234], nanocrystal quantum dots have been carried as cargo by molecular motors along microtubules. The idea is to use the molecular motors to carry the quantum dots into desired locations, effectively creating a scaffold for the quantum dots. Potentially, this can be an active and dynamic scaffold, in which the quantum dots can be moved on-demand. Linear chains of quantum dots are easily viewed by means of fluorescence microscopy. One of the interesting communication-related uses is to position the nanocrystal quantum dots so that fluorescence resonance transfer can be controlled given quantum dots of different sizes and with different emission spectra. One might think of the molecular motors as workers laying optical fiber.

We have discussed amplitude and frequency modulation for molecular transmission, by simply changing either the total amount of concentration or the rate of change of concentration. We have also mentioned the difference between concentration encoding, which as we have just described, involves modulating the concentration, and molecular encoding, in which different types of molecules are transmitted in order to increase the size of the "alphabet." The hope is that using different types of molecules allows the receiver to easily distinguish between them and a molecular division multiple access (MDMA) channel can be constructed. The idea is to have channels separated by different types of odors just as FDMA separates channels by frequency or TDMA separates them by time. While this may be possible with a careful choice of molecules and detectors, as we have seen already with odors, this is not generally true. It is likely that odors and pheromones, even those of different types, will impact one another at the receiver. Thus, simple schemes using multiple types of molecules, that is molecular encoding, for transmission in order to increase the bandwidth of a molecular channel may be complex to implement.

Here, let us make the simplifying assumption that the receiver is able to perfectly distinguish among multiple molecular types. While previously we had discussed the continuous emission of a single molecular signal, let us consider the case of instantaneous emission. Utilizing molecular diversity, it is not necessary to maintain a continuous AM or FM signal. Instead, each molecule acts as a distinct signal by its very nature. From [228], assuming still air and no molecular interference, the pheromone density $U(x, y, z, t)$ in units of mol/cm^3 can be obtained as shown in Equation 8.44. Here we assume Q molecules are released at the origin of a Cartesian coordinate system at time $t = 0$ and that there is a reflecting ground plane, from which molecules bounce back into the space above the plane. D is the diffusion constant in cm^2/sec, which depends upon the transmission medium, and r is the radius of a sphere centered around the emitter such that $r^2 = x^2 + y^2 + z^2$. The concentration decreases at an exponential rate away from the emitter.

$$U(x, y, z, t) = \frac{2Q}{(4\pi Dt)e^{\frac{-r^2}{4Dt}}} \tag{8.44}$$

We assume again that there is some threshold density of the molecular signal K in mol^3/sec required to trigger the receiver's detector. The concentration must exceed K in order for a detection to occur. Recall that we are modeling a single "puff" being emitted from the transmitter and expanding through space. Thus, there is a surface of a sphere expanding outwards from the source that is of concentration K. Setting $U(x, y, z) = K$ and solving for r as a function of time results in Equation 8.45.

$$R(t) = \begin{cases} \sqrt{4Dt \ln\left(\frac{2Q}{K(4\pi Dt)^{3/2}}\right)} & 0 \le t \le \frac{1}{4\pi D}\left(\frac{2Q}{K}\right)^{3/2} \\ 0 & \text{otherwise} \end{cases} \tag{8.45}$$

The concentration is above the threshold within the sphere, however, as the sphere expands, the concentration decreases since there is only a single instantaneous emission of molecules. Thus, there is a maximum size of the sphere, beyond which the density falls below the threshold K. This is shown in Equation 8.46. Notice that this value is independent of the coefficient of diffusion.

$$R_{max} = \sqrt{\left(\frac{2Q}{K}\right)^{2/3} \times \frac{3}{2\pi e}} = 0.527\left(\frac{Q}{K}\right)^{1/3} \tag{8.46}$$

The time at which the maximum range is reached for a given concentration K is shown in Equation 8.47.

$$t_{R_{max}} = \frac{1}{4\pi De}\left(\frac{2Q}{K}\right)^{2/3} = \frac{0.464}{D}\left(\frac{Q}{K}\right)^{2/3} \tag{8.47}$$

Once the maximum range is reached, the concentration within the sphere of signal molecules will decrease, decreasing below the reception threshold K and eventually reaching zero. The time for the concentration to fall below the threshold is shown in Equation 8.48.

$$t_{fadeout} = \frac{1}{4\pi D}\left(\frac{2Q}{K}\right)^{2/3} = et_{R_{max}} = \frac{0.126}{D}\left(\frac{Q}{K}\right)^{2/3} \tag{8.48}$$

A receiver located near the transmitter will sense a high threshold for a longer duration than a receiver located farther from the transmitter. In other words, a receiver located farther away will sense a faster on/off signal pulse. The bandwidth for a modulated signal concentration will depend upon being able to rapidly pulse the signal concentration. A large diffusion coefficient should allow a faster pulse transmission and a faster fadeout.

The concentration gradient is shown in Equation 8.49. This is useful information for a system in which we would like to tune the transmitter to the rate of change of concentration rather than an absolute concentration level.

$$\frac{dU}{dr} = \frac{-Qr}{(4\pi)^{3/2}(Dt)^{5/2}} e^{-r^2/4Dt} \tag{8.49}$$

At the nanoscale, we can imagine analogies with current human-scale technologies. We have seen molecular motors, akin to simple mechanical transport of information. We have discussed the single carbon nanotube radio, an analog of the human-scale wireless system. The diffusion-based systems, such as calcium signaling and olfactory communication, are somewhat unique, perhaps having a closest analog to smoke signals. However, we can also imagine a nanoscale analog of the human-scale fiber-optic system as well.

8.3   OPTICAL TRANSMISSION WITHIN NANOSCALE NETWORKS

The low propagation delay and high bandwidth of optical transmission may enable it to play a similar role at the nanoscale as it does today within networks, namely a long-haul interconnection between shorter-range networks, for example, as an optical channel to connect between smaller networks utilizing molecular motors or calcium signaling. It could also play a role in connections between the human scale and nanoscale, just as, for example, fluorescent marking techniques are used in biology today. Clearly, interfaces would have to be developed to convert a signal from individual molecules to light and back again. One obvious use of light at the molecular level is chemiluminescence, long used by biologists in various ways to illuminate and track cellular activity [235]. Both fluorescent proteins and the molecular organic light-emitting diodes (MOLED) can be included here.

8.3.1   Fluorescence resonance energy transfer

Chromophores and a fluorophores are the parts of a molecule responsible for their color and their potential fluorescence. The light properties at the molecular level are determined by molecular orbitals, which describe the wavelike behavior of an electron in a molecule. This function can be used to calculate chemical and physical properties such as the probability of finding an electron in any specific region or changing an electron from its ground state to an excited state. When a molecule absorbs certain wavelengths of visible light and transmits or reflects others, the molecule has a color. Visible light that hits a chromophore can be absorbed by exciting an electron from its ground state into an excited state. There are biological molecules that capture or detect light energy and the chromophore is the region that causes a conformational change of the molecule when hit by light. A fluorophore, in analogy to a chromophore, is a component of a molecule that causes a molecule to be fluorescent. It is a functional group in a molecule that will absorb energy of a specific wavelength and reemit energy at a different wavelength.

Now imagine that we can line up such fluorescent molecules and use them to transfer an optical signal from one to another. This is known as fluorescence resonance energy transfer (FRET). It is an increasingly popular microscopy technique used to measure the distance between two fluorophores. Resonance energy transfer occurs only over short distances, typically 10 nm, and involves the direct transfer of energy from the donor fluorophore to an acceptor fluorophore. Upon transfer of energy, the acceptor molecule enters an excited state from which it decays through radiation, emitting a longer wavelength than that of the acceptor emission. Thus, by exciting the donor and then monitoring the relative donor and acceptor radiation one can determine when FRET has occurred and at what efficiency. Fluorophores can be used to label biomolecules. The distance condition for FRET is of the order of the diameter of most biomolecules, thus, FRET is often used to determine when and where two or more biomolecules, usually proteins, interact. Since energy transfer occurs over distances of 1 nm to 10 nm, a FRET signal corresponding to a particular location within a microscope image provides a distance accuracy surpassing the optical resolution, which is approximately 0.25 mm, of light microscopes. In addition to spatial proximity, the FRET fluorescent pair must also exhibit overlap of the donor's excitation spectrum with the acceptor's absorption spectrum. The spectrum of the FRET pair cannot be so far apart that there is poor overlap, but at the same time, "cross-talk" between the two imaging channels needs to be avoided. In other words, one wants to sample only the light energy from the donor and none from the acceptor, and vice versa. It is possible to use short bandpass filters that collect light from only the shorter wavelength portion of the donor emission and the longer wavelength portion of the acceptor emission.

A fluorescent protein is a protein, comprised of amino acids, that fluoresces at a certain wavelength when exposed to a different wavelength. As a historical note, these molecules have been developed since the cloning of the green fluorescent protein (GFP) from the jellyfish *Aequoria victoria* (jellyfish) in the 1960s that was sensitive to calcium ion concentration. Since then, fluorescent proteins with different colors and more efficient light output have been developed. The efficiency is measured by the quantum yield. This is the number of times that a defined event occurs per photon absorbed by the system. Thus, the quantum yield is a measure of the efficiency with which absorbed light produces some effect. The quantum yield can be defined by $Q = $ photons emitted/photons absorbed. Green, yellow, and red fluorescent proteins have the highest efficiency at 60%. Genetically encoded fluorescent markers that are sensitive to cell signaling molecules, such as calcium or glutamate, protein phosphorylation state, protein complementation, receptor dimerization, and other processes provide highly specific optical readouts of cell signaling activity in real time. Thus, conversion of molecular signaling information to the optical domain is already a common occurance [236]. In addition to stimulation by radiation and light, it is, of course, possible to create light with electrical stimulation, even at the nanoscale as we discuss next.

8.3.2   Electroluminescence

Electroluminescence is the generation of light from electrical excitation [237]. A molecular organic light-emitting diode (MOLED) is a nanoscale semiconductor that may be used to convert nanoscale information into optical information. Past research on OLEDs has focused on efficient lighting as well as flat panel displays. They involve the injection, transport, capture, and radiative recombination of oppositely charged carriers inside an organic material. However, they can serve as light sources at the molecular level as well. The oppositely charged carriers recombine within the organic material and form an exciton, a form of radiation, which then decays after transmission. The oppositely charged carriers, electrons and holes, propagate in a diffusive manner through the organic film following an Langevin type of model that we have described in Chapter 2 for molecular motors. [238] describes the use of a thin-film OLED sheet for medical implants. The low-power OLED generates light in vivo to examine the impact of light on cells in vivo.

Certainly, we can imagine that optical transmission occurring at the nanoscale will draw lessons from today's knowledge of human-scale free space optical communication in sensor networks [239]. Since light is an electromagnetic wave, we can deduce the received power from Maxwell's equations. Let P_T be the transmitted power, G_T be the transmitter antenna gain, G_R be the antenna gain in reception, k the frequency wavelength, and d the distance between transmitter and receiver. Equation 8.50 shows the receive power.

$$P_R = P_T G_T G_R \left(\frac{\lambda}{4\pi d} \right)^2 \tag{8.50}$$

In order to determine the range, d, we can solve for the distance d as shown in Equation 8.51.

$$d = \frac{\lambda}{4\pi} \sqrt{\frac{P_T}{P_R} G_T G_R} \tag{8.51}$$

Ranges on the order of millimeters can be achieved. As we have seen, while objects at the nanoscale are small, their frequencies tend to be high. Increasing the wavelength, that is, reducing the frequency, in some manner could help improve the power and range. Additionally, typical directional and cooperative techniques might be employed such as beamforming.

There are many other techniques for capturing light and converting it to other forms of energy, including biologically inspired techniques used in photosynthesis and the use of nanowires and other nanostructures in solar cells.

8.3.3 Molecular switches

The molecular switch is yet another biologically inspired nanoscale technique [240]. Molecular switches have been used in high density optical data storage, in which light is used to flip the state of a molecule, thus recording a one or a zero. However, the concept is broadly applicable to other activities besides simple storage of information.

The heart of a molecular switch is a molecule whose state can be changed in a stable and reliable manner by an external stimulus. This is known a bistability. There are a variety of stimuli that have been known to cause state changes in bistable molecules including electron transfer, light, heat, pressure, magnetic or electric fields, pH change, or chemical reactions. Here we focus on light-activated switching. Chiral optical switches are based upon molecules whose state is the chirality, or direction of the spiral shape of the molecule. The chirality, or direction of rotation of the spiral shape, can be switched, in a reversible manner, when exposed to certain wavelengths of light. A nice feature of chiral optical switches is that their chirality can be also be read, in a nondestructive manner, by light of a different wavelength from the one used to write, that is, to change the chirality. Optical switches could clearly provide low-power molecular memory to a nanoscale communication system, allowing more sophisticated signal processing.

Now that we have considered a variety of different techniques for nanoscale optical transmission, let us recap all the channel techniques that have been discussed and consider how they might be interconnected. The next section takes a speculative look at how all the nanoscale channels we have discussed may one day seamlessly connect to one another.

8.4 INTERNETWORKING NANOSCALE NETWORKS

It should be understood that the topic of nanonetworks is still in its infancy; the topic is being defined as this book is written. Thus, there are almost certainly many other nanoscale networking technologies that have not been thought of yet. In fact, one of the purposes of this book is to help stimulate further ideas in this area. For those nanoscale network technologies that are known at this time, we consider how they might interconnect with one another. For example, how can information being transported by a molecular motor be converted, or transduced, into information carried by a calcium signal, or a quantum dot? What is the minimal amount of computation required to perform this transduction to another channel's technology? Is there a "complexity difference" between technologies that indicates the level of processing required?

Table 8.1 shows a matrix of nanonetwork technologies and their potential interconnection mechanisms. It is not exhaustive, since there are other techniques we have discussed, such as neural transmission,
that have been omitted from the table in order to avoid a combinatorial explosion. The techniques included
are a representative sample of the most researched mechanisms. First, the matrix is symmetric, so we only
discuss the upper triangular portion. Second, the diagonal of the matrix below, which forms an identity
matrix, is a transfer of information from one nanoscale mechanism to the same nanoscale mechanism, thus
those interfaces are marked as not applicable (N/A). Third, references to papers that hint at research leading
in a direction that would enable development of that interface are indicated within the matrix whenever
possible.

Proceeding from the top row, molecular motors have been used to carry quantum dots, as discussed
previously. Quantum dots of the proper size absorb and emit optical energy. A direct interface between
molecular motors and calcium signaling does not appear to be well-defined at this point; however, the
reference points to work involved in understanding how calcium signaling regulates molecular motor
operation. Moving to the pheromone/odorant column, one could also imagine molecular motors transporting
and/or releasing pheromones or odorant molecules. With regard to nanostructures and nanotubes, molecular
motors have been used to "sort" microtubules. Also work has been done to understand the relationship
between molecular motors and another form of biological nanotube structure that forms between cells,
allowing for the communication of cytoplasmic material between cells and organelles of cells. Molecular
motors appear to be actively involved in the construction of such biological microtubules.

In the quantum channel row, fluorescent proteins are offered as a potential connection between a
nanoscale quantum network and calcium signaling, as well as between the quantum domain and the optical
domain. The field of quantum biology is positing many connections between quantum mechanics, quantum
computation, and biological processes. This includes such concepts as quantum communication through
microtubules [70] and even how our sense of smell operates [241]. Finally, nanostructures such as carbon
nanotubes have been proposed numerous times for use in quantum computing [242].

Next is the calcium signaling channel row. Here we have already mentioned that fluorescent proteins
have been used to study calcium ions. [243] looks at the regulation of calcium signaling systems by
pheromones. [244] discusses calcium flow through biological membrane nanotubes and [245] discusses
the use of carbon nanotube probes for transferring calcium directly into the cell.

Moving on to the optical channel row, it is not surprising that fluorescent techniques have been
used to study pheromones, which is the topic of [246]. This could provide a low-power interface between
detection of an odor and optical transmission at the nanoscale. An optical interface to nanostructures has
been discussed in Chapter 5 based upon plasmonics.

Finally, for the pheromone/odor channel interface with nanostructures, we have already discussed the
random carbon nanotube network as an electronic nose. Nanotubes have also been proposed as containers
for drug delivery; in this sense they may contain and release odorants as well. Again, it should be
emphasized that this is a new and rapidly developing field, with a combinatorial explosion in the number
of different types of nanoscale technologies that might be interfaced with one another. In the next section,
we look at a proposed in vivo nanoscale communication system that is designed to interconnect different
channel technologies.

8.4.1   The design of an in vivo nanoscale network

A future high-payoff application of nanoscale networking will be in vivo network communication. It is
anticipated that DNA coding various components of the molecular motor, particularly the neck that attaches
to cargo, will be modified to create new cellular transport systems. The motor may be engineered to carry

Table 8.1
Interconnection of Nanonetwork Channel Types

Channel	Mol. mot.	Quantum	Ca sig.	Optical	Pher./olf.	Nano str.
Mol. mot.	N/A	[234]	[247]	Q. Dots	Odorant	[248]
Quantum		N/A	Fl. Pro.	Fl. Pro.	[241]	[242]
Ca sig.			N/A	Fl. Pro.	[243]	[244, 245]
Optical				N/A	[246]	Plasmon
Pher./olf.					N/A	eNose
Nano str.						N/A

new types of cargo, namely, information packets, and also engineered to operate in a manner more desirable for controlled information transport, such as to construct intercellular membrane nanotubes. We take one last look at an in vivo network design that uses many of the concepts covered in previous chapters. This nanoscale network, discussed in [249], proposes an overlapping ring network topology, both DNA and enzymatic computing for biocompatibility, and a network stack that includes error correction. A form of DNA computing has been described in Chapter 3, where the idea is to construct a state machine with a sequence of DNA that is processed into shorter units as computation proceeds by shorter DNA "rules," which match the current state, that break off portions of the longer DNA sequence and eventually leave a final result. Enzymatic computing is used for shorter, faster processing lower in the protocol stack; it is used for computation related to switching and routing. DNA computing is reserved for higher-layer operations such as forward error correction (FEC), addressing, encoding, and decoding of a message.

Briefly, information transmission works as follows. Data is encoded as DNA using the automaton described in Chapter 3. Then, an address is also attached using DNA in a similar manner. One can imagine a portmapper that encodes the proper output onto the end of the information DNA sequence. As previously mentioned, there are overlapping ring networks. Thus, the DNA packet is switched onto the interface for the correct ring that leads to the proper destination. At each node, the address portion is cleaved from the DNA packet and is released into the cell's cytosol. The address portion provides input to the molecular interface control function of the network layer.

In the specific example of a gap junction interface, the output of the enzyme-based circuit will control the permeability of gap junction channels as discussed in Chapter 3. Gap junction permeability is affected by connexin phosphorylation via specific phosphorylation reagents. The concentration of the reagents is controlled by the switching circuit. At the physical layer, the DNA packet, minus its last address, is transported using the corresponding channel mechanism, which may be transported directly to another cell, or a vesicle. This process is repeated as the DNA packet travels from cell to cell and the address is stripped from the packet until it reaches its final destination. At this point, the DNA payload may be decoded and processed by the DNA computer within the destination cell.

It should be noted that the work described in this section is highly theoretical, but it is apropos for this chapter because it points toward the future of nanoscale and molecular computing, providing a vision that will inspire others to help solve the implementation challenges.

8.5 NANOSCALE NETWORK APPLICATIONS

In the limit, as nanonetworking is taken to the extreme, matter at its most fundamental level becomes part of a human-engineered communication network. Certainly, at different philosophical levels, man and matter are already interconnected.

> No man is an island, entire of itself; every man is a piece of the continent, a part of the main. If a clod be washed away by the sea, Europe is the less, as well as if a promontory were, as well as if a manor of thy friend's or of thine own were: any man's death diminishes me, because I am involved in mankind, and therefore never send to know for whom the bell tolls; it tolls for thee.

John Donne, Meditation XVII

In the case of nanonetworks, every molecule will be either part of a communication channel, represent a piece of data, or be the transmitter or receiver of information. Every molecule has the potential to be a sensor and transmit information about physical events. Scientists will have easy access to peer yet deeper into inner space with ease. Given such high resolution detail of the operation of inner space, we will be able to predict events long before they occur. Perhaps we may be able to delay the bell from tolling, or even prevent it from tolling. Product defects will be detected earlier and with such resolution that they will be unheard of at the human scale. Medical imaging will be performed by collecting and correlating in vivo nanonetwork data; current imaging technology, which requires exposure to potentially damaging radiation will no longer be required. In vivo medical "images" will be routinely taken of individual cells and organelles within cells, rather than crude clumps of contrast agent.

An area that will not improve is information assurance. In fact, information assurance, network security, and network management will become an even greater challenge as networks shrink in scale. Military applications and espionage will escalate in step with the reduction in network size.

In the nearer future, we can expect that longer-range nanoscale communication will provide pathways from inside the body to human-scale medical devices. Implant communication may be feasible without requiring the inhuman placement of batteries and radiation inside people. An implanted glucose monitor and an insulin pump may be able to communicate as an almost natural part of the human body. Connections from inside the body to a doctor's personal digital assistant will be commonplace. Remote health service checkups and preventive medicine may be possible.

Autonomous part inspection at the nanoscale will be routine; defects that are imperceptible today, but grow into serious problems that lead to later customer dissatisfaction, will be easily detected and repaired. Within the factory, there will be no environment too harsh for communications. Network communication through water, in areas of high electromagnetic interference, and high temperature that would melt current electronics will not be an issue with nanoscale networks. If any molecule can pass through the system, then an information channel can be formed. Nanoscale manufacturing and the dream of self-assembly will become a step closer to being realized with nanoscale networks. Nanonetwork sensing and control will yield the information necessary to understand and control self-assembly.

Lossless power grid transmission and fine-grained power switching will become a reality through nanoscale networks. Recall that quantum wire has ballistic transport, yielding near zero impedance to current flow. A woven "rope" of such quantum wire might serve as an ideal power cable. In addition, a truly "smart grid" would deliver electric power as packets of energy just like the Internet routes packets today.

One might imagine nanoscale devices attached to products that will provide new functionality. Smart materials will become a reality. The entertainment and gaming industry could detect eye movement more conveniently for three-dimensional displays, haptic devices could connect more directly with the human senses, and additional senses such as smell and taste might no longer be neglected in terms of virtual sensory input.

Nanoscale communication will use less energy enabling a greener environment. A single molecular reaction, which may represent multiple computations, consumes 10,000 times less energy than an electronic transistor [28]. Pollution and its effects may be monitored with an extreme level of coverage and resolution. Biohazard sensors in clothing could detect harmful radiation and fluids and communicate the results. The same coverage and resolution could be achieved for homeland defense. Radiation detection and communication of results need not require large batteries and bulky components. The final section provides an outlook on the future of nanonetworks. We discuss the current state as well as what the challenges and needs are for going forward.

8.6 THE FUTURE OF NANONETWORKS

The new field of nanonetwork communications is passing through its necessary, initial theoretical stages. Many different underlying nanoscale technologies and channel models are being proposed, some of which are being modeled, and fewer still are being implemented and tested. Much theoretical work exists, both from a protocol and architectural standpoint and from an information theory standpoint. The relationship between information theory and diffusion seems to be a common theme that is somewhat unique to nanoscale networks. We can expect that nanoscale networks will come full circle in returning information theory back to the underlying physics from which it came as we saw in Chapter 1. Hopefully, it will be returned in better shape than when it was first borrowed by Shannon and others.

As the field of nanonetworks progresses, the need for better simulation tools also increases. As of the date of this book, the tools available focus on specific subcomponents of nanonetworks, such as current flow in carbon nanotubes [250]. While the appendix lists some sites that make tools available, most of these tools are customized to very specific, usually nonnetworking, applications. In the near future a more complete nanonetwork simulation platform, open to the research community for development, would be desirable. Expanding the current network simulator-2 (ns-2) [251], for example, with nanoscale models could be a good start. Nanonetworking is a multidisciplinary field, heavily influenced by biology and physics. While this may initially be a limiting factor to people who are locked into traditional ways of thinking, it provides the traditional fields of electrical engineering and computer science an exciting opportunity to expand into a rich multidisciplinary field with many opportunities for research contributions. However, although there has been recent activity in actual short-range nanocommunication experimental validation, for example, calcium signaling and molecular motors, we are still largely in the realm of initial theoretical analysis and simulation.

From an industry perspective, it's probably wisest to proceed with the better-developed short-range nanoscale channels in conjunction with existing microscale sensors and implants. Standards will need to be created in order to give industry the confidence to proceed. Selected applications with clear, short-term payoffs will have to be defined, which will depend upon the specific industry that commercial ventures wish to target. Simultaneously, longer-range nanoscale communication approaches need to be developed, whether optical or pheromone/odor-based. However, any cutting-edge technology should be prepared for disruption as the next paradigm-shifting discovery may be made en route to developing any well-planned roadmap. The final subsection describes the ongoing efforts of an IEEE subcommittee that is both promoting and guiding research in this fascinating and potentially revolutionary new form of communication.

8.6.1   The IEEE Nano-Scale, Molecular, and Quantum Networking Subcommittee

The scope of the IEEE Nano-Scale, Molecular, and Quantum Networking Subcommittee is communication and networking in which most, or all, of the system resides on the nanoscale. This includes wireless nanoscale media in support of nanorobotics. Example media include carbon nanotubes, quantum dots, biological structures and molecular communication, and harnessing the advantages of quantum and hybrid classical/quantum effects for applications such as security and coding. This includes the goal of aiding the development of smart materials, nanoscale biomedical applications, and nanorobotics. This subcommittee exists to support the emerging community of engineers, academics, scientists, and others who are developing communication networks on the nanoscale. Activities such as special sessions, symposia, tutorials, and workshops in leading conferences will be vastly expanded to help provide the community with continuing updates about the burgeoning field of nanotechnology and nanoscale telecommunications in particular. A goal of this committee will be to provide a network for required diverse specialties, some of which are in fields typically found outside the IEEE, to come together to advance nanoscale communications. Joint events, special journal issues, and dedicated topical meetings are also organized with other relevant IEEE societies.

8.7   EXERCISES

Exercise 62   In Vivo Nanoscale Networking

Recall the in vivo nanoscale network described in Section 8.4.1.

1. From a qualitative standpoint, how are addressing and routing implemented?

Exercise 63   Fick's Laws

Find the flux, J, of oil droplets through water that float from the bottom to the top surface. The depth of the water is 18 cm, and the concentration of oil on the bottom is 0.1 mol/cm3. Assume that there is no oil at the top yet. $D = 7 \times 10^{-7} cm^2/s$.

1. What is the flux?

Exercise 64   Olfactory Communication

Consider the case of the ant *Pogonomyrmex badius* which has a continuous release of a pheromone alarm at a constant rate. The ratio of Q/K was found to be 1,400 cm^3 or 280 cm^3/sec. D is 0.43 cm^2/sec. K mol/cm^3 is the threshold density.

1. What is R_{max} the maximum sphere around the release site that is above the reception threshold?

Exercise 65   Application

Communication through a harsh environment, such as inside a factory furnace, will destroy today's sensitive electronics. Instead consider a chemical sensor that emits instantaneous gaseous signals to communicate with a boron nitride nanotube sensor that can survive extreme temperatures, up to 2973 degrees centigrade. Assume the chemical sensor releases gaseous chemical signals instantaneously.

1. What is the equation that describes the signal channel?

2. How does the extreme temperature impact the signal behavior?

Exercise 66   Electroluminescence

1. How does electroluminescence work?

2. How does electroluminescence relate to a molecular organic light-emitting diode (MOLED)?

3. How might electroluminescence and MOLEDs be used in a nanoscale network?

Exercise 67   Internet of Nanoscale Networks

Consider an Internet comprised of all the nanoscale technologies listed in Table 8.1.

1. Which nanoscale network interfaces might require flow control and queuing because they connect a higher-speed channel to a lower-speed channel?

Exercise 68   Light Transduction

Assume a molecular optical transmitter with an attenuation of 88 dB [252] and blue wavelength excitation (480 nm).

1. What is the optical transmission range?

Exercise 69   Quantum Dots and Molecular Motors

1. What is a useful relationship between quantum dots riding upon molecular motors and fluorescence resonance energy transfer?

Exercise 70   Fluorescent Proteins

1. How does the mechanism of fluorescence work in a fluorescent protein?

2. How are fluorescent proteins used by biologists?

3. How might they serve a nanoscale network transduction mechanism?

Exercise 71   Molecular Switch

1. How does a chiral optical switch work?

2. What uses would optical switches have in a nanoscale network?

Appendix: Nanoscale and Molecular Communication Network Simulation Tools

I find languages that support just one programming paradigm constraining.

Bjarne Stroustrup

Your paradigm is so intrinsic to your mental process that you are hardly aware of its existence, until you try to communicate with someone with a different paradigm.

Donella Meadows

This appendix provides information about simulation tools relevant to nanoscale and molecular communication networks. Since nanoscale and molecular communication networking is a very new concept, there are currently few simulators available targeted for this particular area. Instead, since communication at the molecular level bridges several disciplines, there is a wide spectrum of simulation tools that are relevant and provide partial empirical solutions. We can expect that as the field of nanoscale and molecular communication matures, these simulation tools, or more likely the paradigms and mechanisms implemented by these simulation tools, will merge into traditional communication network simulators. For now, the simulation tools are distinct and have grown out of their own individual fields; there are electro-optical, biological, classical and quantum mechanical, and communication network simulators. Therefore, we review the concepts as well as a few specific examples of simulators from each of these diverse fields. It would be impossible to cover every simulation modeling concept and tool from each of these fields; we can only look at a few representative samples. It is also true that simulation tools are evolving rapidly, thus, it should be understood that any information mentioned in this appendix about *specific* simulation tools will soon be outdated. Thus, the goal is to provide the reader with a clear sense of the simulation tools that exist in each of the fields, the paradigms that they use, and how they can fulfill the needs of nanoscale and molecular networking.

First, we classify the types of simulation tools that we will discuss as they relate to nanoscale networking. Then we look at a molecular communication network simulator that has been implemented in Java. Next, we step back and look briefly at network-on-chip simulators. These are more mature simulators that simulate communication at the nanoscale. However, as mentioned previously in this book, the goal of nanoscale and molecular networking is to move beyond the carefully designed environment that resides on a chip. Following this, we look at carbon nanotube related simulators. Most of these simulators are again related to chip fabrication, namely, they were generally designed to predict how nanotube structures would perform as interconnects between transistors on chips. Next, we look at typical biological simulators. Then we briefly discuss quantum mechanics and quantum computation related simulators. This discussion considers both the impact of quantum effects at the nanoscale and the utilization of quantum computing to implement nanoscale networking. Simulators specifically dedicated to quantum networking are not extant at this time; most quantum network simulations are constructed from general quantum computation simulation packages. Finally, self-assembly is considered. Self-assembly and self-organization, in their logical forms, are the dream and end-goal of many human-scale network researchers. Communication networks should ultimately self-organize to fulfill their mission and then automatically adapt as needed. At the nanoscale, self-organization is more intimately related with the configuration of the physical environment, namely, of individual molecules.

It is not often taken into consideration that the manner in which simulation tools model a problem influences and often constrains the way we think about and explore a given technology domain. Sometimes the models we are required to use in our simulation tools exert an unconscious influence upon us; as we successfully analyze and model the same way repetitively over many years, we reinforce the notions that these models represent. We come to believe that they are not only correct, but that they are the only correct way to represent the underlying mechanics of the system being modeled. Thus, it is important to consider the modeling paradigms. Since, as we have mentioned, nanoscale networking exists at the junction of several vastly different fields, the corresponding paradigms used by these simulators are also very different.

It should also be emphasized that software changes rapidly. Therefore, simply listing simulation packages extant at the time this book was written is of limited value; we point only to a few representative examples of specific simulation tools. However, we may learn something from the philosophy behind the development of the software tools and their modeling paradigm. Because nanoscale networking is at the interface between several distinct disciplines, each with their own simulation tools and, more importantly, their own way of thinking about how modeling should be done, it makes the topic of nanoscale and molecular simulation quite interesting.

At the network layer, almost every simulator assumes that a queuing model is the fundamental underlying representation of the network; queuing theory is the fundamental theory to be applied. These are usually event-driven simulators that assume a flow of packets through a network of nodes interconnected by channels. At a lower level, individual channels within a network are modeled by channel simulators. These are generally probabilistic simulations, with a link budget and probability of error. Channels have been historically either wired or wireless. Below the level of the channel model, there may be a hardware implementation model, namely, a model of the complete electronic circuit as well as electromagnetic wave propagation. Modeling at such a high level of detail comes at a cost; the computational power required to run the simulation in a reasonable amount of time simply may not exist. However, nanoscale network simulation is, by definition, operating at an extremely high level of detail, namely, at the level of individual molecules.

Network communication technology, its corresponding simulation tools, and their conceptual models may inadvertently lock us into one way of thinking, namely, everything is a queue. At the time this book is being written, wireless communication is the popular research topic of the day; therefore, most people think in terms of wireless channel models. We need to be careful not to constrain our thinking in this manner. In fact, it is interesting to consider the fact that Guglielmo Marconi demonstrated the wireless telegraph in 1896 and packet radio was well-established long before the computer was invented. Yet wireless computer networks took a surprisingly long period of time to evolve given that the fundamental technology existed long before. Before the wireless fad of today, Ethernet channels were most likely the dominant way of thinking. Given historical trends, it is likely that researchers will attempt to fit nanoscale communication analysis and simulation into the current dominant modeling framework, which is wireless communication.

However, nanoscale networking is at the crossroad of several disciplines, including physics and biology, which have their own simulation paradigms. Physics models are of course focused upon the fundamentals of matter, energy, diffusion, and both classical and quantum mechanics. Physics-based models would include the Langevin movement models. Biological simulations tend to take the form of biochemical concentration models, which we will discuss later in this appendix. The differential wave models of biology explain and model calcium signaling communication approaches.

Nanoscale networking is at the convergence of all of these disciplines and has unique simulation needs, namely, combining network simulators, electro-optical simulators, biological simulators, and both classical and quantum physics simulators. Unfortunately, this much-needed convergence of simulators has not yet occurred; nanoscale and molecular networking is still being done piecemeal. Each simulation is being handcrafted. A dominant theme underlying nanoscale and molecular communication appears to be diffusion. Thus, simulators based upon both active and passive diffusion are likely to be developed. We discuss one in this appendix.

To recap, the manner in which we build our tools has an impact upon us in ways that we may not be cognizant of. Thus, it is important to consider the manner in which our simulation paradigms are designed, have been designed in the past, and will be designed in the future. We also need to consider that the market has a huge impact on simulation tools. Questions and situations that occur repetitively in manufacturing and production have driven the development of efficient and modular simulation tools. Questions that a researcher believes are unique will result in the development of only a quick and crude simulator, while those that are recurring will encourage a stable simulator to be constructed.

Let us now consider the different aspects of nanoscale and molecular network simulation. Figure A.1 shows a taxonomy of nanoscale and molecular network simulation tools. Nanoscale network simulation is categorized into self-assembly, biological, and electro-optical simulators. Simulators of self-assembled systems are those based upon distributed agents using local rules to achieve global phenomena. Biological simulators are based upon concepts from molecular biology. Electro-optical simulators are those that simulate electrical or electromagnetic wave propagation. Underneath biological simulators we include diffusive simulators. Note that diffusive simulation is not strictly limited to biology (see Chapter 8 for details on odor communication). However, we place diffusive simulation in this location since, at this time, most diffusive nanonetwork approaches appear to be biologically related or biologically inspired. Under the electro-optical simulation category, we have simulators for interconnects, usually related to interconnects on computer chips, optical transmission simulators, wireless simulators, and finally quantum simulators. Interconnect simulators are focused on carbon nanotube chip interconnects. The wireless simulation category is mentioned primarily for simulation of the single carbon nanotube radio, although this may also include general oscillator simulations such as pulse-coupled oscillation. Finally, the quantum simulation category includes both the simulation of quantum behavior at the nanoscale as well as quantum computation simulators used to simulate communication (i.e., teleportation).

MOLECULAR NETWORK SIMULATOR

An early molecular motor communication simulation [253] has been developed. At the time this book was written, it exists only as a relatively simple, stand-alone simulation tool. It may likely be integrated into more commonly used network simulation tools as this new field progresses. Thus, there is some benefit in discussing how this simulation works. The purpose of this simulator is to model the movement of molecular motors along microtubules, with the intent of transporting information. First, the sender, receiver, and all molecules are represented simply as spheres; the microtubules are cylinders. Potential reactions occur when spheres and cylinders come into contact. The time interval is assumed to be milliseconds and the environment is assumed to exist on an enclosed plane containing a fluid whose viscosity is $0.001 Pa \cdot s$, namely, the same viscosity as water.

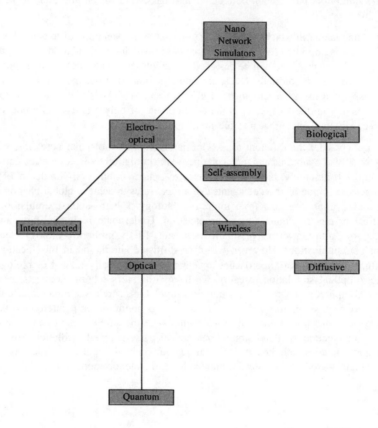

Figure A.1   Nanoscale simulator taxonomy.

In this simulation, the sender releases individual molecules and the receiver reacts to them. It is assumed that multiple molecular motors are attached to the transmitted molecule. It is important to note that the assumed paradigm is that of individual molecules being sent and received, not concentrations of molecules. Thus, the precise location and timing of individual molecules and the time that they arrive at the receiver are important. The molecular spheres are 100 nm in radius and the motors are 50-nm long. A kinesin motor is being modeled, so the motors walk to the positive end of the microtubule at a rate of 800 nm per second. When the molecular motors reach an intersection of microtubules, they switch to the intersecting microtubule probabilistically. That is, with some probability they may jump to the other microtubule and walk along its length. When not bound to a microtubule, the molecule, with its motors attached, is assumed to diffuse with Brownian motion as a sphere with radius 150 nm; the molecule may collide with other objects and surfaces.

The network topology within the simulation assumes a microtubule network with a single sender and a single receiver. The planar container in which the environment resides is assumed to be of infinite length and width and have a height of 10 micrometers. Molecules propagate in three-dimensional space within this area. The microtubule asters have their positive polarized ends in the center of the aster and are assumed to be composed of four 20-nm long microtubules. The receiver is at the center. In its current form, the simulation assumes that the topology remains static throughout the simulation.

There are two cases in which the simulation may take place: with aster- (star) shaped microtubule structures or with no microtubules and only Brownian motion. This provides a convenient comparison of the effect that the microtubules provide on the efficiency of information transfer. This simulator provides a nice example of what a nanoscale network simulation could look like. It simulates at the level of individual molecules and provides both a drift-diffusion (motors on microtubules) and pure diffusion channel model. In the next section, we switch gears and see what we can learn from the more mature network-on-chip simulators.

NETWORK-ON-CHIP SIMULATIONS

At the micro- and nanoscale, electro-optical simulation tools are mature and widely available. The intersection of these tools with network communications accelerated with the advent of network-on-chip (NoC) research. In [254], the benefits of routing packets instead of routing wires is summarized. Instead of laying individual wires as interconnects between components on-chip, a packet network provides general purpose network communication between components, over which packets are routed. Chip interconnects are becoming extremely dense as more components are squeezed into ever-smaller areas. Instead of placing interconnects for all possible data flows, the concept is to include the circuitry to enable packet routing of information inside the chip. While the additional circuitry to support a packet network within the chip is estimated to add approximately 6.6% to the required chip area, it can also provide many benefits, including reduction in the total amount of interconnects, fault tolerance, and better utilization of interconnects. It is estimated that in a traditional nonnetworked chip, an individual interconnect is only used approximately 10% of the time. Nanoscale network simulation of the NoC architecture has been implemented in *network simulator-2* (ns-2) [251, 255]. Thus, instead of simulating an entire network the size of a building, city, or larger portion of the Internet, the entire network fits within a 50- to 60-nm area of silicon. While a NoC architecture is not the goal of ad hoc nanoscale and molecular communication, it points to the use of a traditional network simulator to simulate a nanoscale communication system.

Much work has gone into making the NoC more resilient and adaptive to faults as the component and interconnect sizes continue to decrease. We have already discussed DNA controlled layout of interconnects and the corresponding network used to discover and adapt to misalignment in Chapter 4. Another NoC approach to ease the interconnect bottleneck is optical switching [256]. In this approach, an optical NoC resides in parallel with an electrical NoC. The electrical network carries short control and data messages, while the optical network handles larger data messages. This approach has also been simulated in a commercial networking package. The use of carbon nanotubes as interconnects has been the topic of much research and several simulation tools have been developed to address the performance of individual nanotubes as electrical conductors for the purpose of interconnects on chips. We briefly look at a few of these simulators next.

CARBON NANOTUBE SIMULATION TOOLS

As of the time this book was written, no freely available, fully detailed carbon nanotube network simulator exists. However, several simulators that address various components of nanotube physics exist. For example, *TubeGen* [257] allows one to specify the physical parameters for a nanotube, such as chirality and bond length as input, and the tube structure will be created as output.

Another tool, *CNT Bundle* [258] estimates both conductance and inductance of carbon nanotube bundles, with the specific intent of simulating their performance for interconnect applications. The analysis assumes a carbon nanotube bundle comprised of a mixture of multiwalled and single-walled carbon nanotubes. Such a structure can be a more realistic representation of the of the actual form of interconnects, in some cases, than a single nanotube.

Nano Heatflow [135] simulates the kinetic energy in oscillating carbon nanotubes over the course of a molecular dynamic simulation. Utilizing *TubeGen*, mentioned earlier, *Nano Heatflow* is able to generate nanotubes at the molecular level for simulation purposes. One can simulate varying vibrational modes at different temperatures. Recall that the single carbon nanotube radio is a form of an oscillating nanotube. Also, the flow of heat may someday be explored as a method of information transport.

Nanoscale and molecular communication is intimately related with the movement of matter as information. A field of physics deals with transport phenomena, namely the study of how matter moves from one place to another. This field is concerned with finding the laws that govern such transport and attempt to relate the flux of matter with force, in a variety of forms. This includes diffusion, convection, and radiation. These can all serve as forms of nanoscale and molecular communication. The Boltzmann transport equation is one form of analytical relationship between flux, force applied in the form of a field, and particle interaction; it provides the probability of finding a particle in a particular region at a given time. The derivation is interesting and can be found in [259], however, we leave that to the interested reader to pursue. The *Boltzmann transport simulator* [136] allows one to simulate electron transport in a carbon nanotube using the Boltzmann transport equation. The structure of the nanotube can be modified in this simulation, such as its chirality and length, and the resulting impact on the electron transport can be determined.

These are only a sample of tools, which came from the nanohub Web site. These tools enable the simulation of specific features of carbon nanotubes that are useful for a nanoscale network. They also indicate areas for future improvement and may serve as a starting point for constructing simulation platforms. Now that we have examined a few of the inorganic nanoscale communication related simulators, let us look at some biological ones.

BIOCHEMICAL SIMULATORS

Biochemical simulators are based upon the kinetics, or rate, of a chemical reaction. This is particularly important for molecular transmission and detection. There is at least two general approaches used in biochemical simulation, deterministic networks of coupled ordinary differential equations that describe the chemical reaction rates and their influences upon each other and a more random, stochastic approach that relies upon a single differential equation known as the chemical master equation [260]. Here we consider the more widely used approach of networks of coupled differential equations. Therefore, let us derive a commonly used equation, known as the Michaelis-Menten equation, in order to provide both a feel for how a biochemical simulator works as well as to explain a particular widely used, fundamental biochemical relationship.

Enzyme kinetics have been characterized by the Michaelis-Menten equation. In order to use biochemical simulators, it's important to understand what the Michaelis-Menten equation is. Once you have this understanding, as well as fundamentals of chemical reaction rate equations, you will have a good start towards understanding the modeling of information flow with biochemical simulators. The basic enzyme model is shown in Equation A.1, which we will explain shortly.

$$\mathrm{E} + \mathrm{S} \underset{k_{-1}}{\overset{k_1}{\rightleftharpoons}} \mathrm{ES} \xrightarrow{k_{cat}} \mathrm{E} + \mathrm{P} \tag{A.1}$$

The Michaelis-Menten equation is shown in Equation A.2. Let us derive this equation beginning with the simple enzyme reaction equation above.

$$v_0 = v_{max}[S]/(k_M + [S]) \tag{A.2}$$

Equation A.1 indicates that an enzyme E combines with a substrate S in a reversible manner, hence the double arrow in $\mathrm{E} + \mathrm{S} \underset{k_{-1}}{\overset{k_1}{\rightleftharpoons}} \mathrm{ES}$. Just as we saw in Section 6.4, the rate, sometimes called velocity, of the reaction is designated by k_1. Thus, the enzyme combines with the substrate at rate k_1 to produce the complex ES. Simultaneously, the complex ES is disassociating back into E and S at rate k_{-1}. However, if a catalyst is present, the complex ES will yield the same enzyme E and a product P at rate k_{cat}. This is known as a kinetic model, since rates, or velocities, of chemical reactions are involved. The reactions convey information at the nanoscale.

One way in which to view Equation A.1 is as a reaction controlled by the amount of substrate S. Consider the rate at which the product P is produced. Let this rate be indicated by v_0. Thus, from the right side of Equation A.1, we can see that Equation A.3 results, namely, the product is created at a rate that depends upon the concentration of ES represented as $[ES]$ and the reaction rate k_{cat}.

$$v_0 = k_{cat}[ES] \tag{A.3}$$

A potential problem is that the concentration of ES is generally unknown and, being in the center of Equation A.1, is undergoing complex kinetic changes. Thus, the motivation for the final equation is to represent v_0 in terms of more easily measurable components such as $[E]_{total}$ and $[S]$. The goal, then, is to remove $[ES]$ from the final equation.

In the steady state, one can assume that the rate of formation and consumption of ES is constant. It should be clear from the initial equation that the rate of formation is as shown in Equation A.4 and the rate of consumption is as shown in A.5.

$$[ES] = k_1[E][S] \tag{A.4}$$

$$[ES] = k_{-1}[ES] + k_{cat}[ES] \tag{A.5}$$

So in the steady state Equations A.4 and A.5 must be equal, as shown in Equation A.6.

$$k_{-1}[ES] + k_{cat}[ES] = k_1[ES] \tag{A.6}$$

Next, the rate constants from Equation A.6 can be moved to the left side, as shown in Equations A.7 and A.8.

$$(k_{-1} + k_{cat})[ES] = k_1[E][S] \tag{A.7}$$

$$\frac{k_{-1} + k_{cat}}{k_{-1}} = \frac{[E][S]}{[ES]} \tag{A.8}$$

Now that all the constants are gathered and since they are indeed constant, they can all be replaced by k_M, as shown in Equation A.9.

$$k_M = (k_{-1} + k_{cat})/k_{-1} \tag{A.9}$$

Now, again stating that our goal is to remove $[ES]$ and obtain only measurable quantities, we need to determine the total amount of E, which we will call $[E]_{total}$. This includes all the enzyme in the reaction including the E in the ES complex. Thus, the unbound concentration of E is shown in Equation A.10.

$$[E] = [E]_{total} - [ES] \tag{A.10}$$

Substituting the results of Equation A.9 and Equation A.10 into Equation A.8 yields Equation A.11.

$$k_M = (E_{total} - ES)[S]/[ES] \tag{A.11}$$

Next, solving for $[ES]$ results in Equation A.12.

$$[ES]k_M = [E]_{total}[S] - [ES][S] \tag{A.12}$$

Then, collect the terms with $[ES]$ on the left side, as shown in Equation A.13.

$$[ES]k_M + [ES][S] = [E]_{total}[S] \tag{A.13}$$

Recalling our goal of removing $[ES]$, we can factor it from the left side, yielding Equation A.14.

$$[ES](k_M + [S]) = [E]_{total}[S] \tag{A.14}$$

Then divide both sides by $(k_M + [S])$, leaving $[ES]$ by itself on the left side and yielding Equation A.15.

$$[ES] = [E]_{total}[S]/(k_M + [S]) \tag{A.15}$$

Now, substitute the above Equation A.15 into Equation A.3, yielding Equation A.16.

$$v_0 = k_{cat}[E]_{total}[S]/(k_M + [S]) \tag{A.16}$$

Note that the maximum kinetic rate, or velocity, of the reaction to produce the product P occurs when all of the enzyme is bound with S, forming the complex ES. This means that $[ES] = [E]_{total}$, that is, the entire concentration of E, being bound to S, implies that the total concentration of $[ES]$ is $[E]_{total}$. Thus, Equation A.17 represents this maximum rate of production of P.

$$v_{max} = k_{cat}[E]_{total} \tag{A.17}$$

Now, v_{max} can be substituted into Equation A.16 by replacing $k_{cat}[E]_{total}$ with v_{max}. This yields the final form of the equation shown in Equation A.18.

$$v_0 = v_{max}[S]/(k_M + [S]) \tag{A.18}$$

We can now speak of the rate of change of the product P as shown in Equation A.19.

$$\frac{dP}{dt} = v_{max}[S]/(k_M + [S]) \tag{A.19}$$

The substrate S correspondingly changes as shown in Equation A.20.

$$\frac{dS}{dt} = -v_{max}[S]/(k_M + [S]) \tag{A.20}$$

The biochemical kinetic simulation tools solve systems of such differential equations and are able to plot the change in concentration of substances over time. From a nanoscale and molecular communication network viewpoint, we would generally like to know the information-carrying capacity of a channel implemented by such rates of chemical concentration changes.

A specific example of such a simulator is Cellware, an easy-to-use and freely available simulation platform for building mathematical models of cellular processes [261]. It is capable of analyzing biochemical network properties, kinetic parameters, and network evolution. Cellware's basic interaction is point and click. The user can select a blank species template, a blank reaction node, and links, to couple the reactions and species. After the basic setup, knowledge of biology is a requirement. Fine-tuning the chemical species requires knowledge of the species and their various biological components. Reactions are flexible and allow the user to input reaction equations. Knowledge of mathematical formulas for modeling biological reactions is required.

Another specific example is Copasi (Complex Pathways Simulator), another freely available simulator [262]. Copasi is a biochemical network simulator and therefore may be a way to begin testing and simulating the functionality of biological-based nanonetworks. Model development with Copasi is mostly done through scripting, using its own script or the Systems' Biological Markup Language (SBML). Knowledge of SBML is also important, given that many models are described in this format.

Clearly, the interconnected chemical pathways form molecular communication channels. The ability to simulate these channels allows us to see to what degree these channels can be manipulated without causing harm to the host in which they reside while at the same time providing a reasonably efficient communication channel. Now that we have had a very brief overview of biochemical simulation, let us look at various aspects of quantum simulation.

QUANTUM SIMULATION FOR NANOSCALE AND MOLECULAR NETWORKS

There are two general aspects to quantum simulation for nanoscale networks. These involve whether one wishes to simulate quantum mechanical phenomena at the nanoscale and its impact on the nanoscale network or whether one wishes to simulate quantum computation in order to implement a quantum network.

As an example of quantum mechanical phenomena, as we discussed in Chapter 4, if a wire is considered to be a cylinder, as the radius decreases, quantum confinement modifies the states of the conduction and valence bands, and thus its properties as a semiconductor. This can be seen by direct manipulation of the fundamental Schrödinger equation. Thus, classical approaches to modeling the electrical properties of the nanowire become inaccurate and must begin to take quantum mechanical effects into account. The challenge with this approach is not only the additional computational complexity of implementing the Schrödinger equation for every particle and its interaction, but also the challenge of implementing multiscale modeling, that is, modeling physical phenomena at multiple levels of detail, including both the classical and quantum levels, and knowing when each level of detail is best applied [263]. From a communication networking perspective, there is a loose analogy to protocol layering, that is, each scale level of a multiscale physics simulation may assume the services of a more detailed simulation beneath it.

On the other hand, assuming that some form of limited quantum computation capability is in place, one can design communication protocols entirely within the quantum domain. In this case, one can work entirely with quantum computation concepts as explained in Chapter 5. The challenge with this approach is that the vast potential power that quantum computation can provide also makes it a challenge to simulate on classical computers. For example, recall from Chapter 5 the outer product and the density matrix; the size of the density matrix grows exponentially with the number of qubits making both storage and direct computation of such matrix operations intractable. Recall how the partial trace could be used to extract quantum states from mixed states; this is a common operation necessary for dealing with the unwanted entanglement of qubits with their environment. Thus, incorporating more realistic noise into such quantum simulations adds to the explosive growth in the amount of information that must be stored and manipulated on a classical computer. Therefore, simulating in the quantum computation domain, as of the date this book was written, is limited to only a small number of qubits.

There are freely available quantum simulators for common mathematical packages such as *QDensity* for Mathematica [264] and QUBIT4MATLAB [265] for MATLAB. These are primarily symbolic and functional mathematical tools, allowing the user to conveniently manipulate common quantum computational operators and equations. There are many other quantum computation simulation packages, including some that are graphical and allow convenient viewing and manipulation of quantum circuits [266].

Now that we have covered a selected set of simulators from specific domains, such as electronic, biochemical, and quantum mechanical, let us consider what may be one of the most interesting of all, the simulation of self-assembly and self-organization, the ability, like "God," to create something from nothing, to create or induce order from disorder. Unfortunately, while the topic of self-assembly and self-organization, and the corresponding simulation tools, have been around for a long time, progress in this area appears to be slow. Even attempts at agreement on the precise definitions of "order" and "disorder" can lead to endless philosophical debates. However, we mention some tangible progress on simulators and simulation relevant to nanoscale and molecular networking in the next section.

SELF-ASSEMBLY SIMULATORS

As often mentioned with regard to nanotechnology, understanding self-assembly and self-organization is critical for any realistic engineering at this scale. Because we cannot reach down into this space with our hands, it is just as futile as reaching into outer space; we need to understand how to induce fundamental particles to act, and interact, in the desired manner. Self-assembly and self-organization are generally simulated using agent-based simulation tools. There are many freely available general purpose agent-based simulation tools, including simulators such as Swarm and RePast [267, 268]. The approach is to assign simple rules to each autonomous agent and then examine what kind of global behavior emerges as they freely interact. Generally speaking, in our case, each agent represents a fundamental particle [269]. However, the simulators mentioned do not include realistic physical models; such models would have to be manually coded by the user.

There are simulators utilizing a similar paradigm focused upon the interaction of sequences of DNA. As we saw in Chapter 4, a common approach for implementing self-assembly is to attach DNA sequences to nanoparticles and allow complementary DNA strands to combine, or hybridize, which causes the nanoparticles to join together in a predefined manner [270]. There are several such "kinetic tile" simulators available; one particular example is called Xgrow. The simulated tiles are simply two-dimensional shapes comprised of DNA strands. The tiles will self-assemble as determined by the attraction of the complementary strands to one another. The DNA strands can be considered implementations of simple encoded rules in a form of algorithmic self-assembly. Finally, there are agent-based simulators that attempt to include fundamental physics as an inherent part of the simulator [271].

Hopefully, this appendix has provided enough of an overview of the various simulation components that exist, and their operational paradigms, that the reader will be encouraged to explore their utility in the simulation of nanoscale and molecular networks. If the reader is motivated to adapt, improve, and expand these, and other, tools for the purpose of nanoscale and molecular network simulation, this appendix will have served its purpose. Perhaps one day concepts from all of the aforementioned simulators will be standard elements within nanoscale and molecular communication network simulators.

References

[1] S. F. Bush and A. B. Kulkarni, *Active Networks and Active Network Management: A Proactive Management Framework*. Kluwer Academic/Plenum Publishers, 2001. ISBN 0-306-46560-4.

[2] S. F. Bush, "Third international conference on nano-networks." http://nanonets.org/cfp.shtml, Sept 2008.

[3] S. F. Bush and Y. Li, "Nano-communications: A new field? An exploration into a carbon nanotube communication network," Tech. Rep. 2006GRC066, GE Global Research, February 2006.

[4] M. Chang, V. Roychowdhury, L. Zhang, H. Shin, and Y. Qian, "Rf/wireless interconnect for inter- and intra-chip communications," *Proc. of the IEEE*, vol. 89, pp. 456–466, Apr 2001.

[5] C. Miller, "Electromagnetically coupled interconnect system," U.S. Patent 6882239, Nov 3, 2009.

[6] M. Ghoneima, Y. Ismail, M. Khellah, and V. De, "Variation-tolerant and low-power source-synchronous multicycle on-chip interconnection scheme," *Networks-on-Chip, special issue of VLSI Design*, vol. 2007, pp. 1–12, 2007. Hindawi Publishing Corporation.

[7] T. Lehtonen, P. Liljeberg, and J. Plosila, "Online reconfigurable self-timed links for fault tolerant NOC," *VLSI Design*, vol. 2007, pp. 1–13, 2007.

[8] T. Bjerregaard, *The MANGO Clockless Network-on-Chip: Concepts and Implementation*. PhD thesis, Technical University of Denmark, 2006.

[9] A. Hansson, K. Goossens, and A. Radulescu, "Avoiding message-dependent deadlock in network-based systems on chip," *Networks-on-Chip, special issue of VLSI Design*, vol. 2007, pp. 1–10, 2007. Hindawi Publishing Corporation.

[10] S. Murali, D. Atienza, L. Benini, and G. D. Micheli, "A method for routing packets across multiple paths in NOCs with in-order delivery and fault-tolerance guarantees," *Networks-on-Chip, special issue of VLSI Design*, vol. 2007, pp. 1–11, 2007. Hindawi Publishing Corporation.

[11] P. Bogdan, T. Dumitras, and R. Marculescu, "Stochastic communication: A new paradigm for fault-tolerance networks-on-chip," *Networks-on-Chip, special issue of VLSI Design*, vol. 2007, pp. 1–17, 2007. Hindawi Publishing Corporation.

[12] M. Sitti, "Microscale and nanoscale robotics systems characteristics, state of the art, and grand challenges," *IEEE Robotics and Automation Magazine*, vol. 14, pp. 53–60, Mar 2007.

[13] M. Sitti, "Micro- and nano-scale robotics," in *Proc. American Control Conf*, (Boston, USA), pp. 1–8, Jun 2004.

[14] G. Alfano and D. Miorandi, "On information transmission among nanomachines," *Nano-Networks*, vol. 38, 2006.

[15] N. A. Weir, D. P. Sierra, and J. F. Jones, "A review of research in the field of nanorobotics," Tech. Rep. SAND2005-6808, System Technologies, Intelligent Systems and Robotics Center, Sandia National Laboratories, October 2005.

[16] J. Jackson, "The interplanetary internet [networked space communications]," *Spectrum, IEEE*, vol. 42, pp. 30–35, Aug. 2005.

[17] A. T. Y. Lui, "Tutorial on geomagnetic storms and substorms," *IEEE Trans. Plasma Sci.*, vol. 28, no. 6, pp. 1854–1866, 2000.

[18] N. Kami-Ike, S. Kudo, Y. Magariyama, S. Aizawa, and H. Hotani, "Characteristics of an ultra-small biomotor," in *Proc. An Investigation of Micro Structures, Sensors, Actuators IEEE Machines and Robots Micro Electro Mechanical Systems, MEMS '91*, pp. 245–246, 1991.

[19] Wikipedia, "Semiconductor fabrication." Website, November 2009.

[20] T. A. Hilder and J. M. Hill, "Encapsulation of the anticancer drug cisplatin into nanotubes," in *Proc. International Conference on Nanoscience and Nanotechnology ICONN 2008*, pp. 109–112, 2008.

[21] K. Teker, E. Wickstrom, and B. Panchapakesan, "Biomolecular tuning of electronic transport properties of carbon nanotubes via antibody functionalization," *IEEE Sensors J.*, vol. 6, no. 6, pp. 1422–1428, 2006.

[22] Q. Tang, N. Tummala, S. K. S. Gupta, and L. Schwiebert, "Communication scheduling to minimize thermal effects of implanted biosensor networks in homogeneous tissue," *IEEE Trans. Biomed. Eng.*, vol. 52, no. 7, pp. 1285–1294, 2005.

[23] K. Y. Yazdandoost and R. Kohno, "An antenna for medical implant communications system," in *Proc. European Microwave Conference*, pp. 968–971, 2007.

[24] L. Schwiebert, S. K. Gupta, and J. Weinmann, "Research challenges in wireless networks of biomedical sensors," in *MobiCom '01: Proceedings of the 7th annual international conference on mobile computing and networking*, (New York, NY, USA), pp. 151–165, ACM, 2001.

[25] A. Cavalcanti, T. Hogg, B. Shirinzadeh, and H. Liaw, "Nanorobot communication techniques: A comprehensive tutorial," in *ICARCV '06. 9th International Conference on Control, Automation, Robotics and Vision*, pp. 1–6, Dec. 2006.

[26] A. Cavalcanti, W. W. Wood, L. C. Kretly, and B. Shirinzadeh, "Computational nanomechatronics: A pathway for control and manufacturing nanorobots," in *Proc. International Conference on Computational Intelligence for Modelling, Control and Automation and International Conference on Intelligent Agents, Web Technologies and Internet Commerce*, pp. 185–185, 2006.

[27] S. F. Bush and Y. Li, "Characteristics of carbon nanotube networks: The impact of a metallic nanotube on a CNT network," Tech. Rep. 2006GRC397, GE Global Research, June 2006.

[28] T. Suda, M. Moore, T. Nakano, R. Egashira, and A. Enomoto, "Exploratory research in molecular communication between nanomachines," *Proceedings of Genetic and Evolutionary Computation*, 2005.

[29] T. Suda, M. Moore, T. Nakano, R. Egashira, A. Enomoto, S. Hiyama, and Y. Moritani, "Exploratory research in molecular communication between nanomachines," Tech. Rep. 05-03, School of Information and Computer Science, UC Irvine, 2005.

[30] M. Moore, A. Enomoto, T. Nakano, R. Egashira, T. Suda, A. Kayasuga, H. Kojima, H. Sakakibara, and K. Oiwa, "A design of a molecular communication system for nanomachines using molecular motors," in *Pervasive Computing and Communications Workshops*, 2006.

[31] M. Moore, "Molecular communication: Simulation of a molecular motor communication system, presentation slides," in *Project Presentation Slides*, February 2006.

[32] M. Moore, A. Enomoto, T. Nakano, and T. Suda, "Simulation of a molecular motor based communication network," in *Proceedings of the 1st international conference on Bio inspired models of network, information and computing systems*, vol. 1, 2006.

[33] M. Moore, A. Enomoto, T. Nakano, R. Egashira, T. Suda, A. Kayasuga, H. Kojima, H. Sakakibara, and K. Oiwa, "A design of a molecular communication system for nanomachines using molecular motors," in *Proc. Fourth Annual IEEE International Conference on Pervasive Computing and Communications Workshops PerCom Workshops 2006*, pp. 6–559, 2006.

[34] M. Moore, A. Enomoto, T. Nakano, T. Suda, A. Kayasuga, H. Kojima, H. Sakakibara, and K. Oiwa, "Simulation of a molecular motor based communication network," in *Proc. 1st Bio-Inspired Models of Network, Information and Computing Systems*, pp. 1–1, 2006.

[35] M. Moore, A. Enomoto, T. Nakano, R. Egashira, T. Suda, A. Kayasuga, H. Kojima, H. Sakakibara, and K. Oiwa, "A design of a molecular communication system for nanomachines using molecular motors," in *Proc. Fourth Annual IEEE International Conference on Pervasive Computing and Communications Workshops PerCom Workshops 2006*, pp. 6–559, 2006.

[36] M. J. Moore, A. Enomoto, T. Nakano, A. Kayasuga, H. Kojima, H. Sakakibara, K. Oiwa, and T. Suda, "Molecular communication: Simulation of microtubule topology," in *Natural Computing, Proceedings in Information and Communications Technology*, Volume 1. ISBN 978-4-431-88980-9. Springer Japan, 2008, p. 134 (Y. Suzuki, M. Hagiya, H. Umeo, and A. Adamatzky, eds.), pp. 134–144, 2008.

[37] C. H. Bennett and G. Brassard, "Quantum cryptography: Public key distribution and coin tossing," in *Proceedings of International Conference on Computers, Systems and Signal Processing*, (New York), 1984.

[38] B. Warneke, M. Last, B. Liebowitz, and K. S. J. Pister, "Smart dust: communicating with a cubic-millimeter computer," *Computer*, vol. 34, no. 1, pp. 44–51, 2001.

[39] S.-H. Yook, H. Jeong, and A.-L. Barabasi, "Modeling the internet's large-scale topology," *Proceedings of the National Academy of Sciences of the United States of America*, vol. 99, no. 21, pp. 13382–13386, 2002.

[40] M. Bhardwaj, T. Garnett, and A. P. Chandrakasan, "Upper bounds on the lifetime of sensor networks," in *Proc. IEEE International Conference on Communications ICC 2001*, vol. 3, pp. 785–790, 2001.

[41] K. Jensen, J. Weldon, H. Garcia, and A. Zetti, "Nanotube radio," *Nano Letters*, vol. 7, no. 11, pp. 3508–3511, 2007.

[42] V. Ermolov, "Significance of nanotechnology for future wireless devices and communications," in *18th IEEE international symposium on personal, Indoor and Mobile radio Communications*, 2007.

[43] J. She and J. Yeow, "Nanotechnology-enabled wireless sensor networks: From a device perspective," *Sensors Journal, IEEE*, vol. 6, pp. 1331–1339, Oct. 2006.

[44] D. Zhao, S. Upadhyaya, and M. Margala, "Design of a wireless test control network with radio-on-chip technology for nanometer system-on-a-chip," *IEEE Trans. Comput.-Aided Design Integr. Circuits Syst.*, vol. 25, no. 7, pp. 1411–1418, 2006.

[45] B. A. Floyd, C.-M. Hung, and K. K. O, "Intra-chip wireless interconnect for clock distribution implemented with integrated antennas, receivers, and transmitters," *IEEE J. Solid-State Circuits*, vol. 37, no. 5, pp. 543–552, 2002.

[46] K. Siwiak, *Radiowave Propagation and Antennas for Personal Communications (Artech House Antennas and Propagation Library)*. Norwood, MA, USA: Artech House Publishers, 1998.

[47] F. A. Bais and J. D. Farmer, "The physics of information," 2007.

[48] W. McCarthy, *Hacking Matter: Levitating Chairs, Quantum Mirages, and the Infinite Weirdness of Programmable Atoms*. Basic Books, free multimedia edition ed., 2003.

[49] C. L. Ramos, "Manet routing with chutes and ladders," in *Proc. IEEE Military Communications Conference MILCOM 2008*, pp. 1–7, 2008.

[50] L. E. Doyle, A. C. Kokaram, S. J. Doyle, and T. K. Forde, "Ad hoc networking, Markov random fields, and decision making," *IEEE Signal Process. Mag.*, vol. 23, no. 5, pp. 63–73, 2006.

[51] J. Patwardhan, *Architectures for Nanoscale Devices*. PhD thesis, Department of Computer Science, Duke University, Durham, NC, USA, 2006. Adviser-Lebeck, Alvin R.

[52] J. Patwardhan, C. Dwyer, and A. Lebeck, "Self-assembled networks: Control vs. complexity," in *First International Conference on Nano-Networks*, 2006.

[53] H. Naeimi and A. DeHon, "Fault tolerant nano-memory with fault secure encoder and decoder," in *Proceedings of the 2nd International Conference on Nano-Networks and Workshops*, Sept 2007.

[54] S. F. Bush, "Genetically induced communication network fault tolerance," *Complexity Journal*, vol. 9, pp. 19–33, Nov 2003. Special Issue on Resilient and Adaptive Defense of Computing Networks.

[55] J. P. Patwardhan, C. L. Dwyer, A. R. Lebeck, and D. J. Sorin, "Nana: A nanoscale active network architecture," *ACM Journal on Emerging Technologies in Computing Systems (JETC)*, vol. 3, pp. 1–31, Jan 2006.

[56] S. F. Bush, *Volume III: Distributed Networks, Network Planning, Control, Management, and New Trends and Applications*, vol. 3 of *The Handbook of Computer Networks*, hardcover Active Networking, pp. 985–1011. John Wiley & Sons, 2007. Hossein Bidgoli Ed., ISBN: 0-471-78461-3.

[57] S. F. Bush and N. Smith, "The limits of motion prediction support for ad hoc wireless network performance," in *Proceedings of the 2005 International Conference on Wireless Networks (ICWN-05)*, Jun 2005. Monte Carlo Resort, Las Vegas, Nevada, USA.

[58] P. Gupta and P. R. Kumar, "Capacity of wireless networks," tech. rep., University of Illinois, Urbana-Champaign, 1999.

[59] S. F. Bush, "Wireless ad hoc nanoscale networking - industry perspectives," *IEEE Wireless Commun. Mag.*, vol. 16, no. 5, pp. 6–7, 2009.

[60] R. A. Freitas, "Current status of nanomedicine and medical nanorobotics," *Journal of Computational and Theoretical Nanoscience*, vol. 2, pp. 1–25(25), March 2005.

[61] R. Weiss, S. Basu, S. Hooshangi, A. Kalmbach, D. Karig, R. Mehreja, and I. Netravali, "Genetic circuit building blocks for cellular computations, communications, and signal processing," *Natural Computing*, vol. 2, p. 4784, Mar 2003.

[62] T. Head, M. Yamamura, and S. Gal, "Aqueous computing: writing on molecules," in *Proc. Congress on Evolutionary Computation CEC 99*, vol. 2, pp. 1006–1010, vol. 2, 1999.

[63] H. Wang and T. C. Elston, "Mathematical and computational methods for studying energy transduction in protein motors," *Journal of Statistical Physics*, vol. 128, pp. 35–76, July 2007.

[64] M. Schliwa and G. Woehlke, "Molecular motors," *Nature*, vol. 422, pp. 759–765, 2003.

[65] A. A. Tseng, K. Chen, C. D. Chen, and K. J. Ma, "Electron beam lithography in nanoscale fabrication: recent development," *IEEE Transactions on Electronics Packaging Manufacturing*, vol. 26, pp. 141–149, 2003.

[66] U. Tulu, C. Fagerstorm, N. Ferenz, and P. Wadsworth, "Molecular requirements for kinetochore-associated microtubule formation in mammalian cells," *Current Biology*, vol. 16, pp. 536–541, Mar. 2006.

[67] Y. F. Inclan and E. Nogales, "Structural models for the self-assembly and microtubule interaction of gamma-, delta- and epsilon-tubulin," *Journal of Cell Science*, vol. 114, pp. 413–442, 2000.

[68] H. Hess, "Self-assembly driven by molecular motors," *Journal of the Royal Society of Chemistry*, vol. 2, pp. 669–677, 2006.

[69] N. Glade, "Computing with the cytoskeleton: A problem of scale," *International Journal of Unconventional Computing*, vol. 4, no. 1, pp. 33–44, 2008.

[70] S. Hameroff, A. Nip, M. Porter, and J. Tuszynski, "Conduction pathways in microtubules, biological quantum computation, and consciousness," *Biosystems*, vol. 64, pp. 149–168(20), January 2002.

[71] S. Klumpp, T. M. Nieuwenhuizen, and R. Lipowsky, "Self-organized density patterns of molecular motors in arrays of cytoskeletal filaments," *Biophys. J.*, vol. 88, pp. 3118–3132, May 2005.

[72] S. F. Bush and S. Goel, "The impact of persistence length on the communication efficiency of microtubules and CNTs," in *Nano-Net 4th International ICST Conference, Nano-Net 2009, Lucerne, Switzerland, October 18-20, 2009*, vol. 20, pp. 1–13, Publisher Springer Berlin Heidelberg, 2009.

[73] M. G. L. Van den Heuvel, M. P. de Graaff, and C. Dekker, "Microtubule curvatures under perpendicular electric forces reveal a low persistence length," *Proceedings of the National Academy of Sciences*, vol. 105, no. 23, pp. 7941–7946, 2008.

[74] S. A. Endow, "Determinants of molecular motor directionality," *Nat Cell Biol*, vol. 1, pp. E163–E167, October 1999.

[75] J. L. Ross, H. Shuman, E. L. Holzbaur, and Y. E. Goldman, "Kinesin and dynein-dynactin at intersecting microtubules: Motor density affects dynein function," *Biophys. J.*, vol. 94, pp. 3115–3125, April 2008.

[76] M. G. L. van den Heuvel, M. P. de Graaff, and C. Dekker, "Molecular sorting by electrical steering of microtubules in kinesin-coated channels," *Science*, vol. 312, no. 5775, pp. 910–914, 2006.

[77] H. Hess, J. Clemmens, D. Qin, J. Howard, and V. Vogel, "Light-controlled molecular shuttles made from motor proteins carrying cargo on engineered surfaces," *Nano Letters*, vol. 1, no. 5, pp. 235–239, 2001.

[78] S. Diez, C. Reuther, C. Dinu, R. Seidel, M. Mertig, W. Pompe, and J. Howard, "Stretching and transporting DNA molecules using motor proteins," *Nano Letters*, vol. 3, no. 9, pp. 1251–1254, 2003.

[79] K. Kato, R. Goto, K. Katoh, and M. Shibakami, "Microtubule-cyclodextrin conjugate: Functionalization of motile filament with molecular inclusion ability," *Bioscience, Biotechnology, and Biochemistry*, vol. 69, no. 3, pp. 646–648, 2005.

[80] S. Hiyama, Y. Isogawa, T. Suda, Y. Motitani, and K. Sutoh, "A design of an autonomous molecule loading/transporting/unloading system using DNA hybridization and biomolecular linear motors in molecular communication," *European Nano Systems*, Dec. 2005.

[81] S. Hiyama, S. Takeuchi, R. Gojo, T. Shima, and K. Sutoh, "Biomolecular motor-based cargo transporters with loading/unloading mechanisms on a micro-patterned DNA array," in *Proc. IEEE 21st International Conference on Micro Electro Mechanical Systems MEMS 2008*, pp. 144–147, 2008.

[82] Y. Hiratsuka, T. Tada, K. Oiwa, T. Kanayama, and T. Uyeda, "Controlling the direction of kinesin-driven microtubule movements along microlithographic tracks," *Biophysical Journal*, vol. 81, pp. 1555–1561, 2001.

[83] D. A. Mac Donaill, "Digital parity and the composition of the nucleotide alphabet," *IEEE Eng. Med. Biol. Mag.*, vol. 25, no. 1, pp. 54–61, 2006.

[84] J. D. Watson and F. H. C. Crick, "Molecular structure of nucleic acids: A structure for deoxyribose nucleic acid," *Nature*, vol. 171, pp. 737–738, April 1953.

[85] S. F. Bush and S. Goel, "An active model-based prototype for predictive network management," *IEEE Journal on Selected Areas in Communications Journal*, vol. 23, pp. 2040–2057, Nov 2005.

[86] Y. Benenson, R. Adar, T. Paz-Elizur, Z. Livneh, and E. Shapiro, "DNA molecule provides a computing machine with both data and fuel," *Proc Natl Acad Sci U S A*, vol. 100, pp. 2191–2196, March 2003.

[87] J. Bath and A. J. Turberfield, "DNA nanomachines," *Nature Nanotechnology*, vol. 2, pp. 275–284, 2007.

[88] R. T. Pomerantz, R. Ramjit, Z. Gueroui, C. Place, M. Anikin, S. Leuba, J. Zlatanova, and W. T. Mcallister, "A tightly regulated molecular motor based upon t7 RNA polymerase," *Nano Letters*, vol. 5, pp. 1698–1703, September 2005.

[89] S. Hiyama, Y. Moritani, T. Suda, R. Egashira, A. Enomoto, M. Moore, and T. Nakano, "Molecular communication," in *Proc. of the 2005 NSTI Nanotechnology Conference*, (Network Laboratories, NTT DoCoMo, Inc. and Information and Computer Science, University of California, Irvine), 2005.

[90] B. Albert, D. Bray, J. Lewis, M. Raff, K. Roberts, and J. Watson, *Molecular Biology of The Cell*, vol. 1. Garland, 1 ed., 1983.

[91] T. Nakano, T. Suda, M. Moore, R. Egashira, A. Enomoto, and K. Arima, "Molecular communication for nanomachines using intercellular calcium signaling," *IEEE NANO*, June 2005.

[92] M. Berridge, "The am and fm of calcium signalling," *Nature*, vol. 386, pp. 759–780, Apr. 1997.

[93] D. Selmeczi, S. Mosler, P. Hagedorn, N. Larson, and H. Flyvbjerg, "Cell motility as persistent random motion: Theories from experiments," *Biophysical Journal*, vol. 89, pp. 912–931, 2005.

[94] Y. Moritani, S. Nomura, S. Hiyama, K. Akiyoshi, and T. Suda, "Molecular communication interface using liposomes with gap junction proteins," *Bio-Inspired Models of Network, Information and Computing Systems*, vol. 1, 2006.

[95] J. Rottingen and J. Jversen, "Ruled by waves? Intracellular and intercellular calcium signaling," *Acta physiologica Scandinavica*, vol. 169, pp. 203–219, 2000.

[96] Y. Sasaki, M. Hashizume, K. Maruo, N. Yamasaki, J. Kikuchi, Y. Moritani, S. Hiyama, and T. Suda, "Controlled propagation in molecular communication using tagged liposome containers," in *Proc. 1st Bio-Inspired Models of Network, Information and Computing Systems*, pp. 1–1, 2006.

[97] Y. Moritani, S. Hiyama, S. M. Nomura, K. Akiyoshi, and T. Suda, "A communication interface using vesicles embedded with channel forming proteins in molecular communication," in *Proc. 2nd Bio-Inspired Models of Network, Information and Computing Systems Bionetics 2007*, pp. 147–149, 2007.

[98] J.-Q. Liu and T. Nakano, "An information theoretic model of molecular communication based on cellular signaling," in *Bio-Inspired Models of Network, Information, and Computing Systems 2007. Bionetics 2007. 2nd*, pp. 316–321, Dec. 2007.

[99] R. Ahlswede, N. Cai, S. Y. R. Li, and R. W. Yeung, "Network information flow," *Information Theory, IEEE Transactions on*, vol. 46, no. 4, pp. 1204–1216, 2000.

[100] T. Nakano, Y.-H. Hsu, W. C. Tang, T. Suda, D. Lin, T. Koujin, T. Haraguchi, and Y. Hiraoka, "Microplatform for intercellular communication," in *Proc. 3rd IEEE International Conference on Nano/Micro Engineered and Molecular Systems NEMS 2008*, pp. 476–479, 2008.

[101] K. Francis and B. O. Palsson, "Effective intercellular communication distances are determined by the relative time constants for cyto/chemokine secretion and diffusion," *Proceedings of the National Academy of Sciences of the United States of America*, vol. 94, no. 23, pp. 12258–12262, 1997.

[102] N. Jorgenson, S. Geist, R. Civitelli, and T. Steinberg, "ATP- and gap junction-dependent intercellular calcium signaling in osteoblastic cells," *Journal of Cell Biology*, vol. 139, pp. 497–506, Nov. 1997.

[103] T. Hofer, "Model of intercellular calcium oscillations in hepatocytes: Synchronization of heterogeneous cells," *Biophysical Journal*, vol. 77, pp. 1244–1256, 1999.

[104] A. B. Parekh and J. Putney, James W., "Store-operated calcium channels," *Physiol. Rev.*, vol. 85, no. 2, pp. 757–810, 2005.

[105] A. Goldbcter, G. Dupont, and M. J. Berridge, "Minimal model for signal-induced Ca2+ oscillations and for their frequency encoding through protein phosphorylation," *Proceedings of the National Academy of Sciences of the United States of America*, vol. 87, no. 4, pp. 1461–1465, 1990.

[106] C. Salazar, "Decoding of calcium oscillations by phosphorylation cycles: Analytic results," *Biophysical Journal*, vol. 94, pp. 1203–1215, Feb. 2008.

[107] R. Shetty, D. Endy, and T. Knight, "Engineering biobrick vectors from biobrick parts," *Journal of Biological Engineering*, vol. 2, no. 1, p. 5, 2008.

[108] J. Rospars, V. Krivan, and P. Lansky, "Perireceptor and receptor events in olfaction. Comparison of concentration and flux detectors: a modeling study," *Chem Senses*, vol. 25, pp. 293–311, 2000.

[109] D. B. Strukov, G. S. Snider, D. R. Stewart, and S. R. Williams, "The missing memristor found," *Nature*, vol. 453, no. 7191, pp. 80–83, 2008.

[110] S. P. Brown and R. A. Johnstone, "Cooperation in the dark: signaling and collective action in quorum-sensing bacteria," *Proc Biol Sci.*, vol. 268, pp. 961–965, May 2001.

[111] Y. V. Pershin, S. La Fontaine, and M. Di Ventra, "Memristive model of amoeba's learning," 2008.

[112] V. Krivana, P. Lansky, and J. P. Rospars, "Coding of periodic pulse stimulation in chemoreceptors," *Biosystems*, vol. 67, pp. 121–128, October-December 2002. Issues 1-3.

[113] A. A. Faisal, L. P. Selen, and D. M. Wolpert, "Noise in the nervous system," *Nat Rev Neurosci*, vol. 9, pp. 292–303, April 2008.

[114] N. Wakamiya, K. Leibnitz, and M. Murata, "Noise-assisted control in information networks," in *Proc. Frontiers in the Convergence of Bioscience and Information Technologies FBIT 2007*, pp. 833–838, 2007.

[115] S. Balasubramaniam, D. Botvich, F. Walsh, W. Donnelly, S. Sergeyev, and S. F. Bush, "Applying compartmentalisation techniques to communication protocols of biological nano and mems devices," in *Proc. Fourth International Conference on Broadband Communications, Networks and Systems BROADNETS 2007*, pp. 323–325, 2007.

[116] J.-Q. Liu and K. Shimohara, "A biomolecular computing method based on Rho family GTPases," *NanoBioscience, IEEE Transactions on*, vol. 2, pp. 58–62, June 2003.

[117] J.-Q. Liu and K. Shimohara, "On designing error-correctable codes by biomolecular computation," in *Proc. International Symposium on Information Theory ISIT 2005*, pp. 2384–2388, 2005.

[118] J.-Q. Liu and H. Sawai, "A new channel coding algorithm based on phosphorylation/dephosphorylation-proteins and GTPases," in *Proc. 1st Bio-Inspired Models of Network, Information and Computing Systems*, pp. 1–5, 2006.

[119] F. Pampaloni and F. Ernst-Ludwig, "Microtubule architecture: inspiration for novel carbon nanotube-based biomimetic materials," *Trends in biotechnology*, vol. 26, no. 6, pp. 302–310, 2008.

[120] W. H. Goldmann, "Actin: A molecular wire, an electrical cable?" *Cell Biol Int*, vol. 32, pp. 869–870, Jul 2008.

[121] B. C. Bunker, E. D. Spoerke, J. Hendricks, G. D. Bachand, and P. R. Haddon, "Active transport of carbon nanotubes using motor proteins." Physical, Chemical, & Nano Sciences Center Research Briefs, 2006. Sandia National Laboratories.

[122] A. Bianco, K. Kostarelos, and M. Prato, "Applications of carbon nanotubes in drug delivery," *Current Opinion in Chemical Biology*, vol. 9, no. 6, pp. 674–679, 2005.

[123] E. Thostenson, Z. Ren, and T. Chou, "Advances in the science and technology of carbon nanotubes and their composites: a review," *Composite Science and Technology*, vol. 61, pp. 1899–1912, 2001.

[124] C. Dekker, "Carbon nanotubes as molecular quantum wire," *Physics Today*, pp. 22–28, May 1999.

[125] P. McEuen, "Single wall carbon nanotubes," *Physics World*, pp. 31–36, June 2000.

[126] A. Eckford, "Achievable information rates for molecular communication with distinct molecules," in *BIONETICS*, pp. 10–13, 2007.

[127] P. McEuen, M. Fuhrer, and H. Park, "Single-walled carbon nanotube electronics," *IEEE Transactions on Nanotechnology*, vol. 1, no. 1, pp. 78–85, 2002.

[128] M. Huhtala, A. Kuronen, and K. Kaski, "Carbon nanotubes structures: Molecular dynamic simulation at realistic limits," *Computer Physics Communications*, vol. 146, pp. 30–37, 2002.

[129] A. Krishnan, E. Dujardin, T. Ebbesen, P. Yianilos, and M. Treacy, "Youngs modulus of single-walled nanotubes," *Physical Review B*, vol. 58, no. 20, pp. 14013–14019, 1998.

[130] G. Gao, T. Cagin, W. Goddard, and W. A. G. III, "Energetics, structure, mechanical and vibrational properties of single-walled carbon nanotubes," *Nanotechnology*, vol. 9, no. 3, pp. 184–191, 1998.

[131] S. Berber, Y. Kwon, and D. Tomanek, "Unusually high thermal conductivity of carbon nanotubes," *Physical Review Letters*, vol. 84, no. 20, pp. 4613–4616, 2000.

[132] T. Adams, "Physical properties of carbon nanotubes." Website.

[133] H. Rafii-Tabar, "Computational modeling of thermo-mechanical and transport properties of carbon nanotubes," *Physics Reports*, vol. 390, pp. 235–452, 2004.

[134] T. Ebbesen, H. Lezec, H. Hiura, J. Bennett, H. Ghaemi, and T. Thio, "Electrical conductivity of individual carbon nanotubes," *Nature*, vol. 382, pp. 54–56, 1996.

[135] Z. Aksamija and U. Ravaioli, "Boltzmann transport simulator for CNTs," Website: nanoHUB.org, 2008.

[136] Z. Aksamija and U. Ravaioli, "Boltzmann transport simulation of single-walled carbon nanotubes," *Journal of Computational Electronics*, vol. 7, pp. 315–318, 2008.

[137] M. Suvakov and B. Tadic, *Simulation of the Electron Tunneling Paths in networks of Nano-Particle Films*, vol. Lecture Notes on Computer Science. Springer Berlin, 2007.

[138] D. McLachlan, C. Chiteme, C. Park, K. Wise, S. Lowther, P. Lillehei, E. Soichi, and J. Harrison, "AC and DC percolative conductivity of single wall carbon nanotube polymer composites," *Journal of Polymer Science B*, vol. 43, no. 22, pp. 3273–3287, 2005.

[139] K. Takayanagi, "Suspended gold nanowires: Ballistic transport of electrons," *JSAP Journal*, vol. 3, pp. 3–8, 2001.

[140] M. Uplaznik, "Conducting properties of quantum nanosystems." Seminar, 2004.

[141] K. L. Ekinci and M. L. Roukes, "Nanoelectromechanical systems," *Review of Scientific Instruments*, vol. 76, no. 6, pp. 061101, 2005.

[142] C. Rutherglen and P. Burke, "Carbon nanotube radio," *Nano Letters*, vol. 7, no. 11, pp. 3296–3299, 2007.

[143] M. Stadermann, S. J. Papadakis, M. R. Falvo, J. Novak, E. Snow, Q. Fu, J. Liu, Y. Fridman, J. J. Boland, R. Superfine, and S. Washburn, "Nanoscale study of conduction through carbon nanotube networks," *Physical Review B*, vol. 69, pp. 1–3, 2004.

[144] T. Durkop, S. Getty, E. Cobas, and M. Fuhrer, "Extraordinary mobility in semiconducting carbon nanotubes," *Nano Letters*, vol. 4, no. 1, pp. 35–39, 2004.

[145] A. Javey, J. Guo, Q. Wang, M. Lundstrom, and H. Dai, "Ballistic carbon nanotube field effect transistors," *Nature*, vol. 424, pp. 654–657, 2003.

[146] C. Dwyer, M. Cheung, and D. Sorin, "Semi-empirical spice models for carbon nanotube FET logic," in *Proceedings of the Fourth IEEE Conference on Nanotechnology*, 2004.

[147] F. Y. Wu, "Theory of resistor networks: The two-point resistance," *Journal of Physics A*, vol. 37, pp. 6653, 2004.

[148] C. Pistol, A. Lebeck, and C. Dwyer, "Design automation for DNA self-assembled nanostructures," in *Proceedings of the 43rd Annual Conference on Design Automation*, pp. 919–924, 2006.

[149] C. Dwyer, J. Patwardhan, A. Lebeck, and D. Sorin, "Evaluating the connectivity of self-assembled networks of nano-scale processing elements," in *NanoArch*, 2005.

[150] M. Mihail, C. H. Papadimitriou, and A. Saberi, "On certain connectivity properties of the internet topology," in *44th Annual IEEE Symposium on Foundations of Computer Science*, (Cambridge, MA), 2003.

[151] A. Dasgupta, J. E. Hopcroft, and F. McSherry, "Spectral analysis of random graphs with skewed degree distributions," in *45th Annual IEEE Symposium on Foundations of Computer Science*, (Rome, Italy), pp. 602–610, 2004.

[152] E. L. Wolf, *Nanophysics and Nanotechnology*. Wiley-VCH, 2004.

[153] S. Wolfram, *The Mathematica Book*. Wolfram Media, 5th ed., 2003. ISBN: 1-57955-022-3.

[154] S. F. Bush and Y. Li, "Network characteristics of carbon nanotubes: A graph eigenspectrum approach and tool using Mathematica," Tech. Rep. 2006GRC023, GE Global Research, 2006.

[155] G. Kramer and S. Savari, "Edge-cut bounds on network coding rates," *Journal of Network and Systems Management: Special Issue on Management Of Active and Programmable Networks*, vol. 14, pp. 49–67, March 2006.

[156] F. Y. Hong and S. J. Xiong, "Quantum interfaces using nanoscale surface plasmons," 2008.

[157] M. Yamaki, K. Hoki, Y. Ohtsuki, H. Kono, and Y. Fujimura, "Quantum control of a chiral molecular motor driven by laser pulses," *Journal of the American Chemical Society*, vol. 127, no. 20, pp. 7300–7301, 2005.

[158] W. G. Cooper, "Evidence for transcriptase quantum processing implies entanglement and decoherence of superposition proton states," *Biosystems*, vol. 97, pp. 73–89, August 2009.

[159] M. A. Nielsen and I. L. Chuang, *Quantum Computation and Quantum Information*. Cambridge University Press, October 2000.

[160] S. Bose, V. Vedral, and P. L. Knight, "Multiparticle generalization of entanglement swapping," *Phys. Rev. A*, vol. 57, pp. 822–829, Feb 1998.

[161] H. A. Bachor and T. C. Ralph, *A Guide to Experiments in Quantum Optics*. Wiley-VCH ([Weinheim]), 2 ed., 2004. LC: QC446.2 .B32 2004.

[162] M. Hijlkema, B. Weber, H. P. Specht, S. C. Webster, A. Kuhn, and G. Rempe, "A single-photon server with just one atom," in *European Conference on Lasers and Electro-Optics, 2007 and the International Quantum Electronics Conference.*, pp. 1, 2007.

[163] N. Shiokawa, Y. Shimizu, N. Yamamoto, T. Noguchi, M. Kozuma, and T. Kuga, "Real-time detection of single atoms with transverse mode of high-finesse optical micro cavity," in *Proc. Conference Digest Quantum Electronics Conference 2000 International*, pp. 1, 2000.

[164] S. Haroche, "Entanglement experiments with single atoms and photons in a cavity," in *Proc. Technical Digest Quantum Electronics and Laser Science Conference (QELS 2000)*, pp. 178, 2000.

[165] H. Carmichael and L. A. Orozco, "Quantum optics: Single atom lases orderly light," *Nature*, vol. 425, no. 6955, pp. 246–247, 2003.

[166] H. J. Kimble, "The quantum internet," *Nature*, vol. 453, pp. 1023–1030, June 2008.

[167] M. C. Teich and B. E. A. Saleh, "Squeezed state of light," *Quantum Optics: Journal of the European Optical Society Part B*, vol. 1, no. 2, pp. 153–191, 1989.

[168] M. Oskin, F. T. Chong, I. L. Chuang, and J. Kubiatowicz, "Building quantum wires: the long and the short of it," in *ISCA '03: Proceedings of the 30th Annual International Symposium on Computer Architecture*, (New York, NY, USA), pp. 374–387, ACM, 2003.

[169] M. Hayashi, K. Iwama, H. Nishimura, R. R. H. Putra, and S. Yamashita, "Quantum network coding," in *STACS*, pp. 610–621, 2007.

[170] S. F. Bush, "Active virtual network management prediction: Complexity as a framework for prediction, optimization, and assurance," in *Proceedings of the 2002 DARPA Active Networks Conference and Exposition (DANCE) May 29-30, 2002*, San Francisco, California, USA, pp. 534–553, IEEE Computer Society Press, Santa Fe Institute, Santa Fe, NM, May 2002. ISBN 0-7695-1564-9.

[171] S. F. Bush and T. Hughes, "On the effectiveness of Kolmogorov complexity estimation to discriminate semantic types," in *Proceedings of the SFI Workshop on Resilient and Adaptive Defense of Computing Networks 2003*, Nov 2003. Santa Fe Institute, Santa Fe, NM.

[172] N. G. Anderson, "Information acquisition at the nanoscale: Fundamental considerations," *IEEE Trans. Nanotechnol.*, vol. 7, no. 5, pp. 521–526, 2008.

[173] N. G. Anderson, "Accessible information from molecular-scale volumes in electronic systems: Fundamental physical limits," *Journal of Applied Physics*, vol. 99, no. 4, pp. 043706, 2006.

[174] P. W. K. Rothemund and E. Winfree, "The program-size complexity of self-assembled squares (extended abstract)," in *STOC '00: Proceedings of the Thirty-Second Annual ACM Symposium on Theory of Computing*, (New York, NY, USA), pp. 459–468, ACM, 2000.

[175] D. Polani, "Measuring self-organization via observers," in *ECAL*, pp. 667–675, 2003.

[176] R. Cilibrasi and P. Vitanyi, "Clustering by compression," 2003.

[177] C. E. Shannon, "A mathematical theory of communication," *Bell system Technical Journal*, vol. 27, 1948.

[178] A. W. Eckford, "Nanoscale communication with Brownian motion," in *Proc. 41st Annual Conference on Information Sciences and Systems CISS '07*, pp. 160–165, 2007.

[179] A. W. Eckford, "Achievable information rates for molecular communication with distinct molecules," in *Proc. 2nd Bio-Inspired Models of Network, Information and Computing Systems Bionetics 2007*, pp. 313–315, 2007.

[180] B. Atakan and O. B. Akan, "An information theoretical approach for molecular communication," in *Proc. 2nd Bio-Inspired Models of Network, Information and Computing Systems Bionetics 2007*, pp. 33–40, 2007.

[181] M. Long, S. L, and G. Sun, "Kinetics of receptor-ligand interactions in immune responses," *Cellular and Molecular Immunology*, vol. 3, pp. 79–86, Apr 2006.

[182] B. Atakan and O. B. Akan, "On molecular multiple-access, broadcast and relay channels in nanonetworks," in *ICST/ACM BIONETICS*, November 2008.

[183] I. Akyildiz, "Nanonetworks: A new communication paradigm," *Computer Networks*, vol. 52, pp. 2260–2279, 2008.

[184] A. Lehninger, D. L. Nelson, and M. M. Cox, *Lehninger Principles of Biochemistry*. W. H. Freeman, fifth edition ed., Jun 2008.

[185] J. Wiedermann and L. Petru, "Communicating mobile nano-machines and their computational power," tech. rep. 1024, Institute of Computer Science, Academy of Science of the Czech Republic, May 2008.

[186] A. Vahdat and D. Becker, "Epidemic routing for partially connected ad hoc networks," tech. rep., Duke University, 2000.

[187] S. Kim and D. D. Eun, "Impact of super-diffusive behavior on routing performance in delay tolerant networks," in *Proc. IEEE International Conference on Communications ICC '08*, pp. 2941–2945, 2008.

[188] P. S. Ray, S. I. Sheikh, P. H. Graven, M. T. Wolff, K. S. Wood, and K. C. Gendreau, "Deep space navigation using celestial x-ray sources," in *Institute of Navigation 2008 National Technical Meeting,*, (San Diego, CA), January 2008.

[189] D. N. Matsakis, J. H. Taylor, and T. M. Eubanks, "A statistic for describing pulsar and clock stabilities," *Astronomy and Astrophysics*, vol. 326, pp. 924–928, Oct. 1997.

[190] S. Devasia, E. Eleftheriou, and S. O. R. Moheimani, "A survey of control issues in nanopositioning," *IEEE Trans. Control Syst. Technol.*, vol. 15, no. 5, pp. 802–823, 2007.

[191] O. Felfoul, J. B. Mathieu, G. Beaudoin, and S. Martel, "In vivo MR-tracking based on magnetic signature selective excitation," *IEEE Trans. Med. Imag.*, vol. 27, no. 1, pp. 28–35, 2008.

[192] S. Martel, M. Mohammadi, O. Felfoul, L. Zhao, and P. Pouponneau, "Flagellated magnetotactic bacteria as controlled MRI-trackable propulsion and steering systems for medical nanorobots operating in the human microvasculature," *Int. J. Rob. Res.*, vol. 28, no. 4, pp. 571–582, 2009.

[193] S. Martel, W. Andre, M. Mohammadi, Z. Lu, and O. Felfoul, "Towards swarms of communication-enabled and intelligent sensotaxis-based bacterial microrobots capable of collective tasks in an aqueous medium," in *Proc. IEEE International Conference on Robotics and Automation ICRA '09*, pp. 2617–2622, 2009.

[194] B. Zhen, H.-B. Li, , and R. Kohno, "Networking issues in medical implant communications," *International Journal of Multimedia and Ubiquitous Engineering*, vol. 4, no. 1, pp. 23–38, 2009.

[195] S. F. Bush, *The Handbook of Computer Networks*, hardcover Active Networking, p. 3008. John Wiley & Sons, 2007. ISBN 0-471-78461-3.

[196] S. F. Bush, "A simple metric for ad hoc network adaptation," *IEEE Journal on Selected Areas in Communications Journal*, vol. 23, pp. 2272–2287, Dec 2005.

[197] S. F. Bush, *Focus on Computer Science Research*, hardcover Complexity as a Framework for Prediction, Optimization, and Assurance, pp. 65–129. Nova Science Publishers, Inc, 2004. 1-59033-993-3.

[198] P. Merkle, H. Brust, K. Dix, K. Muller, and T. Wiegand, "Stereo video compression for mobile 3d services," in *Proc. 3DTV Conference: The True Vision—Capture, Transmission and Display of 3D Video*, pp. 1–4, 2009.

[199] T. Wiegand and G. J. Sullivan, "The h.264/avc video coding standard [standards in a nutshell]," *IEEE Signal Process. Mag.*, vol. 24, no. 2, pp. 148–153, 2007.

[200] H. Schwarz, D. Marpe, and T. Wiegand, "Overview of the scalable video coding extension of the h.264/avc standard," *IEEE Trans. Circuits Syst. Video Technol.*, vol. 17, no. 9, pp. 1103–1120, 2007.

[201] S. da Silva, Y. Yemini, and D. Florissi, "The Netscript active network system," *IEEE J. Sel. Areas Commun.*, vol. 19, no. 3, pp. 538–551, 2001.

[202] D. Wetherall, D. Legedza, and J. Guttag, "Introducing new internet services: why and how," *IEEE Netw.*, vol. 12, no. 3, pp. 12–19, 1998.

[203] M. Hicks, A. D. Keromytis, and J. M. Smith, "A secure plan," *IEEE Trans. Syst., Man, Cybern. C*, vol. 33, no. 3, pp. 413–426, 2003.

[204] F. Pottier and D. Rémy, "The essence of ML type inference," in *Advanced Topics in Types and Programming Languages* (B. C. Pierce, ed.), ch. 10, pp. 389–489, MIT Press, 2005.

[205] S. Bhattacharjee, K. L. Calvert, and E. W. Zegura, "Reasoning about active network protocols," in *Proc. Sixth International Conference on Network Protocols*, pp. 31–40, 1998.

[206] S. da Silva, Y. Yemini, and D. Florissi, "The Netscript active network system," *IEEE Journal on Selected Areas in Communications*, vol. 19, pp. 538–551, Mar 2001.

[207] S. Bhattacharjee, K. Calvert, Y. Chae, S. Merugu, M. Sanders, and E. Zegura, "Canes: an execution environment for composable services," in *Proc. DARPA Active Networks Conference and Exposition*, pp. 255–272, 2002.

[208] J. Weldon, K. Jensen, and A. Zettl, "Nanomechanical radio transmitter," *Phys. Stat. Sol.*, vol. 245, no. 10, pp. 2323–2325, 2008.

[209] A. Ayari, P. Vincent, S. Perisanu, M. Choueib, V. Gouttenoire, M. Bechelany, D. Cornu, and S. T. Purcell, "Self-oscillations in field emission nanowire mechanical resonators: A nanometric DC-AC conversion," *Nano Letters*, vol. 7, pp. 2252–2257, Aug. 2007.

[210] C. Elliott, D. Pearson, and G. Troxel, "Quantum cryptography in practice," in *SIGCOMM '03: Proceedings of the 2003 Conference on Applications, Technologies, Architectures, and Protocols for Computer Communications*, (New York, NY, USA), pp. 227–238, ACM, 2003.

[211] R. Alleaume, J. Bouda, C. Branciard, T. Debuisschert, M. Dianati, N. Gisin, M. Godfrey, P. Grangier, T. Langer, A. Leverrier, N. Lutkenhaus, P. Painchault, M. Peev, A. Poppe, T. Pornin, J. Rarity, R. Renner, G. Ribordy, M. Riguidel, L. Salvail, A. Shields, H. Weinfurter, and A. Zeilinger, "SECOQC white paper on quantum key distribution and cryptography," *CoRR*, vol. abs/quant-ph/0701168, 2007. (informal publication).

[212] Y. Nambu, K. Yoshino, and A. Tomita, "One-way quantum key distribution system based on planar lightwave circuits," Mar 2006.

[213] R. Van Meter, T. D. Ladd, W. J. Munro, and K. Nemoto, "System design for a long-line quantum repeater," *IEEE/ACM Trans. Netw.*, vol. 17, no. 3, pp. 1002–1013, 2009.

[214] L. Childress, J. M. Taylor, A. S. Sørensen, and M. D. Lukin, "Fault-tolerant quantum repeaters with minimal physical resources and implementations based on single-photon emitters," *Physical Review A*, vol. 72, pp. 052330, Nov. 2005.

[215] C. Shannon, "The zero error capacity of a noisy channel," *Information Theory, IRE Transactions on*, vol. 2, pp. 8–19, September 1956.

[216] G. Simonyi, "Perfection, imperfection, and graph entropy," *Electronic Notes in Discrete Mathematics*, vol. 5, pp. 278–280, 2000.

[217] B. Emsenhuber, "Scent marketing: Subliminal advertising messages," in *Proceedings of the 2nd International Workshop on Pervasive Advertising*, (Lbeck, Germany), 2009.

[218] M. Kaitz and A. Eidelman, "Smell-recognition of newborns by women who are not mothers," *Chem. Senses*, vol. 17, no. 2, pp. 225–229, 1992.

[219] R. H. Portera, J. M. Cernocha, and F. McLaughlinb, "Maternal recognition of neonates through olfactory cues," *Physiology & Behavior*, vol. 30, no. 1, pp. 151–154, 1983.

[220] R. Porter, "Olfaction and human kin recognition," *Genetica*, vol. 104, no. 3, pp. 259–263, 1998.

[221] J. N. Kaye, *Symbolic olfactory display*. PhD thesis, Massachusetts Institute of Technology. Dept. of Architecture, 2001.

[222] P. M. Greco, S. D. Hunt, and J. W. Seuck, "Communication device having a scent release feature and method thereof," U.S. Patent 7200363, Apr 3, 2007.

[223] D. Harel, L. Carmel, and D. Lancet, "Towards an odor communication system," *Computational Biology and Chemistry*, vol. 27, no. 2, pp. 121–133, 2003.

[224] H. W. King, "Piezoelectric sorption detector," *Anal. Chem.*, vol. 36, pp. 1735–1739, 1964.

[225] J. Shlens, "A tutorial on principal component analysis," tech. rep., Systems Neurobiology Laboratory, Salk Institute for Biological Studies, December 2005.

[226] J. Li, Y. L. and Qi Ye, M. Cinke, J. Han, and M. Meyyappan, "Carbon nanotube sensors for gas and organic vapor detection," *Nano Letters*, vol. 3, pp. 929–933, 2003.

[227] J. Philibert, "One and a half century of diffusion: Fick, Einstein, before and beyond," *Diffusion Fundamentals*, vol. 2, pp. 1.1–1.10, 2005.

[228] W. H. Bossert and E. O. Wilson, "The analysis of olfactory communication among animals," *Journal of Theoretical Biology*, vol. 5, no. 3, pp. 443–469, 1963.

[229] O. F. T. Roberts, "The theoretical scattering of smoke in a turbulent atmosphere," *Proceedings of the Royal Society of London*, vol. 104, pp. 640–654, 1923.

[230] T. S. Ursell, "The diffusion equation: a multi-dimensional tutorial," tech. rep., Department of Applied Physics, California Institute of Technology Pasadena, CA 91125, October 2007.

[231] R. A. Freitas, *Nanomedicine, volume i: Basic capabilities*, 1999. Chapter 7. "Communication."

[232] B. Wyszynski, T. Yamanaka, and T. Nakamoto, "Recording and reproducing citrus flavors using odor recorder," *Sensors and Actuators B: Chemical*, vol. 106, no. 1, pp. 388–393, 2005. ISOEN 2003 - Selected Papers from the 10th International Symposium on Olfaction and Electronic Noses.

[233] L. Carmel and D. Harel, "Mix-to-mimic odor synthesis for electronic noses," *Sensors and Actuators B: Chemical*, vol. 125, no. 2, pp. 635–643, 2007.

[234] G. D. Bachand, S. B. Rivera, A. K. Boal, J. Gaudioso, J. Liu, , and B. C. Bunker, "Assembly and transport of nanocrystal CDSE quantum dot nanocomposites using microtubules and kinesin motor proteins," *Nano Letters*, vol. 4, no. 5, pp. 817–821, 2004.

[235] M. Lowry, S. O. Fakayode, M. L. Geng, G. A. Baker, L. Wang, M. E. Mccarroll, G. Patonay, and I. M. Warner, "Molecular fluorescence, phosphorescence, and chemiluminescence spectrometry.," *Anal. Chem.*, vol. 80, no. 12, pp. 4551–4574, 2008.

[236] A. Miyawaki, J. Llopis, R. Heim, J. M. McCaffery, J. A. Adams, M. Ikura, and R. Y. Tsien, "Fluorescent indicators for Ca2+based on green fluorescent proteins and calmodulin," *Nature*, vol. 388, pp. 882–887, 1997.

[237] W. Brütting, S. Berleb, and A. Mückl, "Device physics of organic light-emitting diodes based on molecular materials," *Organic Electronics*, vol. 2, pp. 1–36, 2001.

[238] T. Yamamura, M. Kitamura, K. Kuribayashi, Y. Arakawa, and S. Takeuchi, "Flexible organic LEDs with parylene thin films for biological implants," in *Proc. Conference on MEMS Micro Electro Mechanical Systems IEEE 20th International*, pp. 739–742, 2007.

[239] P. Verma, A. K. Ghosh, and A. Venugopalan, "Free-space optics based wireless sensor network design." NATO IST Panel Symposium on Military Communications, April 2008. NATO: Unclassified.

[240] B. Feringa, R. van Delden, N. Koumura, and E. Geertsema, "Chiroptical molecular switches," *Chem Rev*, vol. 100, pp. 1789–1816, May 2000.

[241] J. C. Brookes, F. Hartoutsiou, A. P. Horsfield, and A. M. Stoneham, "Could humans recognize odor by phonon assisted tunneling?" Nov 2006.

[242] M. B. Belonenko and N. G. Lebedev, "Two-qubit cells made of boron nitride nanotubes for a quantum computer," *Journal of Technical Physics*, vol. 54, pp. 338–342, Mar. 2009.

[243] E. M. Muller, E. G. Locke, and K. W. Cunningham, "Differential regulation of two Ca2+ influx systems by pheromone signaling in saccharomyces cerevisiae," *Genetics*, vol. 159, no. 4, pp. 1527–1538, 2001.

[244] N. M. Sherer and W. Mothes, "Cytonemes and tunneling nanotubules in cell–cell communication and viral pathogenesis," *Trends in Cell Biology*, vol. 18, no. 9, pp. 414–420, 2008.

[245] M. G. Schrlau, E. Brailoiu, S. Patel, Y. Gogotsi, N. J. Dun, and H. H. Bau, "Carbon nanopipettes characterize calcium release pathways in breast cancer cells," *Nanotechnology*, vol. 19, no. 32, p. 325102 (5pp), 2008.

[246] C. Sherrill, O. Khouri, S. Zeman, and D. Roise, "Synthesis and biological activity of fluorescent yeast pheromones," *Biochemistry*, vol. 34, pp. 3553–3560, Mar 1995.

[247] M. V. Vinogradova, G. G. Malanina, A. S. N. Reddy, and R. J. Fletterick, "Structure of the complex of a mitotic kinesin with its calcium binding regulator," *Proceedings of the National Academy of Sciences*, vol. 106, no. 20, pp. 8175–8179, 2009.

[248] C. Leduc, O. Campas, K. B. Zeldovich, A. Roux, P. Jolimaitre, L. Bourel-Bonnet, B. Goud, J.-F. Joanny, P. Bassereau, and J. Prost, "Cooperative extraction of membrane nanotubes by molecular motors," *Proceedings of the National Academy of Sciences of the United States of America*, vol. 101, no. 49, pp. 17096–17101, 2004.

[249] F. Walsh, S. Balasubramaniam, D. Botvich, M. O. Foghlu, T. Suda, and S. F. Bush, "Hybrid DNA and enzymatic based computation for address encoding, link switching and error correction in molecular communication," in *Proceedings of 3rd International Conference on Nano-Networks (Nano-Net)*, (Boston, MA), Sept 2008.

[250] S. Bush and Y. Li, "Nano-communications: A new field? An exploration into a carbon nanotube communication network," tech. rep., GE Global Research, 2006.

[251] K. Fall and K. Varadhan, "The network simulator–ns-2," Website, January 2009.

[252] M. K. So, C. Xu, A. M. Loening, S. S. Gambhir, and J. Rao, "Self-illuminating quantum dot conjugates for in vivo imaging," *Nature Biotechnology*, vol. 24, pp. 339–343, February 2006.

[253] M. J. Moore, A. Enomoto, T. Suda, A. Kayasuga, and K. Oiwa, "Molecular communication: Uni-cast communication on a microtubule topology," in *Proc. IEEE International Conference on Systems, Man and Cybernetics SMC 2008*, pp. 18–23, 2008.

[254] W. J. Dally and B. Towles, "Route packets, not wires: on-chip interconnection networks," in *DAC '01: Proceedings of the 38th Annual Design Automation Conference*, (New York, NY, USA), pp. 684–689, ACM, 2001.

[255] M. Ali, M. Welzl, A. Adnan, and F. Nadeem, "Using the ns-2 network simulator for evaluating network on chips (noc)," in *Proc. International Conference on Emerging Technologies ICET '06*, pp. 506–512, 2006.

[256] K. Bergman, "Nanophotonic interconnection networks in multicore embedded computing," in *Proc. IEEE/LEOS Winter Topicals Meeting Series*, pp. 6–7, 2009.

[257] J. Frey and D. Doren, "Tubegen." Website, 2005.

[258] S. Tanachutiwat and W. Wang, "Cnt bundle," 2007.

[259] M. Krems, "The Boltzmann transport equation: Theory and applications." Website, December 2007.

[260] S. V. Savant, "Stochastic simulation techniques in systems biology," in *Proc. American Control Conference ACC '07*, pp. 1311–1316, 2007.

[261] O. Haavisto, "Modeling and simulation in cellware." S-114.500 Basics for biosystems of the cell, November 2004. olli.haavisto@hut.fi.

[262] S. Sahle, R. Gauges, J. Pahle, N. Simus, U. Kummer, S. Hoops, C. Lee, M. Singhal, L. Xu, and P. Mendes, "Simulation of biochemical networks using COPASI: a complex pathway simulator," in *WSC '06: Proceedings of the 38th Conference on Winter Simulation*, pp. 1698–1706, Winter Simulation Conference, 2006.

[263] M. Hamdi and A. Ferreira, "Multiscale design and modeling of nanorobots," in *Proc. IEEE/RSJ International Conference on Intelligent Robots and Systems IROS 2007*, pp. 3821–3827, 2007.

[264] B. Julia-Diaz, J. M. Burdis, and F. Tabakin, "Qdensity—a Mathematica quantum computer simulation," *Computer Physics Communications*, vol. 174, p. 914, 2006.

[265] G. Toth, "QUBIT4MATLAB v3.0: A program package for quantum information science and quantum optics for MATLAB," *Computer Physics Communications*, vol. 179, p. 430, 2008.

[266] A. de A. Barbosa, B. Lula, and A. F. de Lima, "Symbolic and numeric quantum circuit simulation," in *Proc. First International Conference on Quantum, Nano, and Micro Technologies ICQNM '07*, p. 6, 2007.

[267] R. J. Allan, "Survey of agent based modelling and simulation tools," June 2009.

[268] C. Nikolai and G. Madey, "Tools of the trade: A survey of various agent based modeling platforms," *Journal of Artificial Societies and Social Simulation*, vol. 12, p. 2, March 2009.

[269] A. Troisi, V. Wong, and M. A. Ratner, "An agent-based approach for modeling molecular self-organization," *Proceedings of the National Academy of Sciences of the United States of America*, vol. 102, no. 2, pp. 255–260, 2005.

[270] K. Fujibayashi and S. Murata, "Precise simulation model for DNA tile self-assembly," *IEEE Trans. Nanotechnol.*, vol. 8, no. 3, pp. 361–368, 2009.

[271] G. Hu, Y. Guo, and R. Li, "A self-organizing nano-particle simulator and its applications," *NASA/ESA Conference on Adaptive Hardware and Systems*, 2008.

About the Author

Stephen F. Bush graduated from Carnegie Mellon University and worked at General Electric Information Services. From there, he obtained his Ph.D. while working as a researcher at the Information and Telecommunications Technologies Center at the University of Kansas, participating in the design of a self-configuring, rapidly deployable, beamforming wireless radio network.

Stephen currently enjoys his role as senior scientist at the General Electric Global Research Center where he has published numerous conference papers, journal articles, and book chapters, and taught international conference tutorials on novel communication and network related topics. His previous book publication, *Active Networks and Active Network Management: A Proactive Management Framework*, explained the development and operation of the intriguing and controversial active networking paradigm. Dr. Bush was presented with a gold cup trophy awarded by Defense Advanced Research Projects Agency (DARPA) for his work in active network related research. Dr. Bush has been the principal investigator for many DARPA and Lockheed Martin sponsored research projects including: Active Networking (DARPA/ITO), Information Assurance and Survivability Engineering Tools (DARPA/ISO), Fault Tolerant Networking (DARPA/ATO) and Connectionless Networks (DARPA/ATO), involving energy aware sensor networks. Stephen also likes creative interaction with students while teaching Quantum Computation and Communication at the Rensselaer Polytechnic Institute and Computer Communication Networks at the University at Albany.

Index

absolute mutual information, 167, 188
absolute temperature, 41, 181
absorption, 128, 189, 206, 207, 233, 241, 244
    spectrum, 242
abstraction, 212
acceptor, 52, 241, 242
accumulator, 24
acetone, 233
acknowledgment, 50, 51, 61, 201
actin, 4, 33, 46–49, 55
actin-myosin system, 85
action potential, 67, 86, 87
    propagation, 70
activation rate, 80, 81, 90
activation threshold constant, 77
active compression rate, 29
active network, 24, 29, 54, 82, 106, 107, 166, 167, 191, 209–
        217, 223, 225
    application program interface, 210
    architecture, 192
    capsule, 210, 211
    code, 24, 29
    code length, 29
    code-carrying, 210
    discrete, 210
    execution model, 24
    framework, 191, 209
    integrated, 210
    node, 24
    packet, 82
    resource allocation, 210
    security, 210
    service composition, 210
active networking, 166
    innovation, 209
active node, 210
active packet, 29, 107, 166, 167, 210, 213–216
    length, 215
activity
    cellular, 241
actuator, 3, 32, 194, 204
    in vivo, 208
ad hoc network, 26
adapt, 106
    network, 252
adaptive application, 212
adaptive behavior, 83
adaptive system, 212
address, 3, 26, 78, 199, 245
address space, 198
addressing, 53, 183, 185, 186, 198, 245, 248
    content-based, 199
adenine, 52
adenosine, 33, 41, 66

adenosine diphosphate, 33, 34, 41, 44, 60, 66
adenosine triphosphate, 31, 33, 34, 41, 42, 44, 45, 49, 58, 60,
        65, 66, 97
    binding site, 42
adhesion-based, 3
adjacency matrix, 59, 103, 109
adjoint, 132, 133
Aequoria victoria, 242
affinity, 67, 89, 96
affinity constant, 69
agent, 76, 176, 246, 260
    autonomous, 260
agent-based, 176
agonist, 75, 76
air, 31, 35, 48, 86, 227, 229, 233–235, 240
airborne, 64
alarm, 68, 248
algorithmic
    information, 214
algorithmic information theory, 24, 29, 176
alignment, 22
allomone, 68
allotrope, 94
α-tubulin, 48
alphabet, 165
    binary, 164
ambient noise, 25
amino acid, 53, 54, 65, 69, 96, 216, 242
amoeba, 85, 86
    protoplasm, 86
AMP, 33, 41
amplification, 96, 195
amplifier, 101, 219
amplitude, 154, 180
amplitude encoding, 81
amplitude modulation, 70, 72–74, 78, 95, 219, 225, 240
amplitude-modulated, 102
amplitude-quadrature, 153
amplitude-squeezed, 154
analog, 8, 13, 70, 78, 83, 241
analyte-specific, 10
ancilla, 222
angle
    random, 112
angle entropy, 112, 117
    high, 112
angular resonant frequency, 102
angular wave number, 21
anisotropy, 112, 114
    high, 112
    low, 112
annealing, 23, 49
annihilate, 142
ansatz, 79
antagonist, 70, 75
antenna, 11, 12, 14, 15, 28, 101, 204, 207, 219, 224
    beam-forming, 68
    efficiency, 28
    electrically small, 15

277

self-healing, 212
service, 213
short-range, 241
single carbon nanotube radio, 218
subnetwork, 196
teleportation, 152
thin film, 102
topology, 6, 108
traditional, 2, 95, 158
traffic, 209
wireless, 110, 111, 209
network address, 55
network capacity, 28
network coding, 70, 71, 91, 119
network conductance, 47
network engineering, 163
network flow, 59
network graph, 103, 119
network layer, 22, 25, 119, 245
network management, 209, 246
network on chip, 5
network operation, 88
network protocol, 82
    standard, 82
network security, 246
network service, 212
    genetically programmed, 213, 215
network simulator-2, 247, 255
network-on-chip, 2, 255, 256
    optical, 256
networking, 1, 2, 108, 207, 210, 248, 251, 256, 260
    biological, 2
    human-scale, 9
    macroscale, 2
    molecular, 251, 253
    nanoscale, 2, 3, 252, 260
    quantum, xiii
neural, 67
neural coding, 86
neural transmission, 86
neuromorphic, xvii
neuron, 58, 64, 67, 86, 88, 89, 96
    spike, 87
    subchannel, 86
neurotransmitter, 7, 65, 67, 75, 76, 86, 89
neutron, 67
Newton, 34
Newton's laws, 37
Newton's second law, 40
Newtonian physics, 72
nexus, 68
nitric acid, 69
nitrotoluene, 233
no-cloning theorem, 146
node, xiii, 3, 7, 8, 10, 11, 13–15, 22–25, 27, 28, 59, 70, 71, 91,
        102–104, 106–110, 114, 115, 149, 158, 176, 177,
        185, 187, 189, 191, 192, 196–202, 209, 210, 212,
        216, 218, 222, 224, 245, 252
    active, 209, 210, 216

adjacent, 197
body coordinator, 208
complexity, 46, 108
defective, 108
high degree, 110
identifier, 107
intermediate, 211
mobile, 200
placement, 108
reachable, 107, 108
relay, 186
simulated, 114
node degree, 197
noise, 7, 23, 68, 86–88, 91, 110, 154, 157, 165, 166, 169, 174,
        196, 227, 230, 232, 260
    Gaussian, 110
    thermal, 17
noise value, 88
noisy channel, 166
noisy channel coding theorem, 166
nonbiologist, 32
noncovalent, 96
nondestructive, 243
nonequilibrium, 239
nonexpansion work, 17
noninterference protocol, 27
noninvasive, 5
nonlaminar flow, 72
nonlinear system, 87
nonlinearity, 9, 102
nonlocality, 143, 144
nonnucleated, 4
nonrandomness, 215
nontraditional, 6
nonuniformly, 58
noradrenoline, 68
normal diffusion, 202
normal distribution, 41, 72
normal matrix, 134
normalization, 8, 133, 134, 159
nuclear, 206
nuclear magnetic resonance, 206
nuclei, 206, 215
nucleic, 97
nucleic acid, 49, 65, 68
nucleotide, 31, 49, 52, 55, 57, 65, 69
nucleus, 4, 97, 194, 206, 215

object model, 212
Occam's razor, 76
odor, 228, 229, 232, 237, 239, 244, 245, 247
    control, 228
    target, 237, 239
odor molecule, 228
odor transmitter, 236
odor vector
    target, 239
odorant, 83, 229, 237, 238, 244
    target, 238

Ohm's law, 84, 103, 234
olfactory, 88, 228, 236, 237
   display, 228
olfactory communication, 58
olfactory receiver, 236
olfactory receptor, 83
olfactory system, 68
olfactory transmitter, 236
omnidirectional, 111
on-demand, 239
one-dimensional, 36
one-dimensional channel, 98
one-dimensional wire, 100
operands, 24
operator
   partial trace, 172
   projection, 170
   reduce density, 188
   trace, 170
optical, 9, 63, 128, 241, 242, 244, 245, 247, 253
   data storage, 243
   free-space, 9
optical fiber, 239
optical signal, 63
optimization, 68
orbital
   molecular, 241
order, 212
organ
   olfactory, 228
organ development, 67
organelle, 46, 69, 75, 205, 244, 246
organic light-emitting diode, 242
organism, xiii, 3, 57, 66, 67, 195
   multicellular, 63
   one-celled, 85
   single-cell, 66
   single-celled, 63, 67, 85, 204
orientation, 47
orthogonal, 104, 137, 142, 169
orthogonal basis, 149
orthogonal component, 142
orthogonal space, 160
orthogonal state, 142
orthogonal subspace, 142
orthogonality, 153
orthonormal base, 169
orthonormal basis, 135–137, 142, 161, 170
orthonormal subspace, 170
oscillating, 128, 219
   carbon nanotube, 256
oscillating signal, 88
oscillation, 46, 76, 78, 86, 101, 102, 191, 219
   baseline, 74
   pulse-coupled, 253
   sinusoidal, 74
oscillation frequency, 90
oscillator, 203, 219
oscillatory pattern, 72

osteosarcoma, 74
outer core, 94
outer product, 131, 137, 146, 149, 161, 170, 171, 260
outer space, 192
overdamped, 102
overhead, 51
oxide, 229

P frame, 209
pacemaker, 208
packet, 24, 107
Packet Language for Active Networks, 210
packet radio, 209, 252
packet-switched, 2
palette, 238
paracrine, 89
paradigm
   simulation, 253
parallel, 85
paramecium, 194
   learning, 194
parity code, 52
Parkinson's Disease, 208
partial trace, 172
particle, 21
   nanoscale, 202
   subnanoscale, 126
particle interaction, 256
particle-wave duality, 125
partition function, 19, 20, 23
passive data, 24, 29
passive network, 212
path loss, 208
pathway, 71, 246
   chemical, 259
pattern formation, 46
Pauli exclusion principle, 99
Pauli gate, 134
Pauli matrix, 134
payload, 32, 51, 53, 57, 58
peak, 87
peptide, 69, 97
percolation, 116, 117, 123, 177
   threshold, 47, 96, 116
periodic signal, 86
Perlis, Alan, 191
permeability, 65, 66, 75, 245
permutation, 152
   cyclic, 149
perpendicular, 48, 205
persistence length, 47, 48, 57, 60, 96, 123
perturbation, 87, 204, 206
pervasiveness, 13
pH, 83, 243
phase, 33, 57, 154
phase velocity, 100
phase-quadrature, 153
phenomenon
   quantum mechanical, 259